Bird Families of the World

A series of authoritative, illustrated handbooks, of which this is the fourth volume to be published

Series editors

THE AUTHORS Ian Rowley is an Honorary Research Fellow in the Division of Wildlife and Ecology, CSIRO, following his retirement after 39 years with that organization. His main interest has been the social behaviour of animals and he has applied this to research work on rabbits, corvids, White-winged Choughs, cockatoos and six species of fairy-wren. In 1975, he wrote *Bird life* (Collins) and in 1990, *The Galah* (Surrey Beatty). Eleanor Russell is also an Honorary Research Fellow in CSIRO following her marriage to Ian Rowley. Previous to that she was a Senior Lecturer in Zoology at the University of New South Wales with particular interests in the maternal behaviour of marsupials. For the past 20 years, Eleanor and Ian have worked as a team studying five different species of fairy-wren.

THE ARTIST Peter Marsack originally trained as a zoologist in the UK, and has worked on a range of feral animal species in Western Australia. In 1991 he left to study Plant and Wildlife Illustration at Newcastle University, New South Wales. He has a particular interest in birds, and much of his artwork has been for the Royal Australasian Ornithologists Union. He is presently working on identification plates for *The Handbook of Australian, New Zealand and Antartic Birds* (Oxford University Press).

Bird Families of the World

1. The Hornbills
   ALAN KEMP
2. The Penguins
   TONY D. WILLIAMS
3. The Megapodes
   DARRYL N. JONES, RENÉ W. R. J. DEKKER, AND CEES S. ROSELAAR
4. Fairy-wrens and Grasswrens
   IAN ROWLEY AND ELEANOR RUSSELL
5. The Auks
   ANTHONY J. GASTON AND IAN L. JONES

Bird Families of the World

Fairy-wrens and Grasswrens

Maluridae

IAN ROWLEY

ELEANOR RUSSELL

both Honorary Research Fellows, Division of Wildlife and Ecology, CSIRO, Australia

Illustrated by

PETER MARSACK

Oxford New York Tokyo

OXFORD UNIVERSITY PRESS

1997

Oxford University Press, Great Clarendon Street, Oxford OX2 6DP

Oxford New York
Athens Auckland Bangkok Bogota Bombay Buenos Aires
Calcutta Cape Town Dar es Salaam Delhi Florence Hong Kong
Istanbul Karachi Kuala Lumpur Madras Madrid Melbourne
Mexico City Nairobi Paris Singapore Taipei Tokyo Toronto Warsaw
and associated companies in
Berlin Ibadan

Oxford is a trade mark of Oxford University Press

Published in the United States
by Oxford University Press Inc., New York

A catalogue record for this book is available from the British Library

Library of Congress Cataloging-in-Publication Data
Rowley, Ian, 1926–
Fairy-wrens and grasswrens: Maluridae / Ian Rowley, Eleanor Russell; illustrated by Peter Marsack.
(Bird families of the world; 4)
Includes bibliographical references (p.) and index.
1. Maluridae. I. Russell, Eleanor M. II. Title. III. Series
QL696.P2485R68 1997 598.8—dc21 96–52084

ISBN 0 19 854690 4

Typeset by EXPO Holdings, Malaysia

Printed in Hong Kong

Preface

The brilliant plumage of the males, their jaunty, gait and their acceptance of man-made suburban gardens as a suitable habitat make fairy-wrens a favourite with anyone in Australia or New Guinea lucky enough to have a local group to enjoy. Besides the early illustrations in journals and in Gould's *Birds of Australia*, the group drew the attention of the painter/ornithologist Neville Cayley, who in 1949 produced a small book with paintings and distribution maps of fairy- and emu-wrens, *The fairy wrens of Australia*. This was followed in 1982 by Richard Schodde's monograph *The fairy-wrens* which covered the whole family, and was beautifully illustrated by Richard Weatherly. Schodde's work is a careful and invaluable review of the taxonomy and evolution of the five genera now recognized as constituting the family Maluridae. It extended the coverage of the family to New Guinea.

We are not museum ornithologists skilled in the feather by feather description of birds, nor are we concerned with taxonomy for its own sake, except as it provides accepted names for the species we work on, and tries to explain their relationships. We are both interested in the social organization and behaviour of animals, and in pursuit of this theme, we have found that birds are much easier to work on than the marsupials and rabbits we respectively 'cut our teeth on'. We have drawn extensively from Dick Schodde's monograph (with his permission) to provide the necessary background to our discussions.

In the 14 years since Schodde wrote his monograph, interest in the subjects of co-operative breeding and sexual selection has grown and led to widespread theoretical discussions throughout the world. Most of the data fuelling these debates have come from long-term studies involving individually marked birds followed through several generations. It is only by understanding the measure of success achieved by different identifiable individuals during their life that we can begin to appreciate the complexity of the material on which natural selection works and brings about evolution. This has been our approach with six different species of *Malurus* and forms the background to the chapters that follow.

At the International Ornithological Congress in Christchurch in 1990, Chris Perrins asked Ian Rowley if he would prepare this book. Due to retire in two months, he thought he would have plenty of time, but subsequently found how wrong he was, having also undertaken to edit *Emu*. Fortunately Eleanor, his hard-working wife, agreed to co-author the book. She has done most of the literature research and correlated much diverse information. She analysed our field data and compiled the section on vocalizations, since, sadly, Ian can no longer hear malurid song.

June 1997 — I.R.
Helena Valley — E.R.
Western Australia

Acknowledgements

This book is the culmination of forty years of field studies by Ian Rowley, the last twenty of them shared with Eleanor. During this time, we have received valuable help in the field on our various malurid studies from Steve Blaber (*Malurus melanocephalus*), Dick and Molly Brown (*M. elegans*), Michael, Lesley, and Belinda Brooker, Bob and Laura Payne (*M. splendens*), Graeme Chapman and Joe Leone (*M. splendens*, *M. coronatus*). Besides our own data, we have depended heavily on the work of others. In particular: Richard Schodde in his comprehensive monograph; the long-term studies of Sonia Tidemann, Ray Nias, and Raoul Mulder, and their respective PhD theses; Andrew Cockburn (Australian National University), and Stephen Pruett-Jones and Michael Webster (University of Chicago) for discussion of work in hand; Graeme Chapman, for sharing his experiences with grasswrens; and Peter Fullagar for his advice and patient help in the collection of sound recordings and preparation of sonograms.

In conjunction with Peter Marsack who drew the plates and other illustrations, we thank Ron Johnstone (WAM), Walter Boles (AM), Richard Schodde (ANWC), Les Christidis (MV), Wayne Longmore (QM), and Phillipa Horton (SAM) for access to the collections in their care, for advice on taxonomic matters, and the loan of material for the preparation of the colour plates and black and white artwork.

Many people provided photographs and tapes for the preparation of the illustrations and sonograms; others provided unpublished data from their records, from those of the RAOU and the Australian Bird Banding Scheme, and valued individual comments: Kevin Bartram, Peter Higgins, Danny Rogers, and Jon Starks (RAOU); Ian Mason and John Wombey (ANWC/CSIRO); Belinda Dettman (ABBBS); Andrew Cockburn and David Green (Australian National University); Bruce Beehler, David Bishop, Walter Boles, Graeme Chapman, Mark Clayton, Brian Coates, Jeff Davies, Nic Day, Jared Diamond, Cliff and Dawn Frith, Rob Heinsohn and Sarah Legge, Richard Jordan and Christine Wilder, John Long, Perry and Alma De Rebeira, Rod Smith, Sue Stevens, John Warham, Bert and Babs Wells, and Ron Wooller.

Comments on various parts of the manuscript were made by David Bishop, Les Christidis, Michael Clarke, Hugh Ford, Stephen Garnett, Richard Hobbs, Richard Schodde, and Graeme Smith. Series editors Walter Bock, Jiro Kikkawa, and Chris Perrins read the whole manuscript, and made many suggestions for its improvement.

Ian Rowley would like to acknowledge the encouragement he received from successive Chiefs of the CSIRO Division of Wildlife Research: Francis Ratcliffe, Harry Frith, and Charlie Krebs; their moral support was very important to him. The authors thank Brian Walker, Chief of the Division of Wildlife and Ecology, CSIRO, and Denis Saunders, Officer-in-Charge of the Division's Western Australian Laboratory, for permitting us to use divisional facilities as Honorary Research Fellows; and our colleagues at Helena Valley for their tolerance and good company. Without the patience and helpfulness of the Divisional librarians, in particular Barbara Staples, we could never have completed this project.

We thank the staff of Oxford University Press and the Bird Families series editors for help and advice at every stage.

Dedication

To all those who helped in the production of this book, especially the hundreds of small birds who tolerated our invasion of their private lives and the well-meant shackling of their small legs with colour-bands.

Contents

Colour plates

Colour plates fall between pages 138 and 139.

Abbreviations

ABBBS	Australian Bird and Bat Banding Scheme
ACT	Australian Capital Territory
ad	adult
agl	above ground level
AM	Australian Museum
ANCA	Australian National Conservation Agency
ANU	Australian National University
ANWC	Australian National Wildlife Collection
Apr	April
asl	above sea level
Aug	August
c.	circa or about
CSIRO	Commonwealth Scientific and Industrial Research Organization
d	days
Dec	December
EPC	extra-pair copulation
F	female
Feb	February
g	gram(s)
hr	hour(s)
ha	hectare(s)
imm	immature(s)
in litt.	unpublished information received in writing
Is.	island(s)
Jan	January
juv(s)	juvenile(s)
kg	kilogram
kHz	kilohertz
km	kilometre
m	metre(s)
M	male
Mar	March
min	minute(s)
Mt(s)	mountain(s)
MV	Museum of Victoria
n	number in sample
Nov	November
NRS	Nest Record Scheme
N, S, E, W	north, south, east, west
NSW	New South Wales
NT	Northern Territory
Oct	October
P	primary feather
P1	innermost primary
P10	outermost primary
Qld	Queensland
QM	Queensland Museum
R.	river
Ra.	Range
RAOU	Royal Australasian Ornithologists' Union
S	secondary feather(s)
SA	South Australia
SAM	South Australian Museum
SD	standard deviation
SE	standard error
sec	second
Sept	September
sp., spp.	species (singular/plural)

subsp./spp.	subspecies (singular/plural)
T	tail feather (T1central)
Tas.	Tasmania
Vic.	Victoria
wt	weight
WA	Western Australia
WAM	Western Australian Museum

Plan of the book

This monograph, as others in the series Bird Families of the World published by Oxford University Press, is in two parts.

PART I

Part I attempts an overview of malurid biology, behaviour, ecology, and evolution, in the context of the environment of Australia and New Guinea and the origin and evolution of their faunas. It begins with a chapter introducing the family Maluridae, and is followed by seven chapters dealing with taxonomy, evolution, biogeography, morphology, ecology, behaviour, social organization, co-operative breeding, life histories, and conservation.

In Chapter 2, we discuss the taxonomic history of the family Maluridae; a full taxonomic treatment is given by Schodde (1982*b*).

In Chapter 3, we discuss the formation of the continent, past and present climates, and vegetation, and how they have influenced the evolution of the Maluridae. Because many readers will be unfamiliar with the Australian continent, in this chapter we have described in some detail the different habitats in which malurids live.

Most malurids are insectivores, feeding on or near the ground, and in Chapter 4, we explore how their structure is suited to this way of life, and how some species have specialized in particular habitats, diets, or feeding methods. We look at whether bills, wings, tail, and tarsus have the same relative size in large and small species, or whether in some species bills or legs, for example, are relatively longer or shorter. To do this, we have plotted a linear measurement (e.g. wing length) against body mass (in technical terms, we have calculated the regression of wing length on body mass). If all the points lie in a straight line with little scatter, there is a strong correlation between the linear dimension and body mass. The correlation coefficient r will have a value approaching 1, and r^2 indicates how much of the variation in the linear dimension is explained by the variation in body mass. If the points are scattered, the correlation is low, and only a small part of the variation in the linear dimension is explained by the variation in body mass. In some cases, plotting one or both measurements on a log scale gives a better correlation. If there is an overall close relationship between a linear measurement and body size, with one exceptional species, then we can ask why that species has, for example, a bill length that is relatively long for its body size.

In these calculations, where possible, we have used mean linear measurements of adult males (taken where available from Schodde 1982*b*, to standardize the method of measurement as far as possible); where tail length varies between immature and nuptial adult males, we have used measurements of males in nuptial plumage. Measurements of bill width for grass-wrens (*Amytornis*) are from Keast (1958) or measured by the authors from museum specimens. Body weights were taken from field measurements by the authors, from records of the Australian Bird Banding Scheme, or from published sources (see species accounts for details).

Apart from the brilliant plumage of the fairy-wrens (*Malurus*), the birds in this family are best known for their unusual social organization (territorial group-living) and breeding habits. This aspect of their biology has been studied in depth for three species and is dealt with in Chapters 5 and 6. We have tried to present a framework of the basic scientific theories proposed to explain the evolution of this way of life and its prevalence in Australia. We have drawn on the results of our own work and that of our colleagues on other species, and have used these results to demonstrate the roles played by different individual birds. Although this information has been obtained from only a few species, we feel confident that the principles emerging probably apply throughout the family, and we have discussed this in detail in these chapters rather than in the species accounts.

Similarly, in Chapter 7, where we describe life histories (breeding biology, productivity, and survival), we have compared information from long-term studies for different species in different habitats. We investigated how egg weight changes with body size, estimating egg weight from the measurements given by Schodde (1982*b* and the sources therein). We used the body weights of adult males rather than females as a measure of body size, since many more were available; because males are generally slightly larger than females, our estimates of relative egg size are a little low, but the general trends are clear. Where they are known, the dimensions of eggs are given in our accounts of individual species, later in the book. Throughout this book, we use the following terminology to describe the timing of events and the aging of young birds. A nestling hatches on Day 1, and it is 1 day old as soon as Day 2 starts, which is at midnight following the day of hatching.

In Chapter 8, we review the impact of European settlement on the Australian environment, its effect on the family Maluridae, and the needs of conservation.

PART II

Part II of the book provides detailed accounts of the 25 species in five genera that make up the family. Each species is dealt with in a standard way and illustrated in the colour plates.

Nomenclature

Each of the five genera in the family is described and followed by the relevant species accounts, which start with the English and Latin names for the species. In a second line, the scientific name in its original form is followed by the name of the author who made the description, the year when it was published for the first time, and what that source was.

Universal use of a single English name avoids confusion in handbooks and solves legislative problems where customs and conservation regulations require precise definition of a subject species. We do not give the plethora of English names that have been used in the past; a full list of these is given in *Recommended English names for Australian birds* (RAOU 1978) and in the *Species checklist of the birds of New Guinea* (Beehler and Finch 1985). Australian ornithology is striving hard to standardize usage, and for the Australian species, we use English names as given by Christidis and Boles (1994). For the five species endemic to New Guinea, we follow the standard works for the three *Malurus*, using the generic name 'Fairy-wren', but for *Sipodotus* and *Clytomyias*, two monospecific genera, we use the generic names 'Tree-wren' and 'Russet-wren'. These were suggested by Schodde (1982*b*) when he described the genera, but he did not use them in his species names, using the simple 'wren'—Wallace's Wren and Orange-crowned Wren, as we have done. Beehler and Finch (1985) changed these to 'fairy-wren' and Coates (1990) followed their example. We believe that the five malurid genera are best named differently. By convention, Grasswren is one word; the other names are hyphenated.

Malurus	Fairy-wrens
Stipiturus	Emu-wrens
Amytornis	Grasswrens
Sipodotus	Tree-wrens
Clytomyias	Russet-wrens

In the text we use English names throughout the general chapters, with Latin names given at first mention in each chapter. Subspecies are given their Latin names, since most lack an English name. For example, the three subspecies of the Splendid Fairy-wren (*Malurus splendens*) are given as *M. s. splendens*, *M. s. callainus*, and *M. s. melanotus*.

Description

This presents details of colour and external form of the plumage, bill, legs, and eye, for adult males, adult females, and immatures. Where subspecies have been recognized, the description is of the nominate form, followed by details of the other subspecies: authorship and date of the original description, further taxonomic discussion, details of the differences that distinguish them, and where they occur. We have drawn (with his permission) on the careful and detailed descriptions and measurements provided by Dick Schodde in his 1982 monograph on the family, supplemented by our knowledge of known-age individuals in the field. This section includes information on moult and changes in plumage with age, where known.

History and subspecies

Describes the discovery and naming of taxa now included in each species. We include a summary of the various changes in nomenclature leading to the present treatment of subspecies.

Measurements and weights

Standard measurements of adult wing length (folded, flattened left wing), tail length, bill length (exposed culmen), tarsus length (all in millimetres), and body weight (in grams) are given for adult males and females. We have used measurements of males in nuptial plumage. Linear measurements are taken from the monograph of Schodde (1982*b*), and are in the form: (n = sample size), mean ± standard deviation. Body weights are from field measurements by the authors (*M. coronatus*, *M. elegans*, *M. leucopterus*, *M. pulcherrimus*, *M. splendens*, *S. ruficeps*), from records of the Australian Bird Banding Scheme, or from published sources (see species accounts for details); they are in the form (n = sample size), range (mean ± standard deviation). Data for each subspecies are presented separately. Where Schodde has shown significant variation in size within the range of a species or subspecies, we give measurements for birds from different regions to illustrate the differences.

Field characters

We identify those aspects of a particular species that enable it to be distinguished from relatives that may occur in the same area and might be mistaken for it in the field. There are few other small ground-living passerines in other families, and so we do not attempt to list all the potential 'confusables'.

Voice

We were able to assemble tape-recordings of almost all species in the family, thanks to many colleagues. Sonograms of the characteristic territorial song and other vocalizations have been prepared through the magic of modern technology and the skill and patience of our good friend Peter Fullagar. They are presented here in a standard format at a uniform scale (within genera) to allow direct comparison. The vertical axis shows the frequency in kiloherz (kHz). Where known, contact, alarm, and other calls are described. Copies of all recordings used from Australian species are housed in the Australian Sound Library at the Australian National Wildlife Collection (ANWC) in Canberra. For some species, we have good recordings of several individuals, and can compare individual differences. For other species, we could locate only one poor recording of one individual, but we can at least compare the structure of the song with that of other species. Comparisons between species, and a general discussion of differences between genera in song structure may be found in Chapter 5. Recordings of most Australian malurids are available on Cassette No. 9 of *A field guide to Australian bird song*, compiled by Rex Buckingham and Len Jackson, available from the Bird Observers Club of Australia, P.O. Box 185, Nunawading, Victoria, 3131, Australia.

Range, status, and maps

The present-day distribution of each species (and subspecies where appropriate) is shown on a map prepared from locations taken from museum data, the Australian Bird and Bat Banding Scheme, the literature, RAOU Atlas records, Blakers *et al.* (1984), from correspondence with many ornithologists, and from our own travels. These maps show the range within which a species occurs in suitable habitat. They do not indicate a continuous distribution, nor do they attempt to show the historical distribution in those cases where habitat destruction has reduced the range of a species. On pages xix and 144, a map of Australia and New Guinea attempts to show all places, rivers, mountain ranges, and other geographical features mentioned in the text. It also shows state boundaries. The terminology used for conservation status has recently been stabilized by IUCN (1994), and we have tried to use those categories in this book. However, the major Australian work on this subject (Garnett 1992*a,b*) uses some other terms, and since we quote his evaluations, in places we need to quote his terms; in particular, 'insufficiently known' for the new IUCN Data Deficient.

Habitat and general habits

Often the preferred habitat and characteristic ways of foraging distinguish between two otherwise very similar species that may occur close together. We have tried to describe these differences using our own observations and the literature. Habitats are described in terms of current structural classifications of Australian or New Guinea vegetation types as outlined in Chapter 3. Social organization, habitat use, foraging behaviour, and diet are described in this section.

Displays and breeding behaviour

Certain aspects of display are characteristic of many malurids, such as the Rodent-run as a distraction display, and Petal Carrying in courtship. However, most species live in dense cover and are difficult to observe; mating is rarely seen. We owe much of our knowledge about displays to the careful observations of one aviculturist, the late Rosemary Hutton (Hutton 1991). These displays are described in general in Chapter 5, and their occurrence in particular species recorded in this section. The complex mating system of *Malurus* is described in Chapter 5, based on detailed studies of a few species; there is sufficient evidence to assume that a similar mating system is found in all species of *Malurus*.

Breeding and life cycle

This section summarizes specific information on breeding seasons, nest sites and dimensions, clutch and egg sizes, incubation, nestling and post-fledging periods, and demographic information on productivity and survival, where known. Long-term studies of several *Malurus* species are quoted in detail in Chapter 7, and where appropriate, we have extrapolated these data to less well-studied species in the species accounts.

Colour plates

The colour plates show an adult male and female for each species and, where they are different, the immature and the eclipse male plumages. For polytypic species, it has usually been sufficient to illustrate one sex only to show the differences.

Glossary

The glossary explains technical terms that are unavoidably used in the text to illustrate the detail of particular topics where no commonly used word is adequately descriptive.

Bibliography

References are mentioned in the text as (Author date), with the full citation at the end of the book.

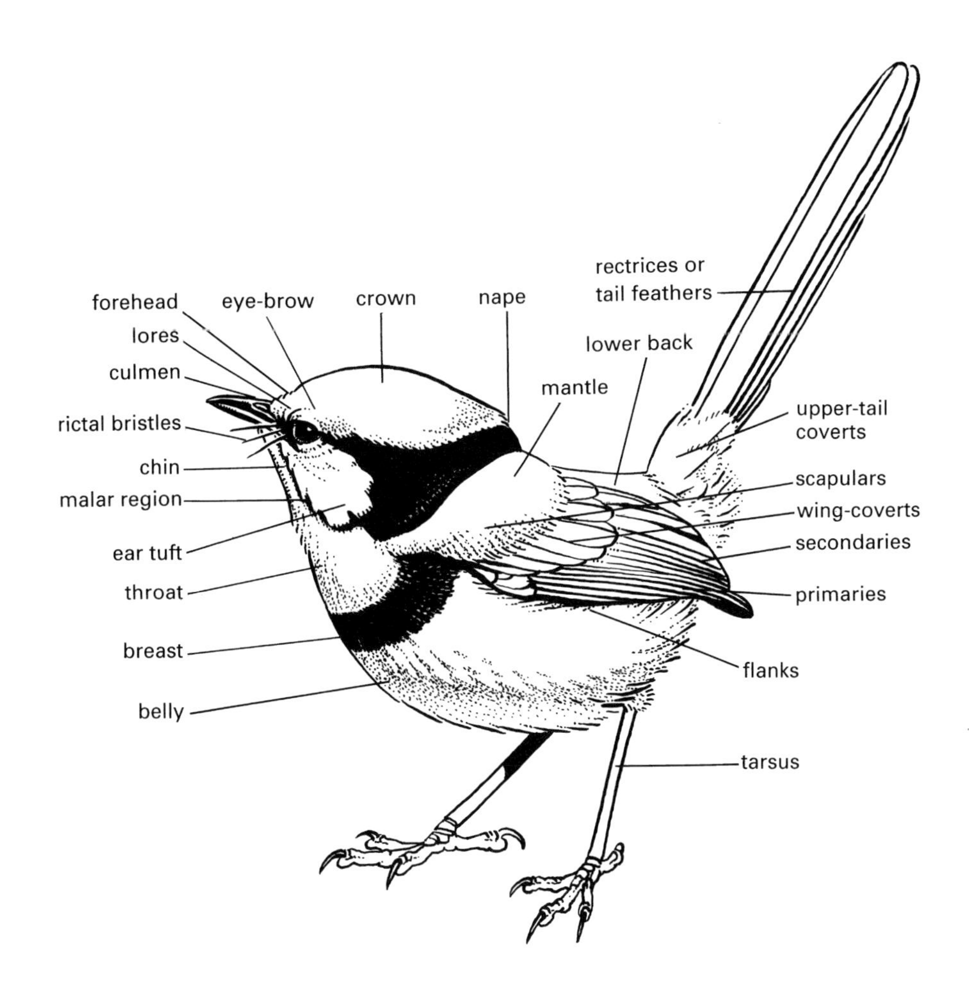
rectrices or
tail feathers
forehead
eye-brow
crown
nape
lower back
lores
culmen
mantle
rictal bristles
upper-tail
coverts
chin
scapulars
malar region
wing-coverts
ear tuft
secondaries
throat
primaries
breast
flanks
belly
tarsus

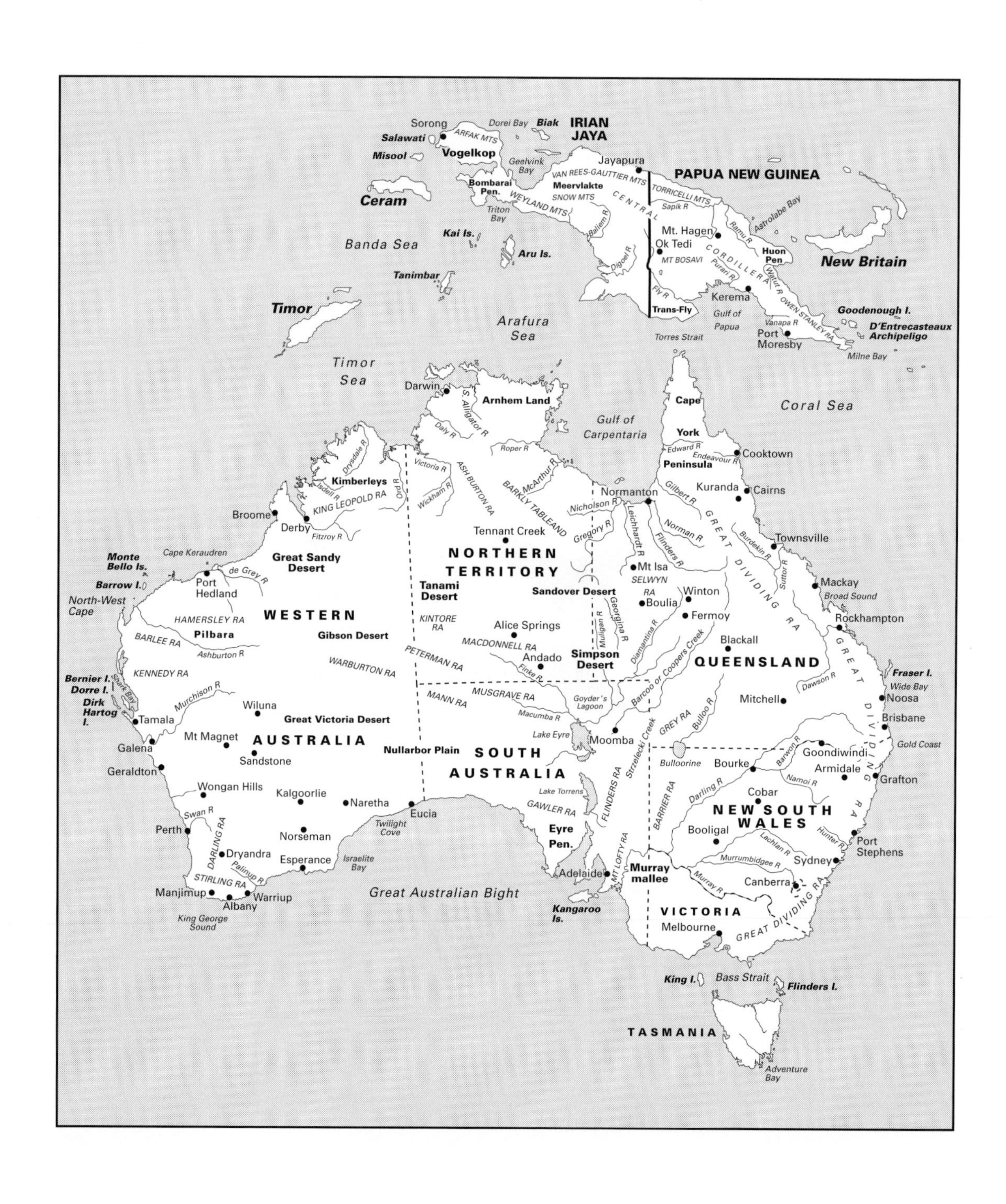
Sorong
Dorei Bay
Biak
IRIAN JAYA
Salawati
ARFAK MTS
Misool
Vogelkop
Geelvink Bay
Jayapura
PAPUA NEW GUINEA
Bombarai Pen.
Meervlakte
VAN REES-GAUTTIER MTS
TORRICELLI MTS
Ceram
SNOW MTS
WEYLAND MTS
CENTRAL
Sapik R
Astrolabe Bay
Triton Bay
Mt. Hagen
Ramu R
Kai Is.
Baliem R
Ok Tedi
CORDILLERA
Banda Sea
Aru Is.
Digoel R
MT BOSAVI
Purari R
Huon Pen
New Britain
Tanimbar
Fly R
Kerema
Watut R
OWEN STANLEY RA
Timor
Trans-Fly
Gulf of Papua
Goodenough I.
Arafura Sea
Vanapa R
Port Moresby
D'Entrecasteaux Archipeligo
Torres Strait
Milne Bay
Timor Sea
Darwin
S. Alligator R
Arnhem Land
Coral Sea
Cape
Gulf of Carpentaria
Daly R
York
Roper R
Edward R
Endeavour R
Cooktown
Peninsula
Drysdale R
Victoria R
ASH BURTON RA
McArthur R
Kimberleys
Isdell R
Ord R
BARKLY TABLEAND
Normanton
Gilbert R
Kuranda
Cairns
KING LEOPOLD RA
Wickham R
Nicholson R
Broome
Derby
Fitzroy R
Tennant Creek
Gregory R
Leichhardt R
Norman R
GREAT
Townsville
Burdekin R
Cape Keraudren
NORTHERN TERRITORY
Flinders R
Monte Bello Is.
Great Sandy Desert
Mt Isa
DIVIDING
Suttor R
Mackay
Barrow I.
de Grey R
SELWYN RA
Port Hedland
Tanami Desert
Sandover Desert
Winton
Broad Sound
North-West Cape
Boulia
Georgina R
WESTERN
Fermoy
RA
Rockhampton
HAMERSLEY RA
KINTORE RA
Alice Springs
Mulligan R
Diamantina R
Blackall
BARLEE RA
Pilbara
Gibson Desert
MACDONNELL RA
Ashburton R
PETERMAN RA
Andado
Simpson Desert
QUEENSLAND
GREAT
WARBURTON RA
Barcoo or Coopers Creek
Fraser I.
KENNEDY RA
Finke R
Dawson R
Wide Bay
Bernier I.
Dorre I.
Shark Bay
MUSGRAVE RA
Goyder's Lagoon
Murchison R
Mitchell
Noosa
Dirk Hartog I.
MANN RA
Wiluna
Macumba R
DIVIDING
Tamala
Great Victoria Desert
GREY RA
Bulloo R
Brisbane
Mt Magnet
Lake Eyre
Strzelecki Creek
Galena
AUSTRALIA
Moomba
Barwon R
Goondiwindi
Gold Coast
Sandstone
Nullarbor Plain
SOUTH
Geraldton
Bulloorine
Bourke
Armidale
AUSTRALIA
Namoi R
Grafton
Wongan Hills
Kalgoorlie
Lake Torrens
FLINDERS RA
BARRIER RA
Cobar
Naretha
Darling R
Swan R
Eucia
GAWLER RA
NEW SOUTH WALES
Perth
Twilight Cove
Eyre Pen.
Booligal
Hunter R
Norseman
Lachlan R
Port Stephens
DARLING RA
Dryandra
Israelite Bay
MT LOFTY RA
Esperance
Murrumbidgee R
Sydney
Palinup R
Adelaide
Murray mallee
STIRLING RA
Murray R
Canberra
Manjimup
Warriup
Great Australian Bight
Albany
Kangaroo Is.
VICTORIA
King George Sound
Melbourne
GREAT DIVIDING RA
King I.
Bass Strait
Flinders I.
TASMANIA
Adventure Bay

PART I

General chapters

1

Introduction to the malurids

Birds of the passerine family Maluridae are found only in Australia and New Guinea. Of the five genera now recognized as making up the family, one, the fairy-wrens (*Malurus*), is found in Australia and New Guinea; two are found only in Australia, the emu-wrens (*Stipiturus*) and the grasswrens (*Amytornis*); and two are found only in New Guinea, the tree-wrens (*Sipodotus*) and the russet-wrens (*Clytomyias*). The largest weigh no more than 40 grams, and most are far smaller, about 10 grams, with the smallest emu-wrens only 5–6 grams. Most species feed predominantly on the ground or in low vegetation, and are not strong fliers; with their long legs and long tail held cocked at right angles to the back, they are a very characteristic element of Australian bird communities.

All species are sexually dimorphic, and this is most extreme in the fairy-wrens, in which the females are plain grey-brown, while the males in breeding plumage are patterned in iridescent contrasting blues and black, and sometimes red. In contrast, the elusive grasswrens, living mostly in red-brown arid habitats, are patterned predominantly in red-browns, streaked with white, as are the tiny, almost invisible emu-wrens, whose most notable feature is their elongated, filamentous tail with only six feathers. No region in Australia is without at least one species of fairy-wren; in more arid areas, two or three fairy-wrens may occur, with an emu-wren and one or more grasswrens.

A little over 200 years ago, when the first European settlers arrived in eastern Australia, four members of the family Maluridae were found near Sydney (or Port Jackson as it was then called). Not surprisingly, they featured in many of the early illustrations from the new colony, but their naming was somewhat haphazard. At the end of the eighteenth century, taxonomy was in its infancy, museum collections were sparse, and the settlers naming new species in Australia relied heavily on similar small birds remembered from Europe. The first specimen of *Malurus* came from Tasmania in 1777 and was given the Latin name *Motacilla cyanea* because it was thought to resemble a wagtail, common in England. A few years later, the same species in Sydney was called the Superb Warbler (Fig. 1.1). Their cocked tail and their liking for dense cover soon led to the name wren or wren-warbler, although the family has no close relationship with the true wrens, Troglodytidae. The unfortunate name fairy-wren came into being in the 1920s, but at least we are spared the Elfin-wren suggested in 1939 by Tom Iredale for the Red-backed Fairy-wren (*Malurus melanocephalus*).

Throughout the nineteenth century, settlements were established at several different points of the Australian mainland. Birds were collected from each of these sites, and from exploratory expeditions into the interior, and because there were no local museums to classify and name them, these specimens were usually sent to a European museum for description. This meant that the type specimens were not available locally for comparisons to confirm identifications, which led to some

1.1 Early illustration of Superb Fairy-wren (*Malurus cyaneus*) by P. Mazell from *The Voyage of Governor Phillip to Botany Bay* (Phillip 1789).

confusing taxonomic tangles, particularly with the grasswrens. Also, the assumption that Australian species were derived from Old World forms was never questioned. The fairy-wrens were assigned at various times to the Old World warblers (Sylviidae), the flycatchers (Muscicapidae), and the babblers (Timaliidae), and it was assumed that, at various times, bird families had reached Australia by island hopping from south-east Asia.

This view has been profoundly altered in the last 30 years by three developments. First, the general acceptance of theories of continental drift led to the realization that the Australian continent was far away from south-east Asia when most vertebrate evolution took place. As a result of this, Australia was not in position to receive immigrants from south-east Asia early enough to have allowed the diverse radiation of endemic species that we see today. The second development was the use of biochemical, molecular methods to assess similarities between species, which has led to a reassessment of avian relationships. This gave support to evidence from morphology of the existence of a diverse Australian radiation of old endemic families only distantly related to those of the northern hemisphere. The third and most recent development has been the discovery of fossil passerines in Australia with an age of *c.* 55 million years. These are the oldest known passerines by about 25 million years, and, consistent with the idea of a southern origin for the passerines, they provide a time base for the Australian radiation.

No matter how their ancestors reached Australia, a major radiation of passerine groups took place, many of which are endemic to Australia or have their greatest diversity there. A radiation of such diversity must have had its origin very early in the history of the passerines, and cannot have occurred after the Australian continent drifted near to south-east Asia. It is strange that our views of Australia's marsupial and avian faunas should have been so different for so long. Since the late nineteenth century, the radiation of marsupials in Australia, evolving in isolation into a range of families and species specialized for different habitats and diets in parallel with mammals elsewhere in the world, has been a textbook example familiar to every student of evolution. The pattern of distribution of marsupials in New Guinea and the islands of Indonesia, Borneo, and the Philippines, with the numbers of marsupial species decreasing with distance from Australia, is mirrored by the distribution of passerine species that belong to distinctively Australian families such as the honeyeaters (Meliphagidae), whistlers (Pachycephalidae), and woodswallows (Artamidae).

Amongst these old endemic families is the Maluridae, which has been isolated from its nearest relatives for many millions of years, at least 20 million and possibly much longer. During this time, species have become adapted to live in almost all habitats from the equator to southern Tasmania, a spread of 43 degrees of latitude, embracing tropical rainforests, desert shrublands, sandhills and rocky gorges, eucalypt forests, and coastal swamplands. The family reached New Guinea at a time when it was still connected to Australia and continuous rainforest covered at least north-east Australia and New Guinea. Of the five genera currently recognized, three occur in Australia (*Malurus*, *Stipiturus*, and *Amytornis*) and three in New Guinea (*Malurus*, *Clytomyias*, and *Sipodotus*).

Some species have a very restricted distribution, some are very widespread. Their evolution and present distribution reflects the cycles of more or less arid conditions experienced by Australia over the last few million years. Australia is an arid continent and not surprisingly several species, including the entire genus *Amytornis*, are arid specialists. Largely because Australia does not have the extremes of seasonal climatic change experienced at higher latitudes, the need to escape very cold winters has not encouraged long distance migration. Many Australian passerines are resident in the same place all year round. Such a resident lifestyle has led to the development of territorial group-living in the Maluridae, with most species breeding co-operatively and retaining young in the family group long after they have learnt to forage for themselves. For such small birds, their survival is remarkably high. They have small clutches, but frequently re-nest both after failure and after raising a brood. This ensures that most pairs produce some young in most years. A long life allows for a delayed onset of breeding and also for abandonment of attempts to breed in particularly unfavourable years (Rowley and Russell 1991).

The 'Stray Feathers' sections in early volumes of *Emu* included many accounts of Superb Blue Wrens, nesting in gardens, raising cuckoos, and endless debate about moult and the acquisition of the brilliant blue plumage—whether they were blue all year round or moulted to a dull brown in the winter. Many accounts referred to groups of Superb Blue Wrens and described a brilliant blue male with his harem—everyone assuming that brown birds were all females; Mormon Wren was a name commonly used.

It was not until 1957 when Ian Rowley used colour-bands to identify individual birds that it was recognized that groups of the Superb Fairy-wren (*Malurus cyaneus*) were an example of 'helpers-at-the-nest', a phenomenon first described by Alexander Skutch in 1935. Members of a group in addition to the mated pair were seen to help feed the young, both in the nest and after they had fledged. Most of these helpers were found to be sexually mature males, at least 1 year old. Since then, intensive studies of several species of *Malurus* have demonstrated the remarkable social organization and mating system of these fascinating birds.

All Australian species of *Malurus* and at least two of the three New Guinea species are now known to breed co-operatively. Strong anecdotal evidence indicates that the grass-wrens (*Amytornis* spp.) are also co-operative breeders (see Glossary), but the other genera, *Stipiturus* in Australia and *Sipodotus* and *Clytomyias* in New Guinea, are not so well known. Nevertheless, so many field observations record groups or family parties in those species that co-operative breeding or at least group-living is probably the most conservative assumption. At first the form of co-operative breeding appeared to be that of a group with a single breeding female, her male partner, and one or more adult helpers. Helpers were usually offspring of the group from previous years who had not dispersed and who took part in territory defence, feeding nestlings, and looking after fledglings.

Such straightforward, typical family groups, however, were not the whole story. After 10 years' study of the Splendid Fairy-wren (*Malurus splendens*) near Perth there was enough information on pedigrees to suggest that many breeding birds were paired with close relatives, and that the level of inbreeding was far higher than was known for any other natural population of birds (Ralls *et al.* 1986, Rowley *et al.* 1986), with no obvious deleterious effects. At about the same time, early accounts recording extra-pair fertilization appeared; the trickle of information soon grew to a flood, showing that the genetic parentage of apparent progeny could not be assumed from field observations alone. Whether the individuals considered to be the breeding pair were genetically compatible with their supposed offspring, or whether the real parents lay outside the group, was investigated using allozyme electrophoresis (see Glossary) (Brooker *et al.* 1990). The extraordinary result was that at least 65 per cent of nestlings, and probably more, could not have been fathered by any male in the group attending their nest, but all were compatible with the bird recognized as the breeding female. This was far higher than the level of extra-pair fertilization reported for any other designated monogamous birds at that time. Subsequently, the more powerful molecular techniques of DNA fingerprinting (see Glossary) have found that the same is true of Superb Fairy-wrens; the majority of eggs laid by any female are not fertilized by her own partner, but by one or more of the males from nearby territories or by an older helper male. These broods of assorted nestlings are reared by the female's social partner and the group's helpers, if any. Current interpretation of this system, based on intensive studies of co-operative breeding in fairy-wrens, is that because opportunities to become a breeding female with her own territory are relatively scarce, a female takes the first territory that presents itself. Later, she has more choice of a male to fertilize her eggs. At the same time, males compete with each other for matings with females, rather than mating with only one female partner. The beautiful male nuptial plumage is part of the process of competing for females, and older males acquire their nuptial plumage well before the start of the breeding season, signalling their age and status to prospective female partners.

Field observations of other Australian fairy-wrens suggest that they share the same sort of mating system, but a lot of work remains to be done before we can attempt to explain the evolutionary pathway leading to this state of affairs, and whether it occurs in other genera of the family.

The family Maluridae has endured substantial loss of habitat following European settlement in Australia and New Guinea, with the development of agriculture and the introduction of exotic predators, pests, and weeds. However, most species have managed to survive these changes and to stabilize their reduced populations since clearing of vegetation has virtually ceased in all states except Queensland. No species or subspecies has become extinct in Australia during the past 200 years, although currently 10 taxa are considered threatened: one subspecies of the Purple-crowned Fairy-wren (*Malurus coronatus*), the Mallee Emu-wren (*Stipiturus mallee*), two subspecies of the Southern Emu-wren (*S. malachurus*), the Carpentarian Grasswren (*Amytornis dorotheae*), two subspecies of the Striated Grasswren (*A. striatus*), and all three subspecies of the Thick-billed Grasswren (*A. textilis*).

2

Taxonomy of the family Maluridae

Introduction

This chapter gives a chronological review of the confused malurid classification over the past 200 years, concluding with the most recent classification based on that of Schodde in his 1982 *Monograph of the Maluridae*, with a few changes based on more recent studies. It provides a broad historical account, which mirrors the history of avian taxonomy in Australia, with reference to the most important classifications and authors. We do not give a detailed evaluation of the characters used in these classifications; access to these is through the literature cited. We discuss the many systems of division into genera that have been proposed. New Guinean species have only recently been included in discussions of Australian species, and so we discuss them separately.

The family Maluridae

The continents with which we are so familiar on world maps today have not always been the same shape nor in the same position on the surface of the earth. It is now generally accepted that the time when most vertebrate evolution took place coincided with major shifts in the continental masses following the break-up of the southern super-continent Gondwana. Before continental drift became widely accepted, it was the general view that the bird fauna of Australia had its origins in the Old World, colonizing in a series of waves of immigration through the island chain of south-east Asia, and undergoing secondary radiation in Australia (Mayr 1944*a*; Darlington 1957; Keast 1959). The most diverse groups, such as honeyeaters and parrots, were supposed to have arrived first.

In the last 30 years, evidence for continental drift, the discovery in Australia of very early fossil passerines, and biochemical evidence of relationships between passerine species, genera, and families combine to support the idea of a massive endemic radiation amongst the Australo-Papuan passerines, similar to the well-known radiation of marsupial mammals in Australia. Biochemical evidence supports the recognition of the Maluridae as a distinct group at the family level, belonging to the old endemic Australo-Papuan lineage. No wonder, then, that systems of avian classification firmly rooted in the idea of colonization of Australia from Asia by species derived from Eurasian families should have been so uncertain in their assignment of the malurids.

The first Australian malurid was collected by William Anderson, surgeon and naturalist on Captain James Cook's third voyage, when the ships *Resolution* and *Discovery* called at Adventure Bay, Tasmania, in 1777. Its long, cocked tail suggested to him that it was a wagtail, and he named it *Motacilla cyanea*, as recorded in Cook's account of that voyage (1784). Anderson died, and his assistant William Ellis published an account of the voyage, with a brief description of *Motacilla cyanea* (Ellis 1782). The genus *Malurus* was established by Vieillot (1816) in his *Analyse d'une nouvelle ornithologie elementaire*, based on Ellis' *Motacilla cyanea* as described in Latham's *A general synopsis of birds* (Vol. 3, 1783). John Gould, in his *Handbook to the Birds of Australia* (1865) listed *Malurus* and *Amytis* (=*Amytornis*, the grasswrens) as 'Family ———-?', calling them all warblers. Clearly, the relationships of the malurids to other groups of passerines were not obvious, and several arrangements have been proposed.

In his *Catalogue of the birds in the British Museum* (1879*b*, 1883), Sharpe placed *Malurus* in the family Muscicapidae, while his 1903 *Handlist of the genera and species of birds* (Vol. 4) placed them in the family Sylviidae. This was followed by the 1926 RAOU Checklist Committee, by G. M. Mathews in *The birds of Australia* (1922–3) and his *Systema avium australasianarum* (Part 2, 1930), and by Berlioz (1950), Beecher (1953), and Wetmore in the 1960 version of his classification of birds of the world. Mayr and Amadon (1951) placed the malurids with the acanthizid thornbills in a group of 85 species of 'Australian Warblers' as the subfamily Malurinae within the family Muscicapidae, a large heterogeneous assemblage of insect-eaters, an arrangement followed by Keast (1959, 1961).

Harrison and Parker (1965) proposed that *Malurus*, *Stipiturus*, and the New Guinean malurids belonged to the family Timaliidae, the Eurasian babblers (at that time including the Australian babblers *Pomatostomus*) on the basis of behavioural similarities (largely co-operative breeding). Subsequently, Harrison (1969) determined that the interscapular gap (apterium) originally described by Parsons (1968) in the Superb Fairy-wren (*M. cyaneus*) was common to all Australian *Malurus*, *Stipiturus*, and New Guinean malurids and was shared by *Amytornis* (see Fig. 4.9). He recognized these species as a distinctive group, but retained them as a subfamily or tribe of the Timaliidae (babblers).

Although the name Maluridae was first used by Swainson, he was not specific as to what he included in the family (Swainson and Richardson 1831). Rand and Gilliard in their 1967 *Handbook of New Guinea birds* used the family Maluridae for the Wren Warblers: 'a group of some eighty-five species found mostly in Australia' (p. 346). Likewise, they were not clear what other species they considered belonging to the family; only five New Guinean species were listed, and *Acanthiza*, *Gerygone*, and *Sericornis*, included in Mayr and Amadon's 1951 Malurinae, were included with the Sylviidae. The Basel sequence of families as used in Volume 11 of the Peters' *Checklist of birds of the world* (Mayr and Cottrell 1986) did not include a family Maluridae, nor does Cracraft's phylogenetic classification (1981), which placed a subfamily Malurinae within the Sylviidae.

Work by Sibley on egg-white proteins (1970, 1976) raised doubts about the relationships of Australian passerine species with the Old World families in which they were placed, and provided evidence of endemism and common ancestry which suggested that adaptive radiation and not multiple invasion had produced most of the present diversity in Australia and New Guinea. For the first time, the 1975 RAOU Interim Checklist (Schodde 1975) recognized the malurine wrens (including grasswrens) as a family in its own right, without known relatives, distinguished by the characters discussed by Harrison and Parker (1965) and Harrison (1969), and quite distinct from the babblers. In his 1982 monograph, Schodde, in addition to the interscapular gap, listed two other unique characters of the family: the enlarged auditory bulla (see Fig. 4.11) and the tail, with never more than 10 rectrices (see Fig. 4.2). Recent biochemical evidence suggests that the malurids are distinct from the other small Australian insectivorous passerines (the acanthizids) with which they have at times been grouped (Sibley and Ahlquist 1990; Christidis and Schodde 1991).

Clearly, despite the common name of the family, 'wrens' or 'warblers' to the early settlers of Australia, and fairy-wrens to a later generation, they are unrelated to the true wrens, the Troglodytidae of Europe and America. Nor are they closely related to the Acanthisittidae of New Zealand, another group labelled wrens by expatriate colonists, but placed in the suboscines by Sibley and Ahlquist (1990).

The history of malurid taxonomy

Despite the difficulties of collecting, naming, and classifying a previously unknown fauna over so vast a continent, the Australian avifauna was regarded as an entity and treated as such, especially after the federation of the separate states into the Commonwealth of Australia in 1901 and the formation of the Royal Australasian Ornithologists' Union (RAOU) as a national body in 1900—two events which happily coincided. New Guinea on the other hand was subdivided into three by separate colonial powers, Britain, Holland, and Germany, with collections of birds ending up in many different museums from a wide variety of sources, and their descriptions appearing in five languages in many different journals. Salvadori's *Ornitologia della Papuasia e della Molucche* of 1881, although an account of species known at the time, was in Italian and not widely available. In the last 50 years, several attempts to provide a comprehensive taxonomic list of New Guinean birds have been made—Mayr (1941), Iredale (1956), Rand and Gilliard (1967), Beehler and Finch (1985), Beehler *et al.* (1986), Coates (1985, 1990). So far as the Maluridae were concerned, it was not until 1982 when Schodde produced his monograph on the family that the confusion of New Guinean nomenclature was efficiently sorted out. Because of their different histories, we discuss Australia and New Guinea separately.

Australia

Throughout the twentieth century, the taxonomy of the family Maluridae has undergone many changes—so much so that a history of the various genera used and discarded is necessary before the individual species can be considered. Taxonomic history within the family Maluridae has two themes running in parallel. The first is the exploration of Australia, with the settlement of several new colonies scattered round the edge of a very large continent, and the collection and identification of its fauna, by many dedicated amateur and professional collectors. Specimens were collected from widely separate localities, and named independently, so that different names for the same birds frequently occurred. The second theme is inextricably linked with the history of change in species concepts, particularly in ornithology. The last 10 years of the nineteenth century and the first 20 years of the twentieth century were a time of revolution in animal taxonomy, with a fundamental change in species concepts. The history of this revolution is described in considerable detail by Haffer (1992); Serventy (1950) explored its significance in Australian ornithology, particularly the taxonomic contribution of Gregory M. Mathews.

Most nineteenth century museum ornithologists used simple differences in morphology to define species, and applied them strictly. For them, the species comprised one morphologically defined taxon—a subspecies or a monotypic species (see Glossary) in current terms—frequently described on the basis of only one or two specimens. Such concepts of species were current before the publication of Darwin's theories of evolution.

Increased knowledge about the geographical variation of birds came from nineteenth century exploration, and showed intergradation between many of the narrow morphospecies of the museum workers. The concept of the subspecies developed after the study of large series of specimens found that many of the 'narrow' species described by museum workers were members of a chain of geographical forms comprising one wide-ranging polytypic species (see Glossary). The species was seen to be composed of a series of interbreeding or potentially interbreeding populations, which, however, did not interbreed with populations of another species. Such a species concept was a natural progression from Darwin's and Wallace's theories of gradual evolution and speciation through the differentiation of geographical races. Towards the end of the nineteenth century, the system of trinomial nomenclature developed and was adopted by American ornithologists; in Europe, its development by Hartert of the Rothschild Museum at Tring had most significance for Australia. Hartert's report on the collections from northern Australia by J. T. Tunney (1905*a*) was the first use of trinomials in Australian ornithology, and was greeted with great hostility (Serventy 1950).

In Australia, then, the first comprehensive accounts of the avifauna are those of Gould, in his *Birds of Australia* (1840–8) and its Supplement (1851–69), and his *Handbook to the Birds of Australia* (1865). Gould presented all known forms, and, among the Insessores, recognized three related genera *Malurus, Amytis*, and *Stipiturus,* but was uncertain of their affinities, placing them in the 'Family——?'. All the current species of *Malurus* were there, although some inland and northern forms had not yet been described. Grasswrens and emu-wrens from desert areas were the main forms lacking.

Subsequently, the accepted authority on nomenclature and classification was Bowdler Sharpe's *Catalogue of the Birds in the British Museum* Volumes 1–27 (1874–95), which generally followed the Gould nomenclature for Australian genera and species. This catalogue was an orthodox, conservative work, and was followed by Sharpe's *Hand list of the genera and species of birds* (1899–1909). Differences from Gould's *Handbook* were chiefly on account of perceived priority or as a result of more recent research.

It was at this point that Gregory Mathews began a project to publish a set of coloured illustrations of the birds of Australia. His starting point was a *Handlist of the birds of Australia* in 1908, on which he invited comment, aiming to establish a basis of the right names and geographical distributions for his larger work. This led to his becoming a major influence in Australian ornithology until his death in 1948, and not always an influence for good. N. B. Kinnear wrote in a biography after Mathews' death that 'during his career (he) probably named more races than any other ornithologist, frequently on very slender grounds' (1949, p. 522).

Matthews' 1908 *Handlist* published in *Emu* followed the orthodox lines of Sharpe's list, and was approved by the ornithological establishment of the time; it formed the basis for the 1913 RAOU Check-list. Mathews, meanwhile, underwent a conversion to the radical approach, and in 1912 published *A reference-list to the birds of Australia*, which was a complete break with tradition, in that it was the first attempt to determine all the subspecies of Australian birds (Mathews 1912*a*). This increased the number of Australian named forms from 800 to 1500, but as Serventy pointed out, 'later taxonomic revisions.... revealed that many of Mathews' races were described in a most uncritical way and were of no validity...' (1950, p.261). Thus, the 1913 Checklist gives *Malurus longicaudus, M. cyaneus, M. cyanochlamys*, and *M. elizabethae*, whereas Mathews (1912*a*, 1913) lists *Malurus cyaneus* with nine subspecies, the four species of the Checklist, plus another five.

Mathews' earlier lists (1908*a*, 1912*a*) retained the broad genera of Gould; however, his 1913 list adopted a position he explained in 1912 as 'the recognition of many small, compact, easily defined groups usually encompassed by colour, and the recognition of these as being of generic value' (Mathews 1912*c*, p. 105). For the next 20 years, Mathews followed this philosophy of small narrow genera, with broad species made up of many subspecies. Thus in his 1913 list, the previously known genus *Malurus*, as used by everyone from Gould to Mathews himself in 1912, was replaced by six genera: *Malurus* (the present *cyaneus* and *splendens*), *Hallornis* (white-winged with blue body, the present *leucopterus leucono-*

tus), *Leggeornis* (red-shouldered, the present *lamberti*, *elegans*, *amabilis*, and *pulcherrimus*), *Rosina* (Purple-crowned, *coronatus*), *Ryania* (Red-backed, *melanocephalus*), and *Nesomalurus* (white-winged with black body, the present *leucopterus leucopterus* and *l. edouardi*). The emu-wrens remained as *Stipiturus*, and for the grasswrens (now all *Amytornis*), he followed the division into four genera that he had established in his 1912 list: *Diaphorillas* (Thick-billed Grasswren *A. textilis*), *Mytisa* (Striated Grasswren *A. striatus*), *Eyramytis* (Eyrean Grasswren *A. goyderi*) and *Magnamytis* (Black and White-throated Grasswrens *A. housei* and *A. woodwardi*). In his 1865 *Handbook*, Gould had used *Amytis* for the grasswrens; Sharpe, in his 1903 *Handlist*, used Stejneger's *Amytornis* (1885) for them, having established that *Amytis* had already been used for another species (not a bird); yet Mathews (1912*a*) considered that *Diaphorillas* Oberholser, 1899 was the correct name. Milligan's original description of *Amytornis housei* in 1902 was as *Amytis,* before Sharpe's use of *Amytornis* in 1903, but Hartert (1905*b*) followed Sharpe's 1903 usage in describing *Amytornis woodwardi.* The 1913 RAOU Check-list used *Amytornis* for all the grasswrens known at that time.

From 1913 until Mathews' *Birds of Australia* in 1922–3, there was little change in the generic names used; some new species were described, and Mathews named many more subspecies (discussed in more detail in the individual species accounts). However, for the grasswrens, he reverted to *Amytornis* while retaining *Eyramytis* and *Magnamytis.*

During the preparation of the second edition of the *Official checklist of the birds of Australia* which began in 1915, it was agreed that, in principle, ''large genera' should be adopted, as used by some British Ornithologists and advocated in America' (RAOU 1926, Introduction, p. iv). This was in disagreement with Mathews' *Birds of Australia* (1922–3), which continued to use the narrow genera he had adopted in 1912–3. A solution was reached: 'After considerable discussion, a compromise was effected with Mr. Gregory M. Mathews, who offered to accept large genera if the numerous generic names as used in his later lists were included as subgenera. This offer was accepted.' (p. v). e.g. *Malurus (Leggeornis) lamberti.* The 1926 Checklist thus reduced the number of malurid genera to four: it retained *Rosina* for the Purple-crowned Fairy-wren, but otherwise only *Malurus*, *Stipiturus*, and *Amytornis* were used.

Mathews, however, in his *Systema avium australasianarum* published by the British Ornithologists' Union in 1930, adhered to his small genera, with *Malurus*, *Hallornis*, *Leggeornis*, *Rosina*, and *Ryania* as before. *Nesomalurus* was used for black-and-white fairy-wrens *leucopterus leucopterus* and *l. edouardi*, and included the White-shouldered Fairy-wren *N. moretoni* from New Guinea (*Malurus alboscapulatus moretoni* in Schodde's 1982 monograph). Other White-shouldered Fairy-wrens from New Guinea were included in the genus *Musciparus* Reichenow, 1897. Mathews retained three genera of grasswrens: *Eyramytis* and *Magnamytis* as used previously, except that the Carpentarian Grasswren *dorotheae* was recognized as a full species instead of a subspecies of *Magnamytis woodwardi.*

In 1934, George Mack, ornithologist at the National Museum of Victoria, published a revision of the genus *Malurus.* He recognized 13 species in the genus (considering only the Australian forms), accepting three subgenera, *Malurus*, *Hallornis*, and *Rosina.* He grouped the small, blue-and-white, Australian black-and-white, and red-backed forms in the subgenus *Hallornis.* The largest species, *coronatus*, he maintained in the subgenus *Rosina.* Both these groups lack the prominent erectile ear coverts of the remaining species, which he grouped as the subgenus *Malurus.*

In the last 50 years (1945–95), remote parts of Australia have become much more accessible, and more effort has been directed to problems of intergradation or isolation of species and subspecies. The work of Keast (1958), Mees (1961), Harrison (1972) with the British Museum Hall Expedition, Storr (1973), Ford (1975), and Schodde (1982*b*) has helped to resolve much of the debate about specific and subspecific status in the family Maluridae. In 1975, an *Interim list of Australian songbirds* appeared (Schodde 1975), to bring up to date the names being used for birds in Australia, before the publication of the final Part 2 of the Third Edition of the Checklist (which has not yet appeared). S. A. Parker was responsible for the family Maluridae, and recognized three genera, *Malurus*, *Stipiturus*, and *Amytornis*. The arrangement of species and subspecies was approaching its present form, except for the complex of *M. lamberti*, the chestnut-shouldered fairy-wrens. This has subsequently been resolved by Schodde (1982*b*) and Ford and Johnstone (1991). At last, in 1994, an updated taxonomic list has incorporated the many changes that have resulted from advances in taxonomic methods and the use of molecular techniques for determining relationships in birds (Christidis and Boles 1994).

New Guinea

Even more so than in Australia, malurids in New Guinea were collected from many scattered sites and often named in ignorance of previously named species. This was partly because, in the nineteenth century, New Guinea was divided between three colonial powers, Holland, Germany, and Britain, and partly because many species came from collectors of bird of paradise plumes, via traders, and their origin was not always known exactly. Nor were the affinities of some species with *Malurus* recognized until recently. Because New Guinea birds are not included in Australian check-lists, and have only rarely been included in discussions of nomenclature, we discuss them separately.

The first known malurids from New Guinea were specimens of *M. cyanocephalus* collected by Quoy and Gaimard in north-western New Guinea on the voyage of *l'Astrolabe* in 1827. Although they had earlier collected fairy-wrens in Australia (*M. leucopterus* in 1818 on Dirk Hartog Island and *M. splendens* at King George Sound in 1826), they did not recognize the new bird as a fairy-wren, but identified it as one of the South American group of todies, and in 1830 named it *Todus cyanocephalus*. This is surprising in view of its broad similarity with *Malurus splendens*, the brilliant cobalt blue of the male, with black face and cocked tail. Charles Bonaparte recognized that it was not a tody, and erected a new genus *Todopsis* in 1854. The Broad-billed Fairy-wren was first described as *Todopsis grayi* by Wallace in 1862, from a specimen collected by his assistant Charles Allen. Its broad bill prompted the French ornithologist Oustalet in 1878 to place it in a new monotypic genus *Chenorhamphus* (= 'goose-bill'; see Fig. 4.6) and this was followed by many subsequent authors.

The White-shouldered Fairy-wren (*Malurus alboscapulatus*), living in grassland habitats, is the most common and widespread of the New Guinea malurids. Its habitat and distribution are not continuous, and many isolated populations that differ considerably from one another are separated by the central cordillera. This has led to considerable taxonomic confusion, as each new variation that was encountered was given a new name. From the 11 named forms, Schodde recognizes six on the basis of their isolation from each other and plumage differences.

The Orange-crowned Wren (*Clytomyias insignis*), the only species in a monospecific genus, was originally described by Sharpe in 1879 from one specimen collected in the Arfak Mountains in the Vogelkop of western New Guinea (Sharpe 1879*a*). In 1907, Rothschild and Hartert recognized its affinity with other malurids because of its reduced number of rectrices. The genus has always been recognized as distinct, and was usually placed with *Todopsis* and *Malurus* in lists of species. Ernst Mayr, in his *List of New Guinea Birds* placed it within the subfamily Malurinae of the

Muscicapidae (1941). The presence of an interscapular gap in the spinal feather tract reinforced its affinities (Harrison 1969).

Wallace's Wren was collected in 1860 by Wallace's assistant, Charles Allen, on Misool Island, and described as *Todopsis wallacii* by Gray in 1862. In one of his rare excursions into the New Guinea avifauna, Mathews recognized the significantly different shape of the bill and the almost identical plumage of male and female (Plate 4), and established a new genus *Sipodotus* for Wallace's Wren (1928). Although this genus was not recognized by many later workers (Mayr 1941; Rand 1942; Ripley 1964; Rand and Gilliard 1967; Gilliard and Le Croy 1970), Schodde in his 1982 monograph separated Wallace's Wren as a monospecific genus, for which Mathews' name *Sipodotus* was available, and to which he gave the vernacular name tree-wrens.

In his 1982 monograph, Schodde considered the Australian and New Guinea malurids together as part of the one radiation, and attempted to follow the evolution of the family, an approach never before taken. In the course of this he concluded that the Broad-billed *(M. grayi)* and Emperor Fairy-wrens (*M. cyanocephalus*) belong to the genus *Malurus,* along with the White-shouldered Fairy-wren, and included two monospecific genera *Sipodotus* and *Clytomyias* as malurids. This arrangement has been followed by recent authors (Beehler *et al.* 1986; Coates 1990).

Conclusion

The classification of the family Maluridae in Schodde's 1982 monograph and his proposed phylogeny were based on morphological studies. This taxonomic arrangement is largely supported by later, preliminary biochemical studies. As yet there is no complete picture of the relationships between all the species of the Maluridae based on DNA comparisons. Sibley and Ahlquist (1982) compared the DNA of *M. lamberti* only with that of *M. splendens, M. alboscapulatus, Stipiturus malachurus, Amytornis textilis,* and a variety of other Australian passerines. The branching pattern that they determined confirmed the Maluridae as a distinct group, but too few species were investigated to show relationships at the species level, or to support any particular grouping of species into genera. Similarly, the allozyme electrophoresis data of Christidis and Schodde (1991) did not examine relationships between genera or species. More recent biochemical analysis (Christidis and Schodde, in press) has identified species groups within *Malurus* but has not yet clearly defined relationships between groups. It suggests that the present taxonomic arrangement of genera and species is appropriate, but the detail of relationships between species may require amendment when biochemical studies of a complete range of forms has been completed.

For these reasons, and because Schodde (1982*b*) does not use them, we have not discussed further sub-families or sub-genera which were largely confusions raised by Mathews.

3

Environment, biogeography, and evolution

Introduction

The family Maluridae occurs only on the continent of Australia, the island of New Guinea and their offshore islands. In Australia, current taxonomy recognizes three genera: the fairy-wrens (*Malurus*; with nine species), the emu-wrens (*Stipiturus*; three species), and the grasswrens (*Amytornis*; eight species). New Guinea also has three genera, fairy-wrens (*Malurus*; three species), and two monospecific genera, the tree-wren (*Sipodotus*) and the russet-wren (*Clytomyias*). The present distribution of species and the isolation of populations which then diverged sufficiently to be recognized as subspecies are the result of the past geological history of Australia and New Guinea. The drift of the continental plates to their present positions, the formation of the land masses identified as Australia and New Guinea, the changes in sea level and climate that have resulted from northward drift, have all influenced the evolution of the environments, soils, flora, and fauna of the Australasian region.

In the Mesozoic geological period 160 million years ago, Australia was part of Gondwana, a large continent that gave rise to Australia, Antarctica, India, Africa, and South America. During the Cretaceous, 144–66 million years ago, Gondwana gradually broke up and the various plates and their continents drifted apart. India was the first to begin the move northwards, about 140 million years ago. Then Africa began to break away at about 120 million years ago, The New Guinea–Australian plate began to separate from Antarctica about 95 million years ago, and by 65 million years ago, may have separated about 200 km from Antarctica. By the mid-Tertiary (*c*. 35–30 million years ago), New Guinea did not exist in its present form; its present southern lowlands were part of the northern edge of the Australian plate, and what is now the northern coast was a chain of islands.

Australia drifted northwards away from Antarctica, at first slowly, and then more rapidly, colliding with south-east Asia between 15 and 3 million years ago, with the formation of the highlands of New Guinea. Between 5 and 2 million years ago, most of New Guinea had emerged from the sea and probably formed the northern part of the Australian land mass. As the Australian plate moved north, the northern edge began to interact with the southern edge of the Indonesian arc, providing avenues for dispersal of plants and animals (Audley-Charles 1987; Burrett *et al.* 1991; Veevers 1991).

Environment and biogeography

Past climates and vegetation

As Australia drifted north after breaking away from Gondwana, its climate varied with changes in latitude and global climate because of altered atmospheric and oceanic circulations. About 65 million years ago, the climate was temperate, mostly warm and moist, with high rainfall in southern Australia, including the interior. The vegetation was more or less uniform subtropical rainforest of Gondwanan origin, with the Southern Beech (*Nothofagus*) widespread. These conditions continued for about 30 million years, but then the global climate cooled progressively with glaciation at the poles; cold arid glacial periods alternated with warm wet episodes. In Australia, the general cooling trend was partly offset by northward drift; the previously humid climate became drier, especially in the central part of the continent. The vegetation became more open, and characteristic 'Australian' elements such as *Eucalyptus* and *Acacia* became established, although wetter forest persisted peripherally. By 15 20 million years ago, northern Australia had reached tropical latitudes, but this coincided with a rapid expansion of polar ice-caps, so that conditions became colder, rainfall decreased, and Australia became increasingly drier. Large parts of northern and central Australia progressively became desert and semi-desert, and the climate in southern Australia alternated between humid and semi-arid (BMR Palaeogeographic Group 1990; Frakes and Vickers-Rich 1991).

In the last 5 million years, there have been wide fluctuations in climate, with some warmer, wetter periods, but in general, the overall aridity has continued. A major change in vegetation occurred as the older rainforests dominated by Southern Beech were replaced by *Eucalyptus*, *Acacia*, and *Casuarina*, with grasses becoming increasingly important. Extensive aridity developed, and during the last 1 million years, there have been several cycles of extreme aridity, the most recent 15–20 000 years ago. This was a critical period for the development of the modern fauna. About 4 million years ago, fossil faunas of central Australia included grassland species, such as grazing kangaroos; lungfish, crocodiles, and flamingoes indicated reliable water bodies, but by 10 000 years ago, these had disappeared from arid and semi-arid regions (Hope 1982; Truswell *et al.* 1987; Frakes and Vickers-Rich 1991).

For much of its history, New Guinea was more closely connected with Australia than with any other land mass. Up to the time when immigration from the north became possible, they shared the same avifauna. Since then, New Guinea has received a proportion of its present-day avifauna by the immigration of paleotropical species from south-east Asia. Nevertheless, 144 of 174 passerine genera occurring in Australia and New Guinea are either endemic to the region or centred in it. Schodde and Calaby (1972) argue that a large part of the New Guinea avifauna shares its origin with the Australian avifauna; the family Maluridae is part of that original avifauna. In particular, the rainforest fauna of the high mountains of New Guinea (more than 1200–1500 metres above sea level) contains a high proportion of old endemics, including the Orange-crowned Wren (*Clytomyias insignis*). Schodde (1991) argues that this montane avifauna represents the earlier fauna of Australasian origin, in its original Gondwanan rainforest habitat which persists in the high mountains. Although the lowland rainforest is predominantly of Indo-malesian origin, its avifauna is mainly of Australasian origin, with relatively few palaeotropical additions. Within the last 2 million years, Antarctic glaciation has led to lowering of sea levels and the development of another land connection between Australia and New Guinea. The associated drier climate favoured the spread of open eucalyptus woodlands and grasslands into southern New Guinea, with their associated avifauna (Schodde and Calaby 1972).

Refugia and speciation

During arid phases throughout the last 2 million years, certain parts were less arid than the rest, and the surviving bird fauna of the originally widespread rainforest retreated to these refugia. During wetter phases, the coastal forests appear to have expanded inland, and with the return of drier conditions in the next phase of aridity, they retreated towards the coast again. These refugia played an important part in the evolution of species. Isolated populations diverged genetically, with the potential to form new species. If, during an expansive phase, these populations came into contact and were sufficiently distinct to avoid hybridization, they might extend as new species into each other's ranges (in sympatry, see Glossary). With the progressive drying of Australia, the most significant refugia have been habitats that remain relatively wetter in dry periods. In addition to coastal areas, mountain regions have served as refugia (Fig. 3.1), including the ranges of central Australia such as the Macdonnell, Petermann, and Flinders Ranges, and patches of rainforest in the Great Dividing Range of the east coast (Keast 1961; Schodde 1982*a*; Ford 1987).

The isolation of populations requires barriers to movement between them. For small birds such as malurids, barriers have generally been areas of greater aridity—for example, patches of desert separating patches of forest. Other barriers to movement were imposed by changes in sea level, such as the opening and closing of sea barriers between Australia and New Guinea in the north and Tasmania in the south. Populations of malurids on small islands off the west coast of Australia have been separated since the last rise in sea level, about 7000 years ago, and have differentiated sufficiently to be interpreted as subspecies, for example, *Malurus leucopterus leucopterus* on Dirk Hartog Island and *M. l. edouardi* on Barrow Island.

The highly dissected topography of New Guinea, with northern and southern lowlands separated by the main cordillera, isolated mountain ranges, and isolated high mountain areas separated by intervening areas of lower altitude and different habitat, has led to patch-

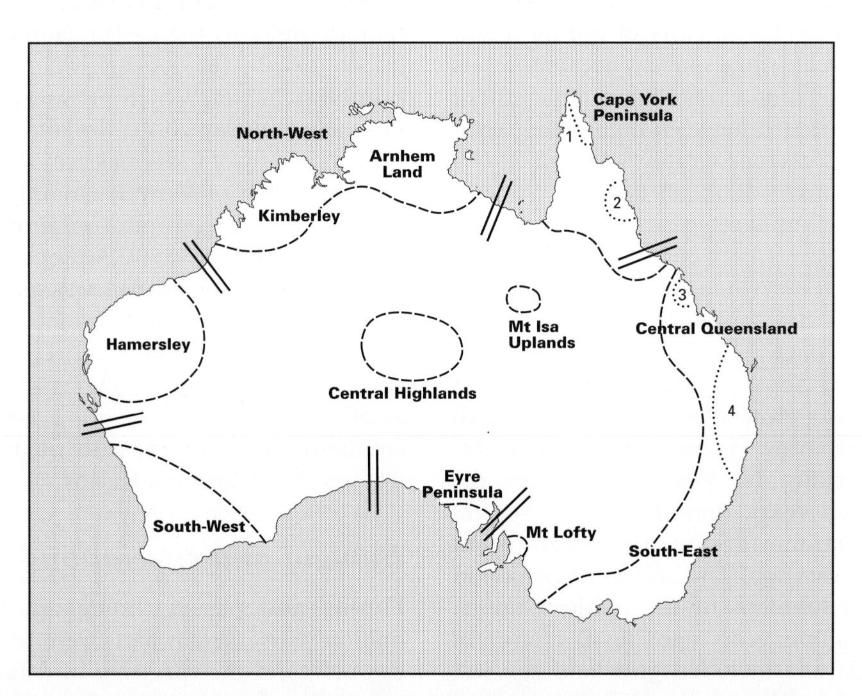

3.1 Major refuge areas and geographical barriers in mainland Australia (after Ford 1987). The numbers 1 to 4 denote rainforest refuges. Parallel lines denote barriers. The extreme highland areas of rainforest refuges 2 and 4 also act as barriers between lowland areas to the north and south, and in the south-east, the Great Dividing Range is a barrier between the east coast and the inland areas to the west.

iness of distribution and isolation of populations with consequent speciation. During the last 5 million years, glaciations further contributed to the development of isolated species and subspecies, with rainforest restricted to higher, moister levels, isolating populations on different mountain peaks. A return to moister conditions allowed rainforests and their faunas to spread (Diamond 1972; Schodde and Hitchcock 1972).

Present climates

In Australia and New Guinea, climate is a major factor affecting the distribution of birds both through direct effects on individuals (extremes of heat, cold, and aridity) and indirect effects on vegetation and insect activity. Australia is a large continental mass spanning a wide range of latitude and longitude with no one generalized climate (Fig. 3.2). The most southerly point of the main Australian land mass is 38 degrees south, with the small area of Tasmania extending to 44 degrees south. Thus southern Australia is at a similar latitude to southern Spain or San Francisco. In the north, about one-third of Australia lies north of the Tropic of Capricorn.

Australia's location in the subtropical high pressure belt results in most of the continent having clear dry air and plentiful sunshine for much of the year. This, along with Australia's low altitude and greatest east–west extent near the Tropic of Capricorn, makes it the hottest continent in terms of duration and intensity of heat. Frosts are common in the cooler months but are rarely severe or prolonged. Only a very small proportion of Australia is cool enough to receive significant snowfalls. About 80 per cent of Australia has a median rainfall of less than 600 millimetres; 50 per cent has less than 300 millimetres a year and is unsuitable for agriculture, but is used for extensive pastoral activities. Variability of rainfall is high over most of the continent, greatest in the centre, with areas of lowest variability in the extreme south-west, south-east, and Tasmania. The arid interior of the continent is bordered by a tropical, monsoon climate in the north and by a predominantly 'Mediterranean' climate (hot, dry summers and cool to mild, wet winters) in the south. Seven main climate types can be distinguished (Fig. 3.2).

New Guinea lies between 10 degrees south and the equator. It is a large island, third in size after Australia and Greenland, about 2400 kilometres long and 680 kilometres wide, with an area of about 800 000 square kilometres. Its main topographical feature is the central mountain range running from one end to the other with many peaks of 4000 metres or more, and a permanent snow line at *c.* 4700 metres on Mt. Carstenz (4884 metres). Much of New Guinea is mountainous, but a large part of the area north and south of the main cordillera is flat and low-lying.

New Guinea lies wholly within the tropics, and its climate is influenced by the south-east tradewinds from May to October and the north-west monsoon from December to April. Although most localities receive some rain in most months, two seasons are recognized, a 'dry' season from June to November and a 'wet' season from November to April. Annual rainfall ranges from less than 1000 millimetres (Port Moresby) to more than 7000 millimetres, with averages of 2500–3000 millimetres most usual. The coastal climate is warm and humid, temperatures ranging from 20 to 34°C; the upland climate is cooler, particularly during the austral 'winter'.

Vegetation—Australia

The distinctive appearance of most Australian landscapes is due to a few widespread genera such as *Eucalyptus* and *Acacia*, whose trees or shrubs dominate the natural vegetation of about 75 per cent of the continent. Most of the *c.* 500 species in the genus *Eucalyptus* are endemic to Australia (Fig. 3.3 (b, c, d); see last page of the colour plates). They generally have sparse foliage with vertically hanging, sclerophyllous (see Glossary), evergreen leaves, and range from trees over 100 metres tall in the high rainfall forests to multi-stemmed shrubs

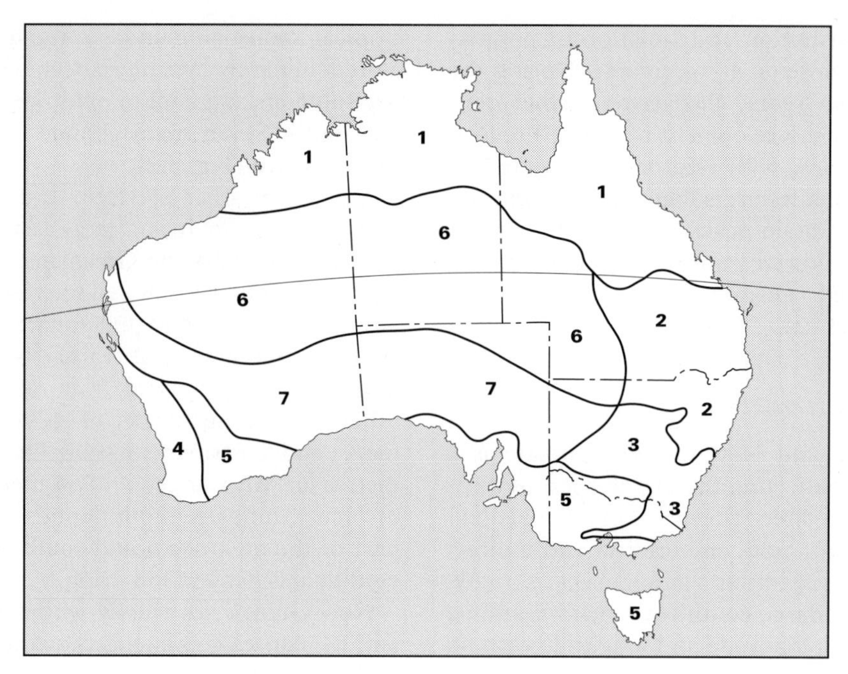

3.2 Australian climate types and their characteristics (from Castles 1994). Temperatures: <10°, cold; 10°–20°, mild; 20°–30°, warm; 30°–40°, hot; >40°, extreme. Arid: less than 300 mm median annual rainfall.

	Seasonal characteristics	
Climate	*Summer*	*Winter*
1. Summer rainfall Tropical	Heavy periodic rains, heavier in coastal and highland areas Hot generally Humid in coastal areas	Generally rainless Mild to warm Dry
2. Summer rainfall Subtropical	Heavy periodic rains, heavier in coastal and highland areas Mainly hot Humid in coastal and highland areas	Some significant rain Mild
3. Uniform rainfall Temperate	Mainly reliable rain Warm to hot	Mainly reliable rain Cool to cold
4. Winter rainfall (moderate to heavy) Temperate	Irregular rain, mostly light Warm to hot	Reliable rain Cool to mild
5. Winter rainfall (mainly moderate) Temperate	Mostly light, irregular rain Warm to hot	Reliable rain Cool to cold
6. Arid (mainly summer rain) Subtropical	Variable rain Hot to extreme Very dry	Mainly irregular light rain Mild to warm Dry
7. Arid (winter or non-seasonal rain) Warm Temperate to Subtropical	Very irregular rain Hot to extreme Very dry	Variable rain, mainly light Cool to mild Dry

1–4 metres tall, the 'mallees' (see Glossary) of harsh coastal or semi-arid environments. Although the genus *Acacia* also occurs elsewhere, in Australia there are more than 700 species, from forest trees 10 metres tall to small heathland shrubs. Acacias form the dominant component of much arid and semi-arid shrubland and many produce large edible seeds. The endemic prickly Hummock Grasses (*Triodia* and *Plectrachne*) form a prominent understorey over more than 25 per cent of the continent and are the primary habitat of several species of malurid (Fig. 3.3(e)); see last page of the colour plates). The mass of repeatedly branching stems bristles with long spine-like leaves and forms a mound up to 1–2 metres high. Hummock Grasses are commonly called 'spinifex' because of a superficial resemblance to the creeping coastal grass *Spinifex* (Anon 1990).

The amount and structure of vegetation depend on the seasonal distribution, quantity, and effectiveness of rainfall, and on soil structure and fertility. Forests and woodlands are restricted to the higher rainfall areas of the periphery, while the arid centre of the continent is sparsely vegetated. Nevertheless, compared with other areas of the world with similar low rainfall, Australian 'deserts' are relatively well vegetated, with a cover of grasses and shrubs that fluctuates with the variable rainfall.

Australian vegetation is floristically very diverse and too complex to describe here in terms of the species present. To describe animal habitats, we use a classification based on the structure of the plant community, the growth form of the tallest layer and its projective foliage cover (the proportion of ground directly beneath the foliage; see Fig. 3.4). For example Closed Forest has the tallest layer in the vegetation with trees 10–30 metres high and a continuous canopy (70–100 per cent cover, generally = rainforest); trees further apart (30–70 per cent cover) are Open Forest; more open still (10–30 per cent cover) is Woodland; and scattered trees (0–10 per cent cover) form Open Woodland. Similar categories are also applied to shrublands and heathlands. This classification is used in describing habitats throughout the accounts of malurid species that follow (Anon 1990).

We have not illustrated in detail the complex mosaic of vegetation types on the Australian continent. Instead, we describe briefly the main types of vegetation found in the eight regions into which Beadle divided the Australian continent in his 1981 analysis of the Australian vegetation (Fig. 3.5).

Changes since human settlement

The effect of humans on the flora and fauna of Australia has been profound. For thousands of years Aborigines manipulated vegetation patterns through the use of fire, and the period since their advent has been one of great

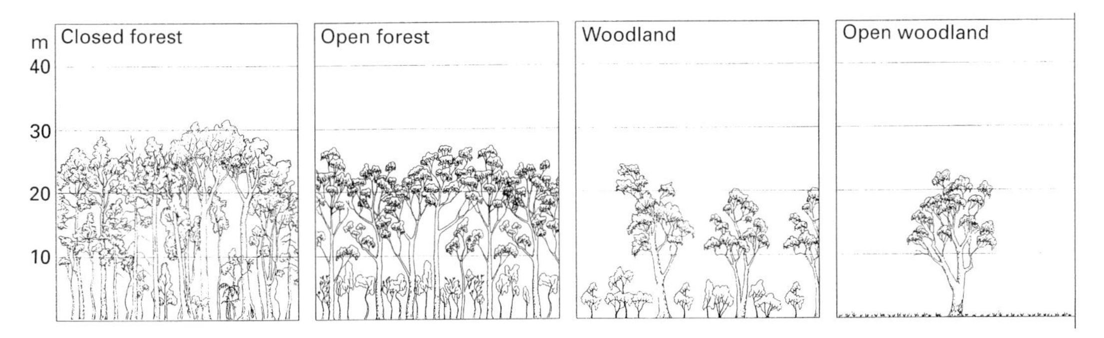

3.4 Vegetation structure—examples of the structural classification used for describing bird habitats in this book, using vegetation where the tallest layer is of trees 10–30 metres tall. (Based on Anon. 1990.)

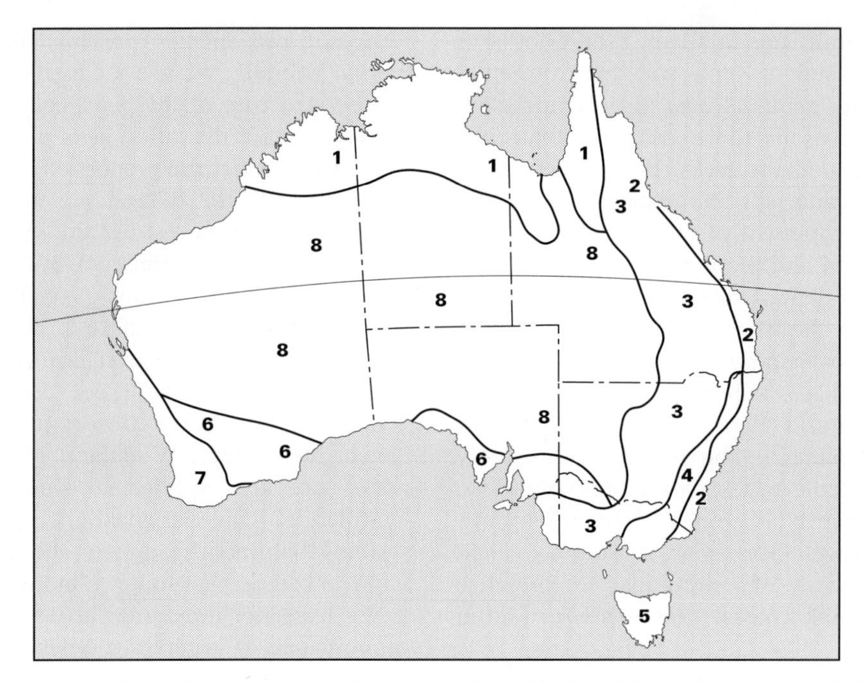

3.5 Australian vegetation. The main types of vegetation found in the eight regions into which Beadle (1981) divided the continent.

1. *Wet–Dry Tropics. Eucalyptus* forests and woodlands, with grassy understorey; small patches of rainforest. Rivers fringed by rainforest, *Eucalyptus* or *Melaleuca* forest, with *Pandanus aquaticus* along the edge (Fig. 3.3(a); see last page of the colour plates). In the drier south, woodlands lower and more open.

2. *Eastern Coastal.* Great Dividing Ra. and E slopes. Most fertile soils support rainforests, largely cleared; a few extensive patches remain, especially in N. Less fertile soils, *Eucalyptus* open forest. Summer rainfall areas in N with grassy understorey, in S, understorey of low trees and shrubs. Least fertile soils support heath or scrub, often swampy.

3. *Eastern Inland.* W slopes of Great Dividing Ra. and inland plains. *Eucalyptus* woodlands, grassy understorey in N, low trees and tall shrubs in S. Mainly cleared for agriculture.

4. *SE Highlands.* Cold winters, with snow; treeless alpine or sub-alpine communities; lower altitudes, open *Eucalyptus* forest or woodland, largely cleared for grazing.

5. *Tasmania.* Gondwanan Southern Beech (*Nothofagus*) cool temperate rainforest in W. Drier E and centre, *Eucalyptus* open forests, some alpine, treeless communities.

6. *The Mallee.* Tall shrubland in S, dominated by many-stemmed 'mallee' eucalypts (Fig. 3.3(c)). Annual rainfall 200–450 mm; much modified for agriculture. Dense low shrubland understorey, in drier areas more open, with spinifex or saltbush.

7. *South-western.* Low coastal shrubland of diverse heath on poor sandy soils. Tall open *Eucalyptus* forests in wettest areas (Fig. 3.3(d)); with lower rainfall, Open Forest gives way to Woodland (Fig. 3.3(b)) and heathy shrubland; understorey low shrubs.

8. *Semi-arid and arid areas.* Not treeless deserts; vegetation determined partly by season of annual rainfall, partly by soil. *Acacia* replaces *Eucalyptus* as dominant tree; *Acacia* woodlands grade into *Acacia* or mixed shrublands; in drier areas replaced by Hummock Grasslands (spinifex, Fig. 3.3(e)). In N and NE

(*Continued on page 21*)

3.5 *(Continued from page 20)*

with summer rainfall, Tussock Grasslands of Mitchell Grass (*Astrebla*). In drier S and SE, low shrublands of saltbush and bluebush (Fig. 3.3(f)), mainly on calcareous and saline soils. Largest area Hummock Grassland, including Great Sandy, Gibson, and Great Victoria Deserts, with sparse trees, particularly Mulga (*Acacia aneura*). Most arid is NE of Lake Eyre; sand dunes, lower slope with spinifex, cane-grass (*Zygochloa*) on dune crests and upper slopes.

change in the Australian fauna. European settlement brought even more rapid, fundamental, and permanent change to vegetation cover. Agriculture and forestry continue to encroach, albeit more slowly, on areas of more or less natural vegetation, and the introduction of exotic flora and fauna has had massive effects on native ecosystems.

Agriculture, forestry, and the grazing of stock now uses nearly 70 per cent of the continent. Although, superficially, maps may suggest that vegetation over much of this area has not been significantly altered apart from urban and cleared arable land, almost all the lands used for grazing have been modified by exotic herbivores—sheep and cattle, rabbits, horses, donkeys, camels, and goats. In particular, understorey vegetation, so important to malurids, is eliminated and there is little regrowth.

Although approximately 5.4 per cent (41.5 million hectares) of Australia is now set aside for flora and fauna conservation in reserves, many ecosystems in areas highly valued for agriculture are inadequately represented.

Vegetation—New Guinea

A large part of New Guinea is still covered with forest, although extensive logging operations are now in progress. The highlands of New Guinea have been inhabited for many thousands of years and the practice of shifting agriculture has resulted in considerable reduction of mid-montane forest. Schodde (1984*b*) considers six principal habitats for birds and we review them briefly here, amplified with reference to Paijman's (1976) classification of the major environments and vegetation types in New Guinea.

Marine and shoreline

Mangrove forests and woodlands are extensive in estuaries and along protected coastlines.

Lowland freshwater swampland

Ranges from open water to swamp forests, chiefly along the major rivers (e.g. the Fly, Sepik, and Mamberamo–Idenberg). Deeper water has floating mats of aquatic vegetation, and extensive areas of grassy swamps may be intermittently dry. Woodlands or forests may grow in permanent or seasonally dry swamps.

Man-made grasslands

Widespread in lowland and mid-montane regions, replacing forest after centuries of repeated gardening and burning.

Eucalyptus-Melaleuca woodlands

An extension of the *Eucalyptus* woodlands in northern Australia, with an understorey of grasses beneath scattered trees. They occur in areas of relatively low, seasonal rainfall, mainly in the lowlands of southern New Guinea.

Alpine meadows

Alpine meadows occur in the central cordillera above the tree line from *c.* 3200 to 4100 metres.

Rainforest

The majority of New Guinea was originally covered with rainforest, which still clothes at least half of the eastern part; 70–75 per cent of the avifauna are essentially rainforest inhabitants. Undisturbed (primary) rainforest is different from secondary rainforest, found wherever rainforest has been disturbed, partly cleared or is regenerating. The strata of a rainforest provide different habitats for birds—some use the crowns of trees, some use the vegetation at lower levels and the forest floor:

1. Lowland forest is very rich in plant species and diverse in structure, with canopy trees usually 30–35 metres high with emergents to 50 metres, many with extensive buttresses at their bases. The understorey is layered with a taller component of palms and gingers, and a sparse lower layer of herbs, ferns, seedlings, forest grasses, and sedges. Climbers, epiphytic ferns, and orchids are common.

2. Forest in foothill country and mountains to 1000 metres is lower, has fewer trees with large buttresses, fewer climbers, but more saplings and treeferns, and a denser herb layer.

3. Lower montane forest (1000–3000 metres) may be mixed forest, or dominated by oaks, Southern Beech, or conifers (above 2400 metres). Canopy height (20–30 metres) is lower, with a large number of low-branched trees and a dense layer of *Pandanus*, shrubs, and bamboos. The ground and fallen logs are covered with moss, and at higher altitudes, epiphytes are abundant, and the forest is referred to as 'moss forest' or 'cloud forest'.

4. Upper montane forest (above 3000 metres) is a low forest with dense trees of relatively few species, including conifers, from about 18 metres tall at 3000 metres to about 6 metres at the tree line (*c.* 3900 metres). The shrub layer has tree ferns, but no *Pandanus* or bamboos.

Evolution

The origin of Australian passerines

The northward drift of the Australian continental plate occurred over the last 100 million years, the period when evolutionary radiation produced the avian families and species that we know today. This raised two major problems for the idea of immigration from south-east Asia that had dominated thinking for the previous 100 years. First, it suggested the possibility of other routes into Australia via Antarctica. Second, Australia began its drift north from more than 60 degrees south, and significant links with south-east Asia were not possible until about 40 million years ago. How much exchange of species occurred over the period from 40 to 15 million years ago is not yet clear (Audley-Charles 1987; Burrett *et al.* 1991).

As the theories of colonization from the north became less tenable, biogeographers began to ask if a southern origin was possible for some modern Australian families (Cracraft 1972; Rich 1975; Schodde 1982*a*, 1991; Boles 1991). Evidence for a Gondwanan origin included the large number, richness and diversity of endemic species, and the occurrence of their closest relatives only in the islands immediately to the north. Fossil evidence reviewed by Olson (1988) suggests that most of the major orders of birds, including some now endemic to the southern continents, are relicts of formerly widespread groups that occurred in the northern hemisphere during the mid-Tertiary. However, Olson identified at least five widespread orders that he believed probably originated in the southern hemisphere, and did not become established in the northern hemisphere until about 25 million years ago: grebes, ducks, pigeons, parrots, and passerines.

In support of a Gondwanan origin for the passerines, Olson cited the greater diversity of primitive passerines such as the Rhinocryptidae and Acanthisittidae in the southern hemisphere. A fossil passerine 15–20 million years old has recently been found in South America, with affinities to the suboscines (Noriega and Chiappe 1993; see Glossary). In the northern hemisphere, the earliest undoubted passerine fossils come from deposits in France *c.* 25 million years old. Until recently, the age of fossil passerines found in Australia was comparable; however, some much earlier passerine material has recently been found dated at *c.* 55 million years old (Boles 1991, 1995, 1997). Two different forms were present, comparable in size to a grassfinch and a thrush, and are the oldest known passerines by almost 25 million years, consistent with a southern origin for the Order Passeriformes. An Australian mid-Miocene specimen (*c.* 15 million years old) is identifiable with the logrunners *Orthonyx*, one of Australasia's most distinctive living passerine genera. This takes the origin of that genus back to a time when an origin from the north is doubtful (Boles 1993), supporting the view that the modern genera of birds were distinct by the Miocene (Feduccia 1995).

Biochemical evidence from molecular biology has also challenged the view of passerine immigration to Australia from the north. The DNA–DNA hybridization studies of Sibley and Ahlquist (1985, 1990) suggested a major dichotomy in the oscines between a group including the major Australo-Papuan lineages that they referred to as the Corvida, and a group centred in Eurasia, Africa, and North America, called the Passerida. The proposed Corvida comprised three primary lineages—lyrebirds and allies (Menuroidea), honeyeaters and allies (Meliphagoidea), and the crows, monarch flycatchers, and allies (Corvoidea). The first two of these superfamilies are all but endemic to Australia and New Guinea and the last, although cosmopolitan, is at its most diverse in Australasia.

Not surprisingly, such a profound reorganization of avian classification has met with criticism from many quarters, and is not accepted by all. However, in its broad outline, the proposed relationships support those previously suggested by morphology and are in turn supported by other investigations carried out using different biochemical methods such as measuring immunological distance (Baverstock *et al.* 1991), allozyme electrophoresis (Christidis and Schodde 1991), and cytochrome *b* (Sheldon and Bledsoe 1993). In consequence, there is enough agreement to support the idea of an endemic radiation amongst the Australo-Papuan passerines, consistent with the available fossil evidence and similar to the well-known radiation of marsupials in Australia.

Analysis of Sibley and Ahlquist's data by Harshman (1994), using different mathematical techniques, has provided support for the distinctness of the Australo-Papuan Corvida from the Passerida. The Passerida includes less than 15 per cent of Australian passerines, mostly members of widespread Eurasian genera. Table 3.1 lists Australian passerine families as currently recognized, divided into two groups, the 'old endemics' (268 species)

Table 3.1 Families of Australian passerines, grouped as those of the older Australian radiations (Parvorder Corvida of Sibley and Ahlquist 1990, 88 genera, 268 species) and families arriving more recently as Australia drifted north (Parvorder Passerida, 23 genera, 42 species).

Old endemic families	*Later arrivals*
Pittidae (1, 3)	Alaudidae (1, 1)
Menuridae (1, 2)	Motacillidae (2, 2)
Atrichornithidae (1, 2)	Passeridae (8, 18)
Climacteridae (2, 6)	Nectarinidae (1, 1)
Maluridae (3, 20)	Dicaeidae (1, 1)
Pardalotidae (15, 49)	Hirundinidae (2, 5)
Meliphagidae (23, 72)	Sylviidae (5, 8)
Petroicidae (8, 20)	Zosteropidae (1, 3)
Orthonychidae (1, 2)	Muscicapidae (1, 2)
Pomatostomidae(1, 4)	Sturnidae (1, 1)
Cinclostomatidae (2, 8)	
Neosittidae (1, 1)	
Pachycephalidae (4, 14)	
Dicruridae (7, 19)	
Campephagidae (2, 8)	
Oriolidae (2, 3)	
Artamidae (4, 14)	
Paradisaeidae (1, 4)	
Ptilonorhynchidae (6, 10)	
Corvidae (1, 5)	
Corcoracidae (2, 2)	

Figures in parentheses after family names are the numbers of genera and species in the families occurring in Australia and Tasmania (not including introduced species or species on Christmas, Lord Howe, or Norfolk Islands). Family names and included genera as in Christidis and Boles (1994).

and the 'new invaders' (42 species). It is now widely accepted that a major radiation of the passerines occurred when Australia and New Guinea were largely isolated from Antarctica and Asia. It is most likely that after Australia came near to south-east Asia, some new families entered Australia, and at the same time, members of old Australian families moved out to the rest of the world (Schodde 1982*a*). Few of the more 'recent' arrivals have invaded the endemic Australian rainforest habitats. Rather, they occupy the comparatively new grassland, savannah, and arid shrubland habitats that developed as Australia dried out in the last 20 million years. These more recently arrived families generally have their major strongholds elsewhere in the world and have relatively few Australian representatives. As New Guinea was closer to Asia, its complement of species belonging to the Passerida is greater.

Our main concern in this section has been to establish the antiquity and uniqueness of the family Maluridae. Although the schemes of relationship based on different biochemical methods do not agree on their position, these methods do confirm the Maluridae as a distinct group separated at the family level, belonging to the Australo-Papuan passerine lineage.

Evolution of the Maluridae

The key to understanding how the Maluridae evolved is the recognition that an early assemblage of Australo-Papuan passerines existed, 'a radiation of such diversity that it must extend deep into the Tertiary' (Schodde 1991, p. 413). The distinct morphological characteristics of the family Maluridae were recognized by the time Schodde wrote his 1982 monograph. For the first time, he considered the evolution of Australian and New Guinea malurids together, not just that of Australian species as was done by earlier authors (Keast 1961; Ford 1974*a*,*b*, 1978*b*).

Both the DNA–DNA hybridization studies by Sibley and Ahlquist (1985, 1990) and the allozyme studies by Christidis and Schodde (1991) suggest that the Maluridae are part of an assemblage that includes the Australian thornbills, pardalotes, honeyeaters, and chats (Sibley and Ahlquist's superfamily Meliphagoidea). Christidis and Schodde suggest that the Australian robins (Petroicidae = Eopsaltridae of Sibley and Ahlquist), and logrunners and chowchillas (Orthonychidae) are also part of this group, which formed part of the radiation of 'old endemics' in Australia before the continent moved far enough north to receive a substantial input from Asian sources. Because the fossil record earlier than 20 million years ago is so sparse, our only ways of tracing the evolution of the Maluridae are from morphological similarities, biochemical evidence showing which species are most closely related, and biogeographical inference from present patterns of distribution and past environmental changes which separated or joined populations. Much of this discussion has to be somewhat speculative, since fossils are few, and detailed comparative anatomical and molecular studies of the relationships between Australian and New Guinea forms have not been done. Sibley and Ahlquist's 1982 study of the Maluridae included only *Malurus lamberti*, *M. splendens*, *M. alboscapulatus*, *Stipiturus malachurus*, and *Amytornis textilis*. Schodde's 1982 monograph proposed an outline of evolution and relationships within the Maluridae, based on morphological and biogeographical evidence. This was largely confirmed by a later study, based on allozyme electrophoresis (Christidis and Schodde, in press), that clarifies relationships between most genera of the Maluridae and within *Malurus* and *Stipiturus* (Fig. 3.6).

The ancestral habitats of these Australian passerines appear to have been the temperate and sub-tropical Gondwanan rainforests that covered much of Australia through the early and mid-Tertiary. As Australia became progressively drier, the rainforest and its avifauna withdrew to the north and east, finally to pockets on the east coast and in the newly formed mountains of New Guinea. The much better known fossil record of Australian mar-

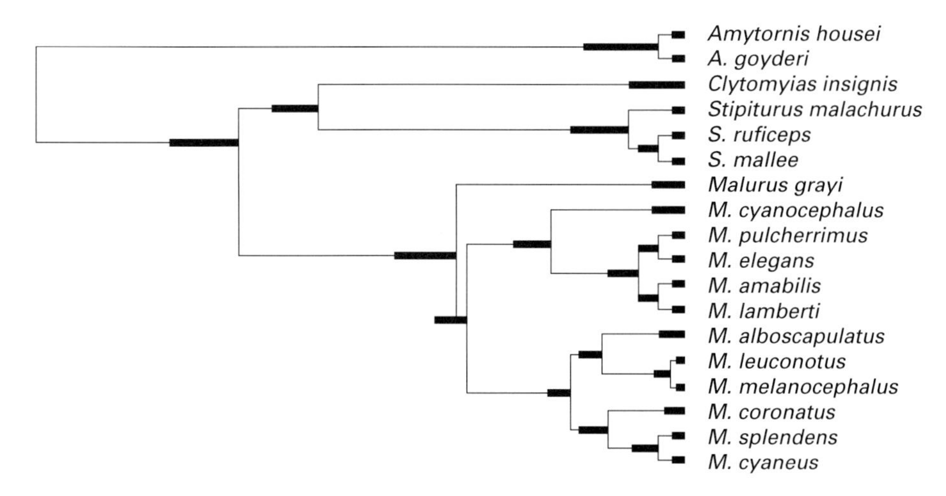

3.6 The relationships between genera and species of the family Maluridae, based on currently known allozyme differences; no *Sipodotus* material available. (From Christidis and Schodde in press.)

supials from the middle Miocene (15 million years ago) shows that there were many 'forest-adapted' forms which closely resemble species present now in the highlands of New Guinea, so that 'a walk up a New Guinea mountain is like a walk back into time' (Archer and Fox 1984, p. 13). Schodde and Faith (1991) draw parallels between the marsupial and bird faunas, showing that avifaunas of New Guinea rainforests and the smaller remnant sub-tropical rainforests of Australia are closely related. Schodde (1982*b*) speculates that the present day *Clytomyias insignis* of New Guinea montane forests comes closest to the ancestral malurid, although it is a specialized form. Diversification from the ancestral form into different niches in its rainforest habitat was presumably the origin of the five different lineages found today (*Clytomyias*, *Sipodotus*, *Malurus*, *Stipiturus*, *Amytornis*) (Fig. 3.6).

Meanwhile, forms adapted to a drier climate and its associated vegetation remained in the rest of Australia. Subsequent speciation leading to the presence of modern species in their current ranges is generally explained by processes of allopatric speciation (see Glossary). This involves the disruption of one continuous range by some climatic or geological event, creating two separate populations that become differentiated to such an extent that they remain distinct if they meet again. Changes in sea level, increase or decrease in effective rainfall, and, in New Guinea, the rising of the mountain spine, can lead to such isolation of populations. Many hypotheses invoking such events (known and unknown) have been proposed to explain present-day patterns of species and subspecies, and we summarize them here (Keast 1957, 1961; Ford 1974*a*,*b*; Schodde 1982*a*,*b*; Cracraft 1986).

Origins of present-day species

Currently, five genera are recognized.

Clytomyias—*the russet-wrens*

The russet-wrens live in the sub-canopy of high montane rainforest in New Guinea (Plate 4). Schodde (1982*b*) suggests that they are an off-shoot from near the root of divergence of emu-wrens and fairy-wrens and since then have persisted in their original habitat. This is largely corroborated by molecular data (Christidis and Schodde, in press)

Sipodotus—*the tree-wrens*

The tree-wrens live an arboreal life in the tangles of undergrowth and vines of the rainforested foot-slopes of mountain ranges throughout New Guinea (Plate 4). Schodde suggests they are another early diverging branch of the malurid line which became isolated in New Guinea rainforests.

Malurus—*the fairy-wrens*

Within *Malurus*, the 12 species fall into five groups, the broad-billed *grayi* group, the emperor *cyanocephalus* group, the blue *cyaneus* group, the bi-coloured *leucopterus* group, and the chestnut-shouldered *lamberti* group, which Schodde suggests radiated during the later Tertiary, 5–20 million years ago (Fig. 3.7). Ancestral Broad-billed and Emperor Fairy-wrens became isolated in northern forest habitats as drier woodland habitats replaced them further south. Ultimately they were isolated in New Guinea when sea levels rose, and evolved in isolation to their present forms. Later, as the mountain spine of New Guinea rose during the late Miocene-Pliocene times, it isolated forms to the north and south which eventually became the subspecies of *M. cyanocephalus* and *M. grayi* (Plate 4).

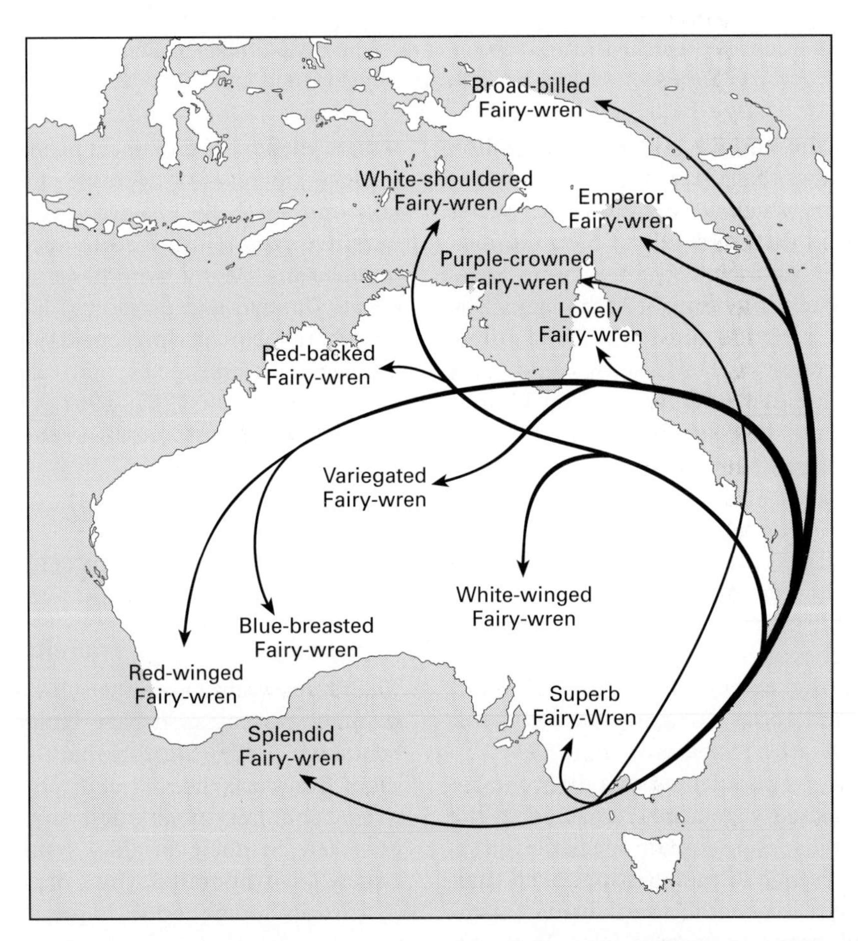

3.7 The radiation of the five main lineages of the fairy-wrens *Malurus* in Australia and New Guinea as suggested by Schodde (1982*b*) and Christidis and Schodde (in press): blue (Superb, Splendid, and Purple-crowned), bi-coloured (White-winged, Red-backed, and White-shouldered), chestnut-shouldered (Variegated, Red-winged, Blue-breasted, and Lovely), emperor, and broad-billed lineages. (Adapted from Schodde 1982*b*, p. 31.)

The other 'blue wrens' now have a mainly southern distribution and presumably evolved from ancestors isolated in southern Australia by the arid centre. This trans-Australian, southern stock was presumably split into south-east and south-west populations, perhaps by the Eyrean Barrier, a shallow trench of sea extending inland from the head of Spencer Gulf to the Lake Eyre Basin. Seas also encroached into the Eucla Basin (at the head of the Great Australian Bight) and the Murray Basin. The eastern enclave gave rise to the Superb Fairy-wren (*M. cyaneus*) and the western to the Splendid Fairy-wren (*M. splendens*) (Plate 1). Schodde (1982*a*) argues that the south-west was always drier than the south-east, and that birds from that area were better adapted to drier conditions and were able to disperse or spread east when cyclical changes allowed, either across the Nullarbor Plain itself, along a coastal corridor of mallee south of the Nullarbor Plain, or through mallee vegetation to the north (Ford 1974*b*). Since the three present subspecies now hybridize freely in areas of overlap, Schodde (1982*b*) suggests that they have diverged only recently, probably when their ancestral stock was last split by an arid glacial period into three—the south-west, the central Australian ranges, and the Murray–Darling Basin, where they evolved in isolation. The present somewhat wetter conditions have allowed the isolates to make contact again and to hybridize.

In the south-east, the Superb Fairy-wrens reached Tasmania at a time when it was continuous with the rest of Australia. When the sea rose at the close of the last glacial period and flooded Bass Strait, fairy-wrens in Tasmania and its islands were isolated from mainland Australia. King and Flinders Island have been separated from Tasmania for not quite so long, and although their fairy-wrens have diverged from the Tasmanian form, Schodde placed them all in the one subspecies *M. c. cyaneus*. Mainland forms have diverged along their range, with the most northerly becoming paler and smaller, but they all remain in contact and intergrade.

An offshoot of the blue wrens was isolated in the forests fringing the northern rivers of Australia as the drier open woodlands replaced rainforests, giving rise to the Purple-crowned Fairy-wren (*M. coronatus*) (Christidis and Schodde, in press). Increasing aridity split the population that was once continuous across northern Australia into two isolates which gave rise to the eastern and western subspecies (Plate 1).

The many attempts to interpret the evolution of the chestnut-shouldered group of fairy-wrens (Plate 2) all take the view that the many species and subspecies must have become differentiated during geographical isolation from each other, and that the present pattern of distribution, with some forms in contact, is secondary (Serventy 1951; Harrison 1972; Ford 1974*b*). Each of the various explanations is coloured by the taxonomic view taken by its author, and for that reason, we prefer the explanation of Schodde (1982*b*), based as it is on the most comprehensive and recent review of the group.

The absence of any chestnut-shouldered fairy-wrens in south-east Australia and the variety of representatives in northern Australia suggest a northern origin. Schodde argues that blue plumage in females is ancestral, and that the browner females of central and southern forms are an adaptation to arid conditions. He suggests that during a warm humid period at the end of the Pliocene or early Pleistocene (about 2 million years ago), ancestral chestnut-shouldered fairy-wrens spread south and became established in the south-west, but not in the south-east, which was colder. When a cold dry glacial period again rendered central Australia inhospitable, it fragmented the ancestral form, isolating a population in the south-west which diverged and eventually gave rise to the Red-winged Fairy-wren (*M. elegans*). Another population isolated in the extreme north-east gave rise to the Lovely Fairy-wren (*M. amabilis*), and a third, in the north-west, was the ancestor of the Variegated Fairy-wrens (*M. lamberti*). Later, another warm, humid cycle allowed the north-west ancestral Variegateds to expand south; in the south-west they met the earlier offshoot *elegans*. Another dry cycle isolated this second wave of chestnut-shouldereds in the south, between Red-winged Fairy-wrens in the south-west and the Eyrean Barrier in the east, and they subsequently gave rise to the Blue-breasted Fairy-wren (*M. pulcherrimus*), whose distribution across southern Australia as far east as the Eyre Peninsula was interrupted so recently that there is little difference between the two populations.

During the last Pleistocene glaciations the Variegated Fairy-wrens were isolated in various refuges where they diverged from each other. In the Kimberley and Arnhem Land Ranges, the ancestral blue females persisted; southern colonies were split by the Great Dividing Range, giving rise to the coastal subspecies *lamberti*, with the inland forms finding refuge in the ranges of central Australia, the Pilbara, and perhaps the Murray–Darling Basin. Since the end of the last glacial period, 12 000–13 000 years ago, these inland purple-backed forms, *assimilis*, have spread out again, making contact with all other forms of chestnut-shouldered fairy-wrens. However, this explanation may need to be modified when more biochemical information becomes available about the timing of the separation between species, particularly with respect to the relationships of Blue-breasted Fairy-wrens (Christidis and Schodde, in press).

The bi-coloured fairy-wrens are quite distinct from the blue and the chestnut-shouldered forms, the males having bi-coloured nuptial plumage of blue or black with red or white on the shoulders, scapulars or back, and plain heads with no ear tufts (Plate 3). Their closest relatives appear to be the 'blue' lineage and their predominantly northern distribution suggests a northern origin. Ancestral bi-coloured fairy-wrens were clearly widespread across the north of the Australian land mass, including the lowlands of New Guinea, where they became isolated during a period when high sea levels separated northern Australia from New Guinea. Later, the New Guinea birds spread along river systems isolated from each other and Schodde suggests that simple genetic mutations controlling female plumage colour occurred more than once to give rise to the complex mosaic now found in the White-shouldered Fairy-wren (*M. alboscapulatus*).

The white-winged member of the bi-coloured group is so clearly arid adapted that it probably evolved in central Australia, perhaps isolated in refuge areas such as the ranges of the Pilbara and central Australia during periods of extreme aridity, and then spread throughout the centre during less arid periods. There is very little variation across its large range, except for two subspecies on islands isolated from the western coast for no more than 8000–12 000 years by rising sea levels.

The tropical grasslands under northern woodlands are the prime habitat of the red-backed member of the bi-coloured group, and Schodde suggests that it evolved there, isolated from ancestral white-winged forms during periods of aridity. Recent aridity during the last glacial periods of the Pleistocene (12 000 years ago or longer) pushed the tropical grasslands to the north and formed arid corridors from the inland to the coast in the north-east (Keast 1961; Schodde and Mason 1980), separating northern and eastern populations of Red-backed Fairy-wrens, which differentiated into the two subspecies now known.

Stipiturus—*the emu-wrens*

Although the emu-wrens *Stipiturus* are widely distributed across a large area of southern and central Australia, they have diversified less than *Malurus* and *Amytornis*, with only three species currently recognized; they have no representatives in northern Australia (Plate 5). Their origin from an ancestral malurid is obscure, although biochemical data indicate that *Stipiturus* and *Malurus* are more closely related to each other than either is to *Amytornis*, and places *Clytomyias* as the closest relative of *Stipiturus* (Fig. 3.6). The Southern Emu-wren (*S. malachurus*) represents the ancestral form from which both inland forms, the

Rufous-crowned (*S. ruficeps*) and Mallee Emu-wren (*S. mallee*) arose (Schodde 1982*b*). In the most likely sequence of events, ancestral emu-wrens were isolated in southern Australia during past glacial periods of the Pleistocene. As the climate improved during the interglacial periods, these ancestral stocks spread outwards, but during later arid cycles populations became isolated in the central Australian ranges and the lower Murray–Darling Basin, and evolved into the Rufous-crowned and Mallee Emu-wrens of today. The lack of variation in the Rufous-crowned Emu-wren across its extensive range suggests that no enclave has been isolated for long.

The alternation of cold, dry, ice-ages with warmer, wetter periods has alternately broken and joined the heathlands across southern Australia, and differences in populations of Southern Emu-wrens may have built up over a number of such cycles, giving rise to the current seven subspecies (Schodde 1982*b*). The populations in southern Australia and Tasmania have been separated since the end of the last glacial period, about 10 000 years ago.

Amytornis—*the grasswrens*

The diversity of grasswrens (see Plates 6–8) in central Australia led Keast in his early study of *Amytornis* (1958, 1961) to suggest a central origin, with a widely distributed parental species the Striated Grasswren (*A. striatus*) budding off a series of species around the periphery. Later events have changed this view. A new species, the Grey Grasswren (*A. barbatus*), was discovered in swamplands in south-west Queensland and northern New South Wales (Favaloro and McEvey 1968). Parker unravelled the history of misidentification that had made the taxonomy of grasswrens so confused (Parker 1972), and the 'lost' Eyrean Grasswren (*A. goyderi*) was rediscovered (Parker *et al.* 1978). Ford's discussions of speciation in birds of arid habitats (1974*b*) also assumed a central origin for grasswrens. However, Schodde (1982*b*) makes a convincing case for a northern origin, proposing that ancestral forms from the north spread into the centre at various times when conditions were favourable, became isolated in refuge areas by periods of increasing aridity, and differentiated there, spreading out from these areas when more favourable conditions returned, and in some cases, making contact with other species (Fig. 3.8).

The strongly marked face of the Grey Grasswren and many fairy-wrens suggests this is an ancestral character. The Grey Grasswren is distinct from all other grasswrens and is perhaps the oldest line (as shown by allozyme differences, Schodde and Christidis 1987), with no obvious close relative. It appears to be a relict species, found only in tussock grassland and Lignum (*Muehlenbeckia cunninghamii*) along rivers and in areas subject to inundation in the Lake Eyre Basin.

The northern grasswrens, Black (*A. housei*), White-throated (*A. woodwardi*), and Carpentarian (*A. dorotheae*), share smaller, unspecialized auditory bullae and slender bills and each occupies the spinifex-clad ranges in its own area (Fig. 3.3(e); see last page of the colour plates). Schodde suggests that a once continuous distribution across northern Australia was interrupted in past times, enabling isolated enclaves of a parental stock to diverge into different species. Spinifex is an endemic Australian group of grasses, probably of Gondwanan origin (Jacobs 1982), whose association with grasswrens may be long-standing. White-throated and Carpentarian Grasswrens also have distinctive white throats and obvious black malar stripes, while the Striated Grasswren has a reduced malar stripe and duller plumage, with far less white on the throat. Schodde suggests that the Striated Grasswren is derived from an eastern offshoot of the northern ancestral forms now represented by White-throated and Carpentarian Grasswrens.

Populations of the Striated Grasswren isolated far apart in the Hamersley Ranges in the Pilbara district of Western Australia (*A. s. whitei*) and in the Flinders Ranges in South Australia (*A. s. merrotsyi*) suggest fragmentation of a more extensive past range

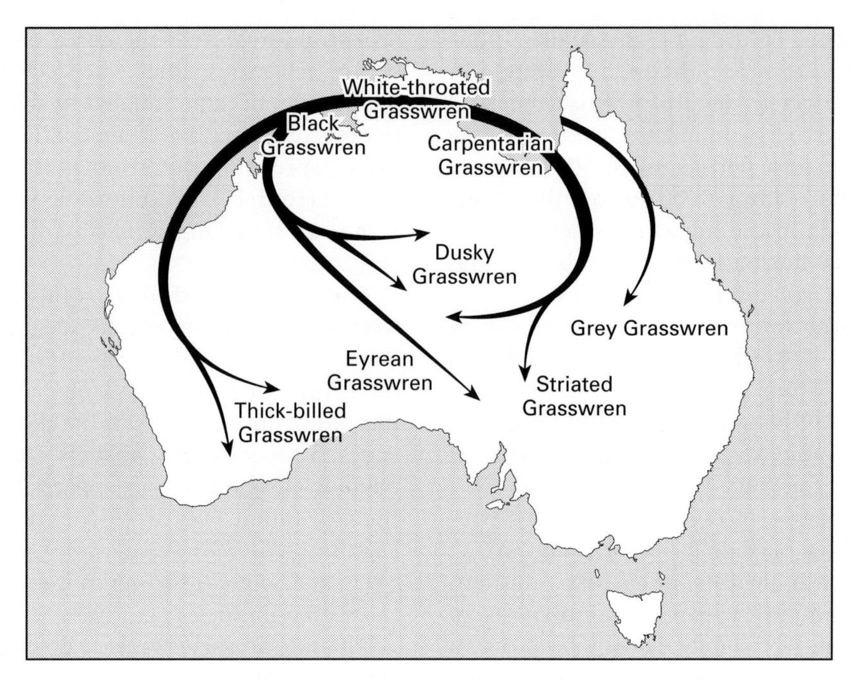

3.8 The radiation of the grasswrens *Amytornis* (after Schodde 1982*b*, p. 143 and Christidis and Schodde, in press): Schodde (1982*b*) suggests that a once continuous distribution across northern Australia has been interrupted and isolated enclaves of the parental stock have diverged into different species, with the *striatus* line (White-throated *A. woodwardi*, Carpentarian *A. dorotheae*, and Striated *A. striatus*) derived from north-east ancestral forms, and the *textilis* line (Black *A. housei*, Thick-billed *A. textilis*, Eyrean *A. goyderi*, and Dusky *A. purnelli*) from the north-west ancestral forms. Association of the *A. goyderi* line with the *textilis* line follows biochemical evidence (L. Christidis, personal communication.)

(Schodde 1982*b*). These two populations live in typical rugged spinifex-clad hills, but the ranges in between, in central Australia, are the home of the Dusky Grasswren (*A. purnelli*) of the *textilis* group. Instead, the habitat of *A. s. striatus* in central Australia is spinifex on sandplain and sand dunes, mostly west of the central ranges. The isolated population of *A. s. striatus* in central Queensland may be a relict of an extension from the western population around the north of the central ranges (Ford and Parker 1974). The origin of *A. s. striatus* populations in western New South Wales is not clear, but the close relationship of these far-flung *striatus* populations is shown by the similarity of song between New South Wales and Pilbara, Western Australia, subspecies (see species account).

Thick-billed (*A. textilis*), Dusky, and Black Grasswrens are darker and more uniformly streaked than *A. striatus*, and the lack of facial patterns and pectoral bands in this *textilis* group led Schodde (1982*a*) to suggest that they are related to the north-western ancestral forms via the Black Grasswren. Biochemical evidence indicates that the Eyrean Grasswren is aligned to this group (L. Christidis, personal communication), contrary to a range of morphological and oological evidence (Schodde 1982*b*); this issue needs further investigation. Populations of the Dusky Grasswren now isolated in the ranges of central Australia and north-west Queensland have diverged into two subspecies *A. p. purnelli* and *A. p. ballarae*. The Thick-billed Grasswren has a predominantly southern distribution, and it was once widely distributed across southern

Australia in chenopod shrubland (see Glossary). Presumably this distribution was once continuous, but Pleistocene cycles of aridity and the Eyrean barrier isolated the eastern subspecies *modestus*. At a later stage, *A. t. myall* of Eyre Peninsula in South Australia was isolated from the main western population (*A. t. textilis*), perhaps by a period of higher sea level that inundated the Nullarbor Plain.

If the biochemical evidence for the affinity of the Eyrean Grasswren with the Thick-billed Grasswren is accepted, birds of the *textilis* line from east of the central ranges moved into the dunes of the Simpson Desert, where another hummock grass (*Zygochloa paradoxa*) occupies and stabilizes the dune crests. Here, they evolved into the Eyrean Grasswren, with its heavy, deep, finch-like bill capable of coping with the large seeds of *Zygochloa*.

The habitat of grasswrens combines interspersed patches of dense cover and open space, either bare rock or open sand. All the species have plumage in tones of red, brown, and black streaked with white, which gives them excellent camouflage. Some subspecies are particularly red, in habitats where the predominant background colour is also red, especially, *A. p. purnelli*, *A. p. ballarae*, *A. s. whitei*, and *A. goyderi*. The diet of all grasswrens includes both insects and seeds; species that live among spinifex and eat its small seeds have relatively long, fine bills. The species that have adopted other habitats—the Thick-billed Grasswren in chenopod shrubland, the Eyrean Grasswren in cane-grass on sand dunes, and the Grey Grasswren in lignum swamp—all have shorter, deeper bills, presumably adapted to cope with the larger seeds they eat.

To convert this very general outline into a detailed account of the evolutionary process will require a more detailed knowledge of biochemical characters, which will enable us to estimate times of divergence of species and relate them to environmental changes.

4

Morphology, locomotion, and feeding behaviour

Introduction

As with other families of Australian birds, studies of malurid morphology have rarely gone beyond what is necessary to distinguish species on the basis of external characters, and to identify them as passerines. Few studies have compared external or internal morphology within the Maluridae or between them and other families. Some comprehensive studies of morphological characters significant in taxonomy, such as the syrinx (Ames 1970) and the palate (Bock 1960), have included a malurid specimen. A few aspects relevant to other studies have been investigated in some detail, such as the cloacal protuberance (see Glossary) and the tympanic region of the skull. In this chapter, we attempt to compare external morphology within the family Maluridae, and illustrate how the typical malurid body form differs from that of other small insectivorous passerines.

Body form

Variation in size and shape

Many studies of bird families have shown that foraging behaviour and external morphology are closely related. Most members of the Maluridae forage mainly on the ground or in low shrubs close to it, and share a general body form with the long tarsi and short rounded wings generally associated with this mode of foraging. Only the New Guinea species, the Broad-billed Fairy-wren (*Malurus grayi*), the Orange-crowned Wren (*Clytomyias insignis*), and Wallace's Wren (*Sipodotus wallacii*) regularly forage further from the ground.

The most obvious difference between species is a difference in size, from the smallest, the Rufous-crowned Emu-wren (*Stipiturus ruficeps*) at 4–5 grams, to the largest, the White-throated Grasswren (*Amytornis woodwardi*) at 33–40 grams. As well as variation in size between species there is variation within species, between individuals, between sexes, and between populations. This is particularly evident in some widespread species with patchy distributions, such as the Variegated Fairy-wren (*M. lamberti*) in Australia or the White-shouldered Fairy-wren (*M. alboscapulatus*) at different altitudes in New Guinea, where individuals of *M. a. naimii* and *M. a. lorentzi* in the highlands are larger than those in the lowlands.

As a general rule in animals, linear dimensions do not increase in direct proportion to body mass or weight, but only as its cube root (Calder 1984; Schmidt-Nielsen 1984). This holds good in the Maluridae: as body mass increases from 10 to 20 grams, wing length does not increase from 50 to 100 millimetres but only to *c.* 60 millimetres. We have compared some dimensions of bills (length, width, and depth), wing, tail, and tarsus across all species of the family Maluridae, using mean linear measurements of adult males (see Plan of the Book, p. xiv). Two species may be the same size but different shape, the same shape but different size, or they may differ in both size and shape. Body mass is the most convenient means of comparing the size of birds that may differ in shape. The size of individual parts tends to be correlated with overall body mass, but the correlations are not always strong, nor similar for different parts. Table 4.1 summarizes the range of sizes. Four figures (Figs 4.1, 4.3, 4.4, and 4.5) show how each of the four external dimensions varies with body mass across the family. These figures show significant correlations between linear dimensions and body mass (the lowest correlation coefficient is $r = 0.55$), although the amount of variation explained by body mass alone varies considerably (as measured by the value of r^2). For some dimensions, especially tail and bill size, much of the variation is not explained by size; for example, species may have longer or shorter tails or bills than we would expect

Table 4.1 Body mass and measurements of wing (folded, flattened, left wing), tail, tarsus, bill length and width for adult males of all species of *Malurus*, *Stipiturus*, *Amytornis*, *Sipodotus*, and *Clytomyias* (nominate subspecies only).

Species		*Body mass*	*Wing*	*Tail*	*Tarsus*	*Bill length*	*Bill width*
Malurus							
M. cyaneus	Superb Fairy-wren	9.4	49.6	58.6	22.8	8.3	3.5
M. splendens	Splendid Fairy-wren	10.6	52.4	60.4	23.5	8.7	3.9
M. lamberti	Variegated Fairy-wren	8.1	48.1	71.9	21.9	9.0	4.1
M. amabilis	Lovely Fairy-wren	8.8	51.6	54.3	21.7	10.1	4.2
M. elegans	Red-winged Fairy-wren	9.9	51.6	75.3	24.5	10.0	4.2
M. pulcherrimus	Blue-breasted Fairy-wren	9.4	50.5	73.0	23.0	9.1	4.1
M. leucopterus	White-winged Fairy-wren	7.6	47.3	57.4	19.8	8.5	3.6
M. alboscapulatus	White-shouldered Fairy-wren	11.2	50.6	41.3	21.2	11.0	4.3
M. melanocephalus	Red-backed Fairy-wren	8.0	44.0	48.9	20.0	8.6	3.9
M. coronatus	Purple-crowned Fairy-wren	11.3	56.2	71.7	24.9	11.3	4.5
M. grayi	Broad-billed Fairy-wren	14.0	60.2	55.4	24.8	14.8	6.7
M. cyanocephalus	Emperor Fairy-wren	13.3	59.4	58.3	24.2	13.1	4.9
Sipodotus							
S. wallacii	Wallace's Wren	7.9	47.8	46.5	19.1	14.8	4.3
Clytomyias							
C. insignis	Orange-crowned Wren	12.3	56.2	65.8	24.7	13.1	5.7
Stipiturus							
S. malachurus	Southern Emu-wren	7.5	43.1	113.3	19.4	8.3	2.2
S. mallee	Mallee Emu-wren	5.5	39.8	85.1	15.5	8.6	2.3
S. ruficeps	Rufous-crowned Emu-wren	5.4	38.9	70.9	14.9	8.3	2.2
Amytornis							
A. barbatus	Grey Grasswren	18.5	60.8	107.8	24.6	10.8	3.5
A. goyderi	Eyrean Grasswren	17.5	59.0	78.4	23.8	10.8	4.2
A. striatus	Striated Grasswren	20.8	63.6	86.4	24.5	10.2	3.2
A. dorotheae	Carpentarian Grasswren	24.5	63.5	86.8	23.8	11.7	4.9
A. woodwardi	White-throated Grasswren	34.1	76.5	104.0	28.4	14.0	5.4
A. housei	Black Grasswren	29.8	74.7	92.4	25.9	14.2	5.5
A. purnelli	Dusky Grasswren	21.0	62.4	76.5	24.9	12.1	4.0
A. textilis	Thick-billed Grasswren	24.2	66.3	91.8	25.2	10.7	5.2

Length measurements (mm) from Schodde (1982*b*), body mass (g) from museum specimens, many literature sources, field measurements by the authors and from records of the Australian Bird and Bat Banding Scheme. Details of sample size and range of measurements are given in the species accounts.

from their overall body size. This shows up in the figures either as a wide scatter of points or as one or two aberrant species, suggesting that the length of the bill or tail may have some extra significance; we have explored such cases further.

Wing

Wing length is very closely related to body size, which explains a high proportion of the variation ($r^2 = 0.94$, Fig. 4.1). Birds from different genera and different geographical localities fit this regression line closely. For a plot of $\log_{10}$ wing length against $\log_{10}$ body mass, the slope is 0.32, very close to the general relationship for birds of 0.33 (Greenewalt 1962, from Schmidt-Nielsen 1984). All members of the family are resident or locally nomadic, and live mostly within dense vegetation, feeding from the ground or not far above it, gleaning from vegetation with occasional sallying to snatch insects from vegetation or the air. All species have short, rounded wings, with 10 primaries, and primaries 1–7 very similar in length (Fig. 4.10, Table 4.2). In *Malurus* spp., primaries 3–6 are the longest, in *Amytornis*, primaries 5–7 are longest, and in *Stipiturus*, primaries 4–7 are the longest. Such short, round, 'elliptical' wings provide good lift at take-off and permit ready mobility in dense vegetation (Keast 1996). In the more arboreal New Guinea species such as Wallace's Wren and the Broad-billed Fairy-wren, the wings are also short and rounded, but the feathers are narrower, and somewhat emarginate.

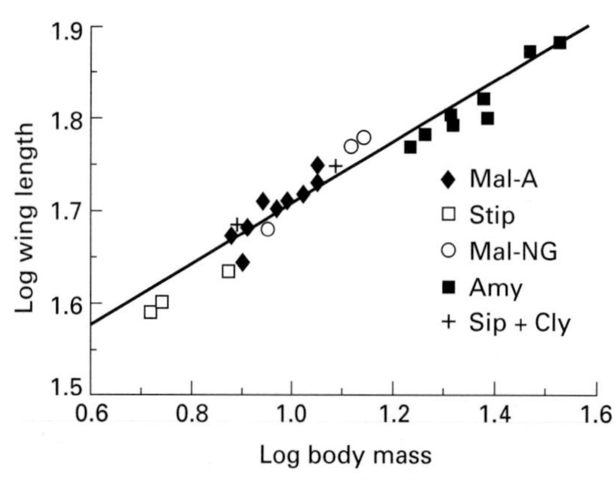

4.1 The relationship between wing length and body size (represented by body mass) for male malurids ($n = 25$, from Table 4.1). Both are plotted on a $\log_{10}$ scale for the closest correlation ($r = 0.97$, SE = 0.017, $r^2 = 0.94$). No species stand out as exceptions to this relationship, indicating that body size is the most important factor affecting wing length. Mal-A = Australian *Malurus*; Stip = *Stipiturus*; Mal-NG = New Guinea *Malurus*; Amy = *Amytornis*; Sip+Cly = *Sipidotus* plus *Clytomyias*.

Tail

Tails, along with the brilliant plumage of male *Malurus*, are one of the trademarks of the family (Fig. 4.2). The 12 rectrices normal in passerines are reduced in the Maluridae; fairy-wrens (*Malurus*) and grasswrens (*Amytornis*) have 10. The innermost rectrices are the longest, and in most species, all but the outer pair are very similar in length, generally rounded, blunt at the tip. The two outer rectrices in *Amytornis* are half the length of the other eight, whereas those in *Malurus* and the

Table 4.2 Wing shape of Superb Fairy-wren (*Malurus cyaneus*): the length of primaries and secondaries (measured when plucked from a single bird, from Parsons 1968).

Primary		*Secondary*	
Number	*Length (mm)*	*Number*	*Length (mm)*
10	20	1	42
9	33	2	41
8	39	3	40
7	42	4	39
6	44	5	37
5	44	6	35
4	44	7	32
3	44	8	30
2	42	9	24
1	42	10	16

P1–7 and S1–3 are similar in length, indicating a rounded wing (see Fig. 4.10).

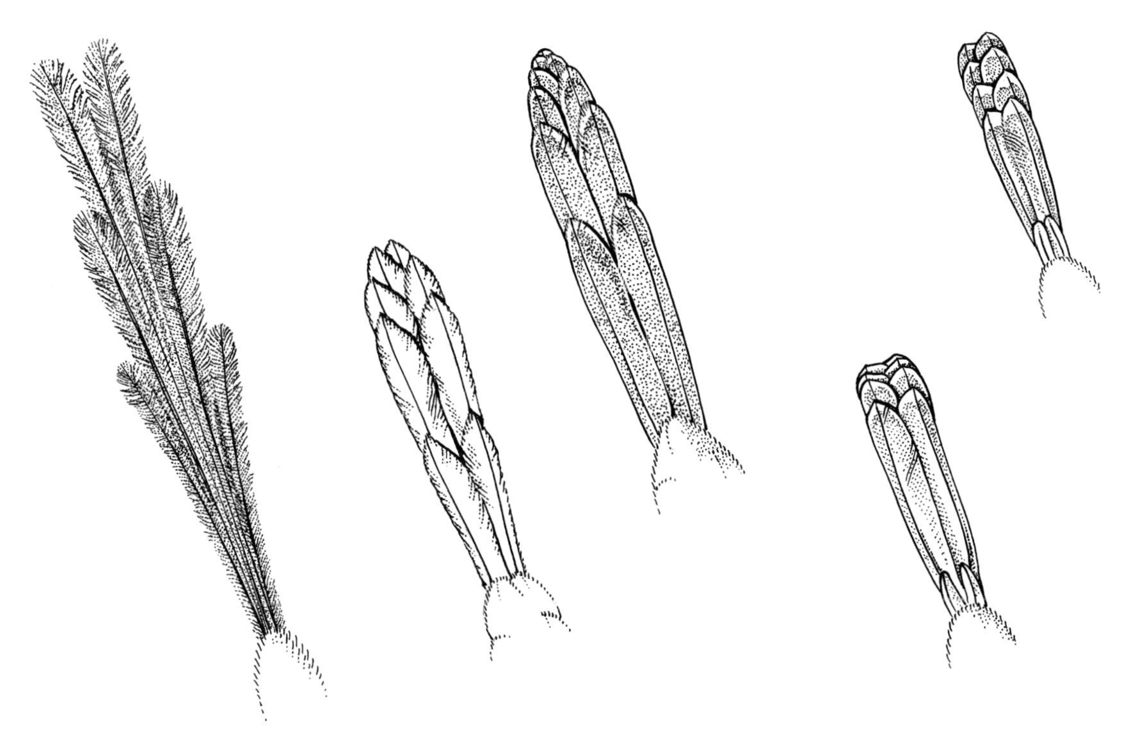

4.2 The tails of malurids (from left to right): Southern Emu-wren (*Stipiturus malachurus*): tail very long, with only three pairs of rectrices; feathers appear filamentous because barbules lack hooks to mesh the sparse barbs together. Orange-crowned Wren (*Clytomyias insignis*): tail with 10 rectrices, but outer two are vestigial. Thick-billed Grasswren (*Amytornis textilis*): 10 rectrices, outer pair about half the length of longest. Superb Fairy-wren (*Malurus cyaneus*): 10 rectrices, outer pair reduced to one-fifth length of longest, tail with little taper at tip, and long rectrices similar in length. Wallace's Wren (*Sipodotus wallacii*): similar to Superb Fairy-wren, but relatively shorter, and more gradation in length of rectrices. (Redrawn from Schodde 1982*b*.)

tree-wren (*Sipodotus*) are only one-fifth as long. The tail of the Grey Grasswren (*Amytornis barbatus*) is exceptional in being relatively long, with more tapered rectrices increasing in length gradually from outer to inner. The russet-wren (*Clytomyias*) also has 10 rectrices, but the outer ones are so reduced as to be almost invisible. In the emu-wrens (*Stipiturus*), the tail is reduced to only three pairs of elongated structurally highly modified feathers, unusually long for such small birds. The shafts of the rectrices are robust, but the barbs are sparse and like true Emu (*Dromaius novaehollandiae*) feathers, the barbules lack hooks to mesh them together, giving the feathers a filamentous appearance. The Southern Emu-wren (*S. malachurus*) has the sparsest barbs and most filamentous appearance, while in the Rufous-crowned Emu-wren, the unmeshed barbs are denser.

The length of the tail is only loosely related to body size (Fig. 4.3). The Southern Emu-wren has the longest tail of all and the two other species of emu-wren also have very long tails. When these species with greatly elongated tails are removed from the analysis, the relationship with size appears closer, but still only 54 per cent of the variation in tail length is explained by body size, and the correlation with wing length is weak, suggesting that the tail is significant for other than aerodynamic functions. Other ground-foraging species such as Scrub Robins (*Drymodes*), Scrub Wrens

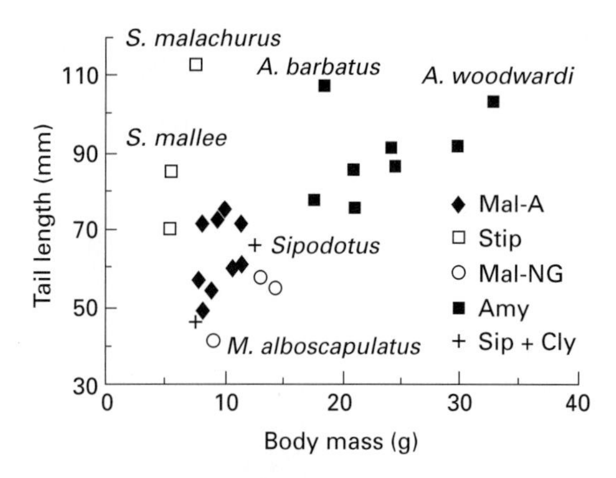

4.3 The relationship between tail length (mm) and body mass (g) for male malurids ($n = 25$, from Table 4.1) The correlation is not strong ($r = 0.55$, SE = 0.43, $r^2 = 0.31$); only 23% of the variation in tail length is explained by body size. The tiny emu-wrens have exceptionally long tails, the more arboreal New Guinea forms have shorter tails than Australian *Malurus* of similar size, and the Grey Grasswren (*Amytornis barbatus*) has a particularly long tail. Even in Australian *Malurus*, tail lengths vary between species of similar size. Most chestnut-shouldered fairy-wrens have relatively long tails and the bi-colored fairy-wrens have short tails. See Fig. 4.1 for key.

(*Sericornis*), and some tyrannid flycatchers (Fitzpatrick 1985) do not have extremely long tails, so these are unlikely to be essential for terrestrial locomotion.

The New Guinea species in general have short tails, in particular the various subspecies of the White-shouldered Fairy-wren, which along with the Red-backed Fairy-wren (*M. melanocephalus*) have the shortest tails relative to body size in the family Maluridae. These two species both live in dense habitats of tropical tall grass. Tails of other New Guinea species, Wallace's Wren, Orange-crowned Wren, Broad-billed Fairy-wren, and Emperor Fairy-wren (*M. cyanocephalus*) are also relatively short, but we could find no direct correlation with habitat type: White-shouldered Fairy-wrens in tall grassy habitats, Orange-crowned Wrens in the sub-canopy of rainforest substage, Wallace's Wren in arboreal tangles of vines, Emperor Fairy-wrens and Broad-billed Fairy-wrens in the dense shrubbery of forest regrowth and edges. The only shared feature of these habitats is their density, without the open spaces between and under shrubs that are so much utilized for ground foraging by Australian *Malurus* and *Amytornis*. In contradiction, however, *Stipiturus*, with the longest tails of all, live in dense swampy heath (Southern Emu-wren) or spinifex (Mallee *S. mallee* and Rufous-crowned Emu-wrens). Of the Australian *Malurus*, some of the chestnut-shouldered fairy-wrens, the Variegated, Blue-breasted (*M. pulcherrimus*), and Red-winged (*M. elegans*) Fairy-wrens have long tails in relation to body size, while the Lovely Fairy-wren (*M. amabilis*) has a relatively short tail. The habitat of the Lovely Fairy-wren at the fringes of northern rainforests is most similar to that of the Emperor Fairy-wren in New Guinea, and both spend more time foraging above the ground. The Splendid (*M. splendens*) and Superb Fairy-wrens (*M. cyaneus*) have relatively short tails, and Red-backed Fairy-wrens have the shortest of all.

Tail length varies between individuals and with wear, as might be expected, but there is another source of variation. In the Purple-crowned Fairy-wren (*M. coronatus*), immature males have longer tails than adult males, pointed out by Schodde (1982*b*). Our field measurements of both subspecies confirmed this. Mean tail lengths of immature and adult males differ by *c.* 4 millimetres, with the extremes of ranges differing by up to 20 millimetres. This same difference was not apparent for most other species for which we have field measurements of males of known ages (Table 4.3). For the Red-backed Fairy-wren, Schodde (1982*b*) reports that the tails of female-plumaged males are longer than those of males in nuptial plumage, with the difference greatest in the northern subspecies *M. m. cruentatus.* It is not clear whether these differences in tail length in Red-backed Fairy-wrens are due to age, since no longitudinal field studies of banded birds have been done,

Table 4.3 Tail length of adult and immature male fairy-wrens (*Malurus*).

Species	*Adult male*	*Immature male*
Purple crowned Fairy-wren		
M. c. coronatus	78.7 ±0.55 (70–87, 41)	82.6 ±0.73* (74–90, 31)
M. c. macgillivrayi	81.4 ±0.97 (73–93, 27)	85.7 ±0.84* (80–93, 15)
Blue-breasted Fairy-wren		
M. pulcherrimus	81.1 ±0.47 (73–88, 38)	80.3 ±0.29 (78–84, 12)
White-winged Fairy-wren		
M. leucopterus leuconotus	62.5 ±0.93 (58–67, 12)	63.9 ±0.60 (59–69, 23)
Red-winged Fairy-wren		
M. elegans	80.4 ±0.26 (74–88, 107)	78.9 ±0.45* (71–85, 48)
Splendid Fairy-wren		
M. s. splendens	61.9 ±0.67 (55–67, 25)	60.2 ±0.67 (55–65, 19)
Red-backed Fairy-wren		
M. m melanocephalus	48.9 ±0.41 (40)	53.9 ±1.36* (12)
M. m. cruentatus	40.8 ±0.43 (63)	51.6 ±0.97* (17)

Data based on field measurements by the authors, except for the Red-backed Fairy-wren (*M. melanocephalus*), which is from Schodde 1982*b*), and is for female-plumaged males of unknown age. Mean ± SE, (range, *n*). Significant differences between adult and immature males ($P<0.001$) are marked *.

and it is impossible to tell the age and sex of brown birds in the non-breeding season (Rowley and Russell, unpublished data).

In most species, both sexes hold the tail cocked, nearly at right angles to the body. In *Sipodotus* and most *Malurus* the rectrices are white tipped on the under surface, visible from behind. The white tip varies from 1 to 2 millimetres in width, and is most obvious on newly grown feathers, becoming frayed or completely worn away with time. It is widest in the Lovely, Variegated, Purple-crowned and female Emperor Fairy-wrens. In White-winged Fairy-wrens (*M. leucopterus*), the male has a whitish fringe on blue tail feathers that becomes worn, while the female has very faint white edges and tip. In species with predominantly black males (White-shouldered and Red-backed Fairy-wrens), there is no white on the tail of the males, but the brown females have a faint white tip that does not contrast with the plain buff tail. The Grey Grasswren has very marked white edges and tips to the rectrices; no other Grasswrens have white in the tail, only light rufous or buff edges to the rectrices.

It is tempting to suggest that white tail tips have some signal function, or help to promote group cohesion between birds in dense habitats. The tail has the potential to convey information about status or motivation, through its angle to the back, and its rate of movement—a slow oscillation or a rapid vibration. The tail is very mobile, but generally not fanned; only the Lovely Fairy-wren of the rainforest fringes of north Queensland fans its tail as it moves, in the manner of *Rhipidura* fantails. (Rowley and Russell, unpublished data). In other species, the closed tail may be 'wagged' when the bird is foraging on the ground or when singing (W. T. Cooper, *in litt.* 3 July 1996; Rowley and Russell, unpublished data).

Tarsus

In many families of birds, species that spend a lot of their time hopping and foraging on the ground generally have longer and thicker legs than species that perch on branches in the typical passerine manner (Storer 1971). Ground-foraging tyrannid flycatchers have significantly longer tarsi than aerial foragers, and the tarsus length of ground-foraging birds in the 10–25 gram mass range is about 25 millimetres (Fitzpatrick 1985). Most of the Maluridae forage predominantly on the ground and the tarsus is longer than in arboreal foragers of similar size but is nevertheless correlated with body size (Fig. 4.4; $r^2 = 0.56$; Table 4.1). The increase in size of tarsus with body size is more gradual in the grasswrens than in the smaller species, and the tarsi of the smaller grasswrens are similar in length to the tarsi of the larger fairy-wrens, such as the Purple-crowned and Red-winged; a tarsus length of about 24 millimetres is common to birds with a range of sizes from 10 to 25 grams. Purple-crowned Fairy-wrens that forage in the spiny fronds of *Pandanus aquaticus* have relatively robust tarsi and feet, as have all the grasswrens. Blue-breasted, Variegated, and White-winged Fairy-wrens have very slender tarsi; we had to trim 1 millimetre from the smallest size of colour-bands to make them fit safely.

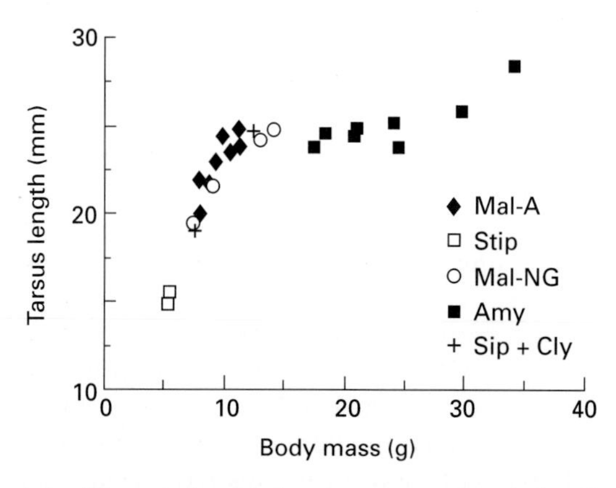

4.4 The relationship between tarsus length (mm) and body mass (g) for male malurids ($n = 25$, from Table 4.1). Tarsus length increases with increasing body size, but not directly in proportion to body size ($r = 0.75$, SE =0.56, $r^2 = 0.56$). The six smaller grasswrens (weighing 18–20 g) have tarsus lengths of *c.* 24 mm, similar in length to that of larger fairy-wrens (10–12 g). See Fig 4.1 for key.

Bill shape

The morphological feature that varies most between malurid species is bill shape, reflecting differences in foraging methods and food choice (Storer 1971; Fitzpatrick 1985; Grant and Grant 1989; Weiner 1994). In general, larger birds have larger bills (Table 4.1), but variation in length, width, and depth produce differences in shape. We plotted bill width against bill length for 25 species; body size effects were removed by dividing both width and length by $\log_{10}$ body mass (Fig. 4.5). Most *Malurus* lie between the extremes of width and depth. Three main bill shapes are distinguishable, which may be related to differences in foraging mode and diet (Figs 4.6, 4.7):

1. Relatively long, narrow, pointed bill, slightly wider at the base, width greater than depth; typical of emu-wrens, Australian fairy-wrens, and the White-shouldered Fairy-wren in New Guinea (Fig. 4.6(b)), and also the range of other Australian gleaning insectivores compared by Wooller (1984). These malurids are all insectivorous, foraging mostly on the ground by gleaning or probing, or gleaning and probing from a perch, flying an occasional sally from the ground or a perch. As in other families, these modes of foraging are associated with poorly developed rictal bristles (see Glossary) (Conover and Miller 1980; Fitzpatrick 1985).

2. Long bills, wide at the base, width much greater than depth, with well developed rictal bristles; typical of the New Guinea arboreal foragers Emperor and Broad-billed Fairy-

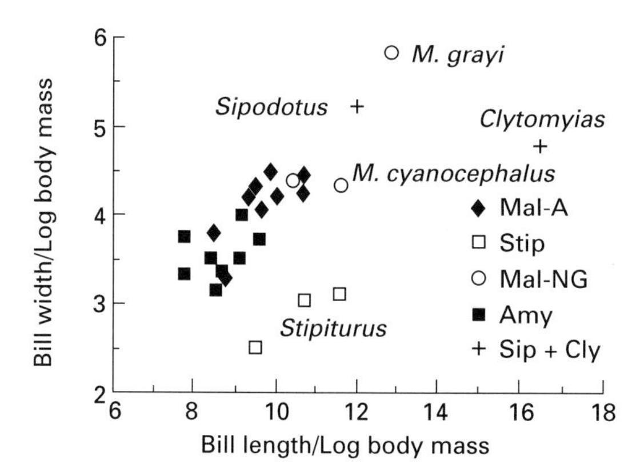

4.5 Bill shape in the Maluridae: the relationship between bill width and bill length. Body size effects are removed by dividing both width and length by $\log_{10}$ body mass. At the lower left of the graph are the grasswrens, with relatively short, narrow bills; some grasswrens have particularly deep, finch-like bills (see Fig. 4.7(d)). At the right of the figure are the long, wide bills of the arboreal foraging New Guinea malurids, the Orange-crowned Wren (*Clytomyias insignis*), the Broad-billed Fairy-wren (*Malurus grayi*), Wallace's Wren (*Sipodotus wallacii*), and the Emperor Fairy-wren (*M. cyanocephalus*) (Fig. 4.6 (a,d)). In between are the typical relatively long, narrow, pointed bills of most *Malurus*. Emu-wrens *Stipiturus* have relatively long pointed bills.

wrens, and Wallace's and Orange-crowned Wrens (Fig. 4.6(a, c, d)). The extreme of this development is in the smallest species, Wallace's Wren, which has the longest bill. These species are all insectivorous, foraging in forest canopy or understorey canopy by gleaning and snatching. This bill shape is very similar to that of upward-striking tyrannid flycatchers as characterized by Fitzpatrick (1980, 1985), who suggested that as well as a protective function, the rictal bristles increase the area swept by the open bill during upward-striking (similar to snatching as defined by Recher *et al.* 1985, p. 403).

3. Relatively shorter, narrower, deeper bills, typical of most grasswrens; the most extreme development is the finch-like bills of Eyrean (*Amytornis goyderi*) Grey, and Thick-billed (*A. textilis*) Grasswrens (Fig. 4.7). Depth is equal to or greater than width, and rictal bristles are well developed. The diet of grasswrens frequently includes seeds (Schodde 1982*b*; Barker and Vestjens 1990). They forage predominantly on the ground, with some gleaning in shrubs, and the large rictal bristles may serve to protect their eyes and face from the spines of spinifex and other spiny arid adapted plants (Schodde 1982*b*).

Variation in the morphology of the tongue associated with differences in diet is not known. Only the tongue of the Superb Fairy-wren has been described, as thin, horny, and slightly frayed at the tip, similar to the tongues of many other small insectivores (McCulloch 1975).

Sexual dimorphism in body size

In most malurids, males are slightly larger than females. To compare the size of males and females, we used wing length, since records of body mass were not available for females of several lesser known species, and wing length is highly correlated with body mass. In Fig. 4.8, wing length for males is plotted against that for females. The line represents equal length of wing in males and females—for all species but one (Orange-crowned Wren), males are slightly larger than females. Wing length of females is 96.7 per cent of the value for males in *Malurus* (12 species), 97.1 per cent in *Stipiturus* (three species), and 97.2 per cent in *Amytornis* (eight species). Thus although differences in plumage between males and females are much greater in *Malurus* than *Amytornis*, there is no difference in the degree of dimorphism in size. In the Orange-crowned Wren, the one species in which females are the same size as males or a little larger, the plumage of males and females is virtually indistinguishable.

Variation within species

In species whose geographical range is small, such as the Red-winged Fairy-wren or the

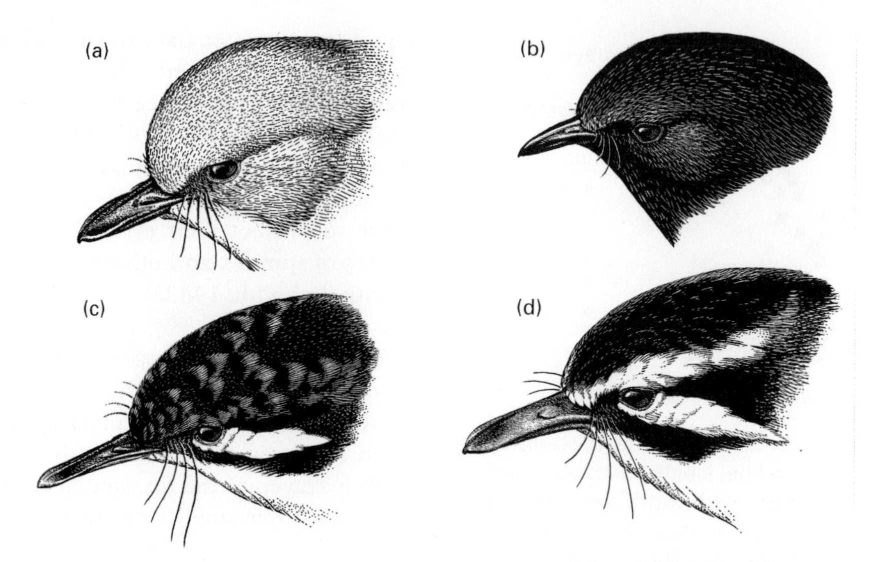

4.6 The bill shape of some New Guinea Malurids. (a) Orange-crowned Wren (*Clytomyias insignis*) (very broad, relatively long), (b) White-shouldered Fairy-wren (*Malurus alboscapulatus*) (narrow, pointed, similar to bill of Australian *Malurus*), (c) Wallace's Wren (*Sipodotus wallacii*) (very long, flattened, narrow), (d) Broad-billed Fairy-wren (*M. grayi*) (long, broad). Longer, broader bills are found in those species that are most arboreal and include upward snatching among their foraging methods, flying upwards from a perch to snatch insects from overhead.

4.7 The bill shape of some grasswrens. (a) Striated Grasswren (*Amytornis striatus*), (b) Black Grasswren (*A. housei*), (c) Grey Grasswren (*A. barbatus*), (d) Eyrean Grasswren (*A. goyderi*). Grasswren bills are short and narrow in relation to body size, but in some species are short, deep, and finch-like. Most species include some seeds in their diets; in Striated and Black Grasswrens, the common seeds available are the small seeds of the Hummock grasses, for example *Triodia* spp., but for the Eyrean Grasswren, the common seeds available are the large, hard seeds of Dune grass *Zygochloa paradoxa*.

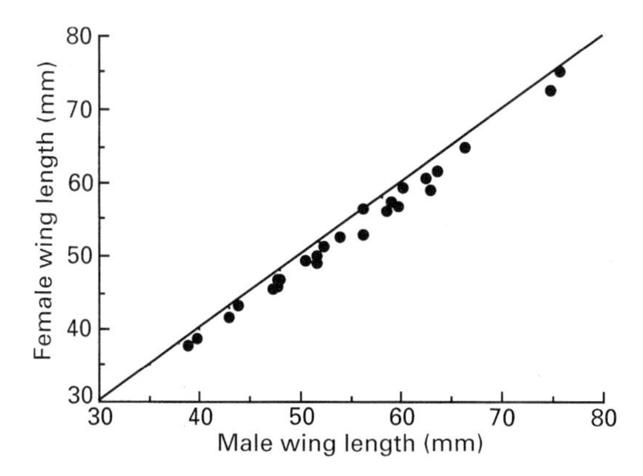

4.8 Sexual dimorphism in size in malurids: wing length of males plotted against that of females. The line represents equal length of wing (and thus body size) in male and female. In all species but one (Orange-crowned Wren *Clytomyias insignis*), males are slightly larger than females.

Mallee Emu-wren, variation in size and colour within species is generally slight. However, many species occur over a large geographical area, living in a range of climatic conditions. In some cases, suitable habitat is patchy or fragmented, and populations in different regions are now isolated, or have been in the past, by changes in sea level or periods of greater aridity. Some of these populations have differentiated to the extent that they are recognized as subspecies. In other cases, species or subspecies vary continuously over a large range. For example, in the Variegated Fairy-wren, three subspecies are identified at the north, north-west, and south-east fringes of the continent. A fourth, *M. l. assimilis*, occurs over most of the rest of Australia, with slight variation in plumage; southern forms are slightly larger, with shorter, narrower bills and relatively the longest tails of any fairy-wren. Similar variation in a widespread species occurs in the Southern Emu-wren, in suitable heathland habitats across southern Australia. Isolated populations have diverged considerably, with seven subspecies differing in size and plumage both within and between subspecies. The range of some subspecies is small, but the two at the south-east and south-west edges of the continent span a considerable geographical and climatic range, and both show similar differences between darker and more rufous plumage with deep blue throat in wet coastal areas and much less intense blue throat and less rufous plumage in drier inland areas.

In New Guinea, changes in sea level during the last 2 million years combined with the mountainous nature of the country have led to the isolation of populations in different mountain ranges or valleys. In widespread species such as the White-shouldered Fairy-wren, a profusion of subspecies has been described, but within subspecies, there is also variation in size with altitude, forms at higher altitudes being considerably larger. Schodde's 1982 monograph gives detailed accounts of variation within species in size and plumage.

Locomotion

Flying

As described earlier in this chapter, most malurids have short rounded wings which, while providing good lift on take-off and for flitting from bush to bush, do not encourage them to indulge in long flights. A few exceptional display flights are quite prolonged, but these are slow and rarely performed. We had a good example of their limitations in the Millstream–Chichester National Park in the Pilbara region of Western Australia. A group of Variegated Fairy-wrens were foraging high in the canopy of riverside paperbarks (*Melaleuca* sp.) when they were scared by the passage of a raptor, and all the group dropped 10 metres to ground level following the alarm call. After a few minutes we watched the birds resume their high level foraging—not by flying up to the canopy, but by hopping along leaning trunks until they were high enough to resume snatching insects among the foliage.

Although one does not often see malurids in the process of dispersing, twice we have encountered groups travelling across open agricultural landscapes by following fencelines, briefly flying from post to post, occasionally seeking protection in the tall weed growth between road and fence, or foraging on the ground. They would be very conspicuous and vulnerable if they flew as a group across such open country.

Hopping

Most movement in search of food is by hopping, jumping with both feet off the ground at once. Whether on the ground or within a tangle of branches, 'hop-search' is the usual tactic—a hop then stop and probe the litter or glean a leaf, then another hop or two and repeat the search. The Eyrean Grasswren hops with one foot slightly ahead of the other, leaving characteristic footprints in sand (Parker *et al.* 1978).

Running

The only time that fairy-wrens run (one foot after the other, Fig. 5.4) is when they are performing the 'Rodent-run' display to distract a predator that is threatening their nest or young. This is described on p. 60, in detail. Grasswrens are even more terrestrial and run both in display and escape.

Plumage

Feather tracts

The feathers of birds are distributed on the body in distinct tracts or pterylae. Most dorsal feathers arise from a distinct spinal tract which runs from the nape of the neck to the base of the tail, with a narrow interscapular region and a broader region on the mid-back. Malurids differ from most species in the distribution of feathers on this spinal tract; they have a break in the interscapular region where feathers are completely absent. This was first illustrated by Parsons, who in 1968 published a series of drawings of the feather tracts of a number of Australian birds, including the Superb Fairy-wren for which he recorded merely 'There is a very well-defined break in Pt. spinalis as shown' (Fig. 4.9). The significance of this break was

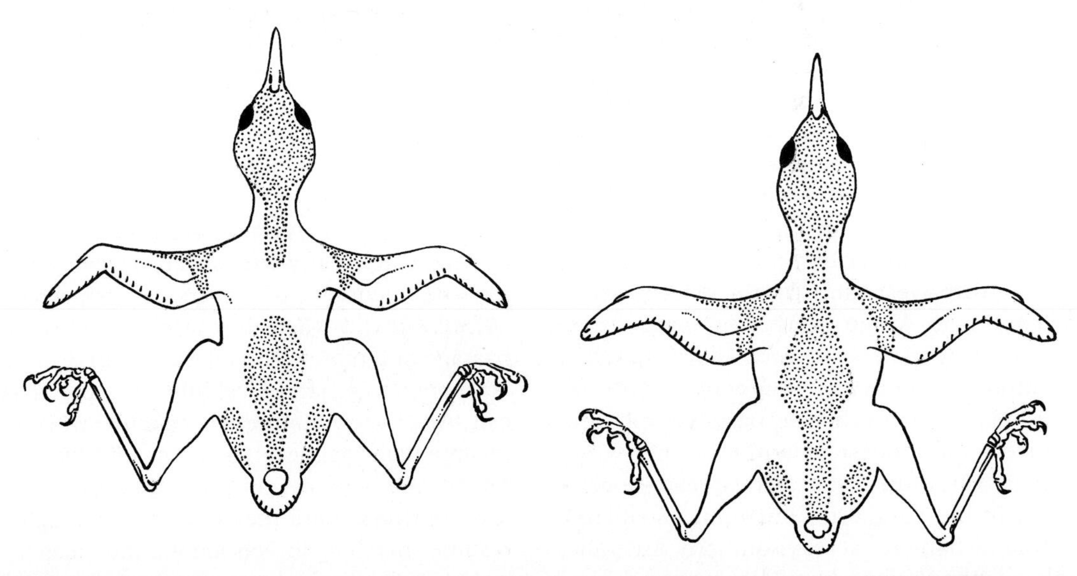

4.9 The interscapular gap or *apterium* characteristic of malurids. The feather tracts of the Superb Fairy-wren (*Malurus cyaneus*) (*left*) and, for comparison, the White-browed Scrub-wren (*Sericornis frontalis*, Pardalotidae) (*right*) indicating the mid-dorsal region in malurids where no feathers originate. (Redrawn from Parsons 1968.)

not appreciated until Harrison also noticed it, and found that it occurred in all species of *Malurus*, *Amytornis*, and *Stipiturus* that he examined, as well as in the New Guinea species, including *Sipodotus* and *Clytomyias* (Harrison 1969). This interscapular gap sets these genera apart from any others with which they have previously been grouped. There is a large bare patch where the mantle would normally be; this area is covered by feathers from the nape and the humeral tracts on either side of the space, which are long, profuse, and often erectile. It is this conspicuous feathering of the humeral tracts that forms the chestnut shoulders of the chestnut-shouldered group. It is interesting that F. Lawson Whitlock, that experienced traveller, collector, and observer of Australian birds, had observed this many years before: 'all the white-winged wrens that I have examined show a parting like that of the human hair, down the centre of the white feathers' (1921, p. 180).

Sexual dimorphism in plumage

The most striking feature of the Maluridae is the brilliant colour of the nuptial males. Most species of the family Maluridae exhibit some degree of sexual dichromatism, but it is most extreme in the Australian *Malurus*, in which adult males assume a brilliant plumage in the breeding season, predominantly of blues, violets, and blacks, with some tinges of red. These brilliant colours are distributed in particular areas where they contrast with plainer backgrounds, and may serve display functions. Significant areas are the crown, the ear tufts, the scapulars, and upper back. In the blue and chestnut-shouldered species, the crowns, upper back, and elongated ear tufts are iridescent blue of various shades from silvery blue to violet. In the bi-coloured species, it is the scapulars and upper back that are contrasting, red or white. In display postures, the ear tufts

4.10 Fairy-wrens in flight: Blue-breasted Fairy-wren (*Malurus pulcherrimus*) (*left*), Red-winged Fairy-wren (*M. elegans*) (*top*), Variegated Fairy-wren (*M. lamberti*) (*right*). The long tail is carried straight behind the body or drooping slightly. The coloured region of the upper back is exposed in flight. Note the short, rounded wings.

and coloured scapulars are erected; the upper back may be exposed when the bird flies or the wings are depressed (Fig. 4.10).

The blue, violet, and black colours of all malurids are structural colours, in which minute air-filled cavities in the feathers are embedded in a layer of highly refractive pigment granules in the surface of the barbules, and scatter incident light. When this pigment is black eumelanin, the feathers appear blue; when chestnut-red erythromelanin is present, the feather appears magenta or shades of violet. Without scattering, erythromelanin appears as its true chestnut-red colour in the scapular feathers of the chestnut-shouldered wrens. The different black and blue-black breast colours of the chestnut-shouldered fairy-wrens depend on the amounts of pigment present (Harrison 1965, 1971). Phaeomelanin pigments produce the various browns, buffs, and reds of female *Malurus*, and male and female emu-wrens and grasswrens.

The blue plumage of *Malurus* males, as well, has a glittering iridescence, from light reflected by the surface of barbules that are flattened and twisted for part of their length, so that if a flattened surface faces the observer who has the light behind him, the feathers appear iridescent. The exaggerated ear tufts of some male *Malurus* are particularly iridescent. More information on iridescence and plumage colours can be found in Greenewalt (1960) and Vevers (1985). It is now recognized that the colour vision of birds extends into the ultra-violet, in which iridescent feathers reflect strongly, so that humans may not fully appreciate the effect and significance of these colours (Bennett and Cuthill 1994).

Sexual dichromatism in the emu-wrens (*Stipiturus*) and the grasswrens (*Amytornis*) is less spectacular but quite distinct. In emu-wrens, males have light blue throats and breasts, and more rufous crowns, while females lack blue in the throat and are much less rufous. Male and female grasswrens are generally very similar in appearance, except for more rufous flanks and sometimes bellies in females. The open environment, terrestrial habit, and drab plumage suggest that the need for camouflage has been the overriding factor affecting the appearance of grasswrens. All are a mixture of browns, reds, and buffs, with the predominant colour matching their habitat, especially the very red forms of the central deserts (*A. striatus whitei* and *A. p. purnelli*), and the dark reds with brown or black markings in the northern red sandstone habitats (White-throated, Carpentarian *A. dorotheae* and Black *A. housei* Grasswrens). In addition, complex disruptive patterns of streaking on the feathers, with white and black shaft-streaks, and paler fringes add to the camouflage (see Plate 6).

Most New Guinea malurids are less sexually dichromatic. The most different are the Emperor Fairy-wren and some White-shouldered Fairy-wrens. Female Wallace's and Orange-crowned Wrens differ from males mainly in their paler breast colour; female Wallace's Wrens also have paler crowns. In the Broad-billed Fairy-wren, females are also paler and less blue ventrally, and a little darker and less blue on the crown. The intense all-over blue male Emperor Fairy-wren is very different from the female, with her blue head, chestnut back, and white belly. White-shouldered Fairy-wrens are the most confusing, with three geographically distinct female plumages: black with white shoulders, almost identical to the male except that they lack iridescence and have brown wings; pied, which are similar to males above, but whiteish ventrally; and brown, very similar to females of the Australian White-winged or Red-backed Fairy-wrens, predominantly brown, grey, and white. Unfortunately, the biology of the New Guinea species is so little known that we have no idea whether differences in the degree of dichromatism are correlated in any way with differences in social organization and mating systems.

Moult

In the Australian *Malurus* with marked sexual dimorphism in plumage, the full nuptial plumage displayed by adult males is replaced at the end of the breeding season by a dull 'eclipse' female-like plumage that is worn during the non-breeding season retaining a black bill and black lores. Only the body feathers are replaced in the moult from eclipse to nuptial plumage. The wing and tail feathers are replaced at the post-nuptial moult, and retained for a year, although tail feathers in all species may be replaced at any time if they are worn or lost. The moult of primaries starts with P1, and progresses regularly (descendantly) outwards to P10. Secondaries 1–6 are moulted ascendantly, starting with S1; the innermost three secondaries (tertials) moult more or less together, starting at the same time as or before S1. Tail moult appears to progress from the central rectrices towards the outer ones, but is often irregular.

In spring-breeding species, young male and female *Malurus* moult some feathers in the autumn/winter after they hatch. This may involve a change in colour of the lores from red or chestnut to black, a change in the colour of primaries and tail (e.g. Splendid Fairy-wren, from brown to turquoise), and acquisition of a white eyering that later becomes blue (e.g. Blue-breasted Fairy-wren). We have recorded young Red-winged Fairy-wrens moulting their inner primaries in March–April, and some young males may acquire a few blue feathers on the back, but immatures hatched late in the breeding season (December or January) generally do not moult their primaries at all (Russell *et al.* 1991). The extent to which immatures moult at this time has not been studied in detail and we can only recognize that changes occur because the appearance of birds changes. Purple-crowned Fairy- wrens may breed in most months, and young birds moult a few months after fledging, rather than at a regular time; the same may occur in other northern species that are not strictly spring breeders.

In the other Australian genera, adult males and females moult only once a year, with no change in the pattern of plumage. In species that breed in the spring, moult generally occurs in late summer or autumn; in desert species that breed irregularly, it probably occurs after breeding. Emu-wren fledglings leave the nest with a duller and less streaked plumage than the adult pattern, and the blue throat of males is paler, but still distinguishes the sexes. When they acquire the full adult plumage is not known (Fletcher 1913, 1915; Schodde 1982*b*). Grasswren fledglings are drabber and less clearly streaked than adults; after a few months they moult into adult plumage, and the females acquire their distinguishing redder flanks.

In the New Guinea species, there is no eclipse plumage, and feathers are probably moulted once a year. Species are insufficiently studied to be certain whether this moult is regular or can occur at any time, or even if it occurs after breeding which itself may occur at any time. There are some indications that moult may be protracted over several months and somewhat irregular (Schodde 1982*b*). Male and female juveniles of White-shouldered and Emperor Fairy-wrens moult first from juvenile to female-like plumage, probably within 6 months of fledging, and males acquire adult male plumage at their next moult. In other New Guinea species, juveniles moult direct to the appropriate full adult plumage within several months of fledging.

Acquisition of nuptial plumage in Fairy-wrens

Australian male *Malurus* vary considerably in the extent to which they acquire full nuptial plumage in their first breeding season. In some species, such as Superb and Splendid Fairy-wrens, all immature males undergo moult of body feathers and most acquire close to full nuptial plumage in the first breeding season after they hatch. This plumage may not be quite perfect: the belly of Splendid Fairy-wrens may not be completely blue, but remain buff, and they may not achieve perfect full nuptial plumage until they are 2 years old. The same probably applies to Purple-crowned Fairy-wrens, but here it is the perfection of the purple crown which is not achieved until a male is 2 or 3 years old. Our experience with Purple-crowned Fairy-wrens in northern

Australia suggests that the main moult occurs in April–June after the wet season, with males moulting into nuptial plumage by August–September when breeding generally starts. However, after a poor wet season, males may not moult into nuptial plumage at all (see plate in Rowley 1988).

In Red-winged and Blue-breasted Fairy-wrens, most 1-year-old males do not get as close to perfect nuptial plumage in their first breeding season as do Superb and Splendid Fairy-wrens. They generally have a 'spotty' appearance, with both blue and grey feathers in the head and face, some blue-black and some grey feathers on the breast, and some chestnut on the shoulders. At 2 years old, most males appear perfect, although some do not attain full plumage until they are 3 years old. In the Red-winged Fairy-wren, males from early broods tend to progress further towards full plumage in their first breeding season, while in some late-hatched males, the only indication of their sex is their black lores, with no sign of blue at all. No marked, known-age individuals of any subspecies of the Variegated Fairy-wren have been studied, but our limited field experience suggests that they are similar to Red-winged and Blue-breasted Fairy-wrens in that not all 1-year-old males achieve full nuptial plumage.

In all the species so far discussed, there may be more than one male in full nuptial plumage in a group. In the Red-winged Fairy-wren, we have encountered groups with five or six males, of which four or five, ranging in age from 2 to more than 6 years old, were in full nuptial plumage. There was no indication that the acquisition of nuptial plumage was influenced by dominance or social position, as has been suggested in many popular accounts. Acquisition of nuptial plumage in the White-winged Fairy-wren is very different. In each clan, there is only one full plumaged blue male in the breeding season, and this blue plumage is not acquired until the fourth year. Younger males are sexually mature at 1 year (Tidemann 1983), but remain brown, indistinguishable from females (except for the presence of a cloacal protruberance in the breeding season) until their third year, when they may acquire some blue feathers, and some white on the wings. If the blue male dies, another male very quickly moults to become blue.

Timing of pre-nuptial moult

In all species in which marked birds have been studied in detail, males undergo the prenuptial moult earlier as they age. Some 1-year-old males may not moult into nuptial plumage until after the breeding season has started, whereas some old males in good seasons may moult from blue to blue without entering an eclipse plumage at all. The timing of moult into nuptial plumage may be delayed in years of more adverse winter weather conditions or extreme drought. Mulder and Magrath (1994) have suggested that females may use variation in the timing of the acquisition of nuptial plumage to assess the relative quality of prospective mates for extra-pair copulations.

Internal morphology

The internal anatomy of malurids has not been formally investigated. We would expect there to be anatomical differences correlated with the progression from the most terrestrial (grasswrens) to the most arboreal species (Wallace's and Orange-crowned Wrens). We have discussed external differences, in bill morphology relative to seed eating in grasswrens and arboreal insectivory in Wallace's and Orange-crowned Wrens, and the differences between long-legged, long-tailed terrestrial species and shorter-legged, shorter-tailed arboreal species. Some differences in internal morphology associated with these different lifestyles are mentioned briefly in Schodde's 1982 monograph. He observes that, although the flight of fairy-wrens is weak, 'it is still stronger than in emu-wrens and grasswrens, and this is reflected in the fairy-wrens' sternums, which are more deeply keeled and have strengthened craniolateralis processes' (p. 30). In the grasswrens, both the keel of the sternum and the sternum itself are very shallow, suggesting a reduction in relative size of the flight muscles. The grasswrens also have 'strongly muscled gizzards and the habit of eating quantities of grit' (p. 42). Richardson and Wooller (1986) compared the size of gizzards and the length of the intestine in a series of insectivorous and nectar-eating birds ranging in size from 5.8 to 62 grams. The size of these structures in Splendid Fairy-wrens was typical of insectivores of their size (9–10 grams) and the gizzards had the thick muscular walls found in insectivores. The syrinx of malurids is the typical complex of the oscines (see Glossary), with several pairs of intrinsic muscles (Ames 1970).

The morphology of the skull of the family Maluridae has not been studied in detail. Schodde (1982*b*) has commented on differences in the structure of the palatine region of the skull in *Malurus*, *Amytornis*, and *Stipiturus*, which he suggested may be correlated with differences in bill morphology, based on Bock's (1960) examination of the palatine process of the premaxilla throughout the passerines.

In one aspect of cranial morphology, the ear region, the malurids differ from other Australian families so far investigated. Shane Parker described structures in the squamosal region at the rear of the skull of some malurids, especially in birds of open, more arid areas, similar in appearance to the tympanic bullae of mammals (Parker 1982*b*). His paper contains detailed drawings of the skulls of *Malurus*, *Stipiturus*, and *Amytornis*. For comparison, there are drawings of skulls of five other species of Australian passerine, and he refers to comparisons with seven other species. As he explains, in most passerines the tympanic cavity is usually a shallow concavity in the skull enclosed at the back by the Tympanic Wing of the exoccipital, and ventrally by the Tympanic Wing of the parasphenoid. In malurids, the tympanic cavity is enclosed in a more inflated bony wall, similar to the tympanic bulla of mammals that encloses the middle ear (Fig. 4.11 (a) & (b)). In the 12 non-malurid species examined by Parker, the auditory region is the typical open cavity with little sign of inflation (Fig. 4.11 (c)). In his 1982 monograph, Schodde also refers to this structure and illustrates it in four more species. The bulla-like effect is slight in Orange-crowned Wrens and Emperor Fairy-wrens of dense forests in New Guinea and Superb Fairy-wrens of dense thickets in moister habitats of eastern Australia. The degree of inflation is greater in the White-winged Fairy-wren, the Southern Emu-wren, the Grey, and Striated Grasswrens, with the most enclosed, inflated tympanic cavities in the Thick-billed, White-throated, and Dusky Grasswrens (Fig. 4.11 (a) & (b)). This inflated tympanic region is an important diagnostic character of the Maluridae.

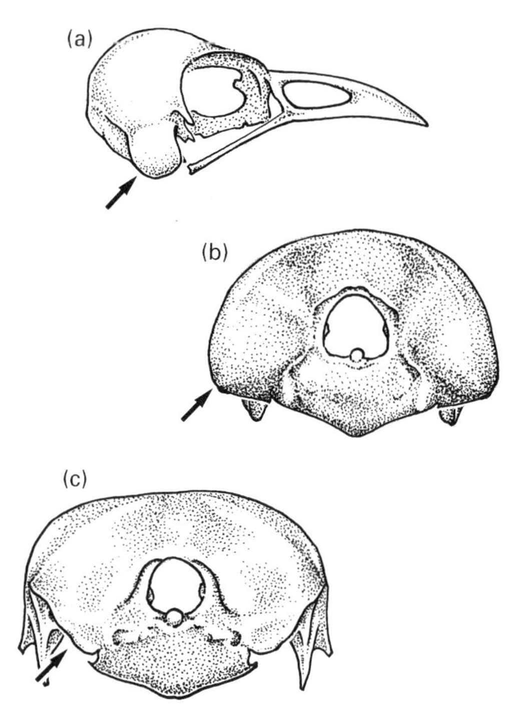

4.11 Views of the skull showing the tympanic region (arrow) inflated in (a) & (b). (a) Lateral view of Dusky Grasswren (*Amytornis purnelli*), (b) posterior view of Thick-billed Grasswren (*A. textilis*), (c) posterior view of the Little Crow (*Corvus bennetti*) of arid and semi-arid areas, with no inflation of the auditory region (Redrawn from Parker 1982*b* and Schodde 1982*b*; not drawn to scale; *C. bennetti* is much bigger than *A. textilis*.)

Both Parker and Schodde draw a parallel with some terrestrial mammals, where there is a correlation between increased relative size of the tympanic bulla and increased aridity of habitat (in rodents, Webster and Webster 1975; in the marsupial *Sminthopsis*, Archer 1981). Similar inflation of the squamosal region of the skull is found in some finches of the family Fringillidae, again more frequently in birds of open habitat (Tordoff 1954). Studies in mammals suggest that enlarged middle ear cavities enhance low frequency sound reception, and hence are adaptive in avoiding predators such as snakes and owls. However, no firm relationship between habitat and the extent of modification to the tympanic bulla has been found. In the Maluridae, as Parker points out, contrary to expectation, the Southern Emu-wren of dense, usually wet, heaths, shows greater development of the tympanic region than the White-winged Fairy-wren, from arid areas. Clearly, there is a long way to go before the relationship between structure, function, and the evolution of auditory structures is understood. Forms that evolved in arid areas may have given rise to forms that now inhabit more mesic areas, but retain the auditory structures appropriate to drier habitats. It may also be significant that the grasswrens have much more varied and complex songs than are found in fairy-wrens and emu-wrens.

Male reproductive anatomy

During the breeding season, most male *Malurus* develop a swelling around the cloaca, the cloacal protuberance, which is a site for sperm storage (Parsons and Cleland 1926; Tidemann 1983) and results from enlargement of the distal end of the ductus deferens (Wolfson 1954). It has also been found in the Thick-billed Grasswren (M. G. Brooker, unpublished observations), but its presence in other species of grasswren, emu-wren, or the New Guinea species is not known.

The cloacal protuberance appears as a large, bulbous structure surrounding the vent, slightly bilobed at the posterior end and with a prominent pointed tip at the anterior end (Fig. 4.12). The average dimensions are: length 7.9 × width 6.3 × height 5.3 millimetres. The protuberance changes seasonally; in winter, it is invisible and in Superb Fairy-wrens in Canberra, it begins to swell 3–4 weeks before the first eggs are laid. It declines after the breeding season, at the onset of the major moult. The moult into nuptial plumage by older males occurs some months before the swelling of the cloacal protuberance; all males develop one and its size does not vary with age (Mulder and Cockburn 1993).

The cloacal protuberance of *Malurus* is among the largest so far reported in passerines; similarly, their estimated relative testis mass (10 per cent of body mass) is larger than that of most other species of comparable size. Many

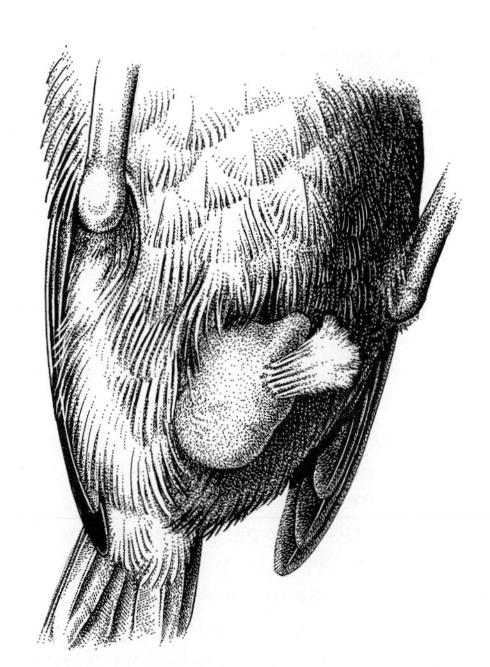

4.12 The cloacal protuberance of the Superb Fairy-wren (*Malurus cyaneus*) when fully developed during the breeding season. The ring of feathers on its ventral surface surrounds the cloaca. The cloacal protuberance serves as a sperm storage organ, probably significant in allowing frequent extra-pair copulations. (Drawn from the photograph in Mulder and Cockburn 1993.)

species with large cloacal protuberances have high copulation rates, but copulation in *Malurus* is rarely observed and rates are unknown. The large cloacal protruberances and testes of fairy-wrens provide large reserves of sperm, presumably for extra-pair copulations (Mulder and Cockburn 1993; Tuttle *et al.* 1996).

Feeding behaviour and food

Foraging

The Maluridae are mainly insectivorous, eating a wide variety of prey captured as they forage over the ground and through the understorey shrubs. Most members of the family spend their lives within 1 metre of the ground, over which they progress rapidly in hops, with both feet together, and from which they spring up to grab prey under leaves and branches as they pass. Several Australian field studies have tried to understand how various species exploit the food resources available in the environment in different ways. They have followed individual birds as they forage through the forest, and recorded foraging acts, scoring the type of foraging behaviour used, and where the food item was located. Many of these studies have included fairy-wrens, and give us a good picture of how they forage and compete with other insectivores.

Nine different methods used by forest birds to catch food were defined by Recher *et al.* (1985), and these categories have been generally used in later studies. Six of these are practised by the Maluridae:

glean—prey is taken from a substrate while the bird remains perched;

probe—bird inserts beak at least partly into a substrate while perched;

snatch—flying bird takes a prey item from a substrate;

hover—similar to snatch, but the bird remains stationary in the air before or while taking the prey;

hawk—flying bird takes a flying insect;

pounce—a bird flies or drops down from a perch to take a prey item on the ground.

Specific feeding locations are classified as ground (whether grassy, bare, or litter), foliage, twigs, branches and trunk, and air.

The feeding of Superb, Splendid, and Red-winged Fairy-wrens has been studied in forest and open woodland habitats. Feeding close to the ground renders them liable to predation from a variety of birds, reptiles, and mammals and so they seldom venture far from cover and tend to forage in family groups that afford added security. The prey taken varies widely with location, time of year, and time of day. As Table 4.4 shows, all the Maluridae so far studied forage predominantly either at ground level or within 2 metres of it, in low vegetation, chiefly understorey shrubs. Most foraging is by gleaning, hopping across the ground, or through shrubs and bracken, in what Rowley (1965) called 'hop-search', seizing whatever insects they encounter. They also spend some time gleaning on branches and twigs and at certain times of year, they probe for insects in the loose bark of eucalypts as it is shed. In fibrous-barked eucalypts such as iron-barks and stringy-barks, insects are present all year round, but for much of the year, the smooth-barked eucalypts provide little habitat for insects. When the bark is shed during the annual cycle, often late summer/autumn, loose bark flaking off the trunk harbours many arthropods, and we have frequently seen Red-winged Fairy-wrens make brief forays onto the trunks of Karri trees (*Eucalyptus diversicolor*) and probe among the loose bark. It is an exposed site, and such forays are generally brief. In years when Marri (*E. calophylla*) flowers in January–February, Splendid Fairy-wrens spend a lot of time more than 5 metres above ground level in the canopy feeding on the insects attracted to the blossoms.

Occasionally, the forager jumps or briefly flies to capture prey a short distance above it—this 'snatching' sometimes develops into hawking when the prey takes off. Most such

Table 4.4 The location of foraging activities in different fairy-wren (*Malurus*) species, and the foraging methods used.

(a) Foraging substrate (percentage of feeding events)

Species	*Ground (bare, litter, grass, logs)*	*Branches/ twigs*	*Trunk/ loose bark*	*Foliage*	*Air*
Superb Fairy-wren[1]	85	2	—	9	4
Superb Fairy-wren[2]	56.1	15.1	2.4	24.2	2.1
Superb Fairy-wren[7]	64.0	7.0	0.0	26.3	2.6
Red-winged Fairy-wren[3]	39	6	—	52	3
Splendid Fairy-wren[4]	—	10	10	45*/35†	—

*Saplings
†Low shrubs

(b) Method of taking prey (percentage of feeding events)

Species	*Glean*	*Snatch*	*Probe*	*Hover*	*Pounce*	*Hawk*
Superb Fairy-wren[1]	92	4	—	—	—	4
Superb Fairy-wren[2]	79.3	17.7	0.2	0.3	0.5	2.1
Superb Fairy-wren[7]	84.2	12.3	0.0	0.8	0.0	2.6

(c) Height (m) at which birds foraged (percentage of feeding events)

Species	*Ground*	*<2*	*0.2–4*	*≥2*	*Season*
Superb Fairy-wren[1]	80	—	20	—	Oct–Jan
Superb Fairy-wren[2]	55	34	—	11	All year
Superb Fairy-wren[6]	16	62	—	22	Aug–Dec
Superb Fairy-wren[7]	43.3	55.0	—	1.7	Oct–Mar
Superb Fairy-wren[7]	63.0	35.2	—	1.9	Apr–Sept
Variegated Fairy-wren[6]	3	74	—	23	Aug–Dec
White-winged Fairy-wren[6]	2	70	—	27	Aug–Dec
Splendid Fairy-wren[5]	56	20	—	24	Apr–Jul
Red-winged Fairy-wren[3]	39	44	—	17	May

Source: [1] Recher *et al.* (1985); [2] Ford *et al.* (1986); [3] Wooller and Calver (1981*a*); [4] Keast (1975); [5] Tullis *et al.* (1982); [6] Tidemann (1983); [7] Cale (1994) and personal communication.

foraging is opportunistic, and involves a range of prey. Sometimes an abundance of food occurs, as when termites swarm, and the focus is on one technique—in the case of flying termites, the birds often fly a towering chase in pursuit, rising several metres vertically to catch such rewarding large prey. At other times, an individual may perch over an ant-nest snatching the winged ants as they emerge (W. T. Cooper, *in litt.* 3 July 1996).

For most species, winter is the time when food is scarcest and prey are sought on the ground, or on stems and twigs. At higher latitudes, ice and snow may preclude such foraging, but in most of Australia, resident species can find food throughout the year. A universally available resource, not always recognized, is the great variety of ants that are active for most of the year. Peter Cale compared summer and winter foraging sites of birds in eucalypt forest in Tasmania (Cale 1994): in winter, 85 per cent of foraging by Superb Fairy-wrens was on the ground or fallen logs, and another 7 per cent was on branches, with very little on foliage. In summer, only 47 per cent of foraging was on the ground, and 45 per cent was on foliage.

In semi-arid shrubland, Sonia Tidemann compared the foraging of three coexisting fairy-wrens (Tidemann 1983). Superb Fairy-wrens spent more time on the ground than the other two species, but they also foraged more and ventured higher into the occasional trees, with *c.* 20 per cent of foraging in trees up to 5 metres from the ground, and their diet included more ants. White-winged and Variegated Fairy-wrens spent almost all their time in shrubs, but whereas the Variegated Fairy-wren gleaned in the inner part of bushes, the smaller White-winged Fairy-wren gleaned in the outer parts.

The Purple-crowned Fairy-wren is found in *Pandanus* fringing tropical rivers, and it uses the basic *Malurus* equipment to feed on that very specific substrate. Its legs are thick and its feet are large, and it can move along the spiny *Pandanus* fronds and grip the spiny edges, probing the composted detritus accumulated in the leaf axils with its relatively long and stout bill. Of the other Australian *Malurus*, the tropical Lovely Fairy-wren probably uses the ground least; it appears to spend most of its time in the thick shrubby vegetation of rainforest margins, gleaning among the shrubs and vines; it also appears to fan its tail while it forages.

Emu-wrens are so hard to watch for more than a few seconds that their foraging ecology is not very well known. What has been recorded suggests that they spend little time feeding on the ground and feed mostly by gleaning within the shrubs from which they so rarely emerge. Most information on the behaviour and breeding of Emu-wrens comes from the patient observation of a Tasmanian schoolteacher nearly 70 years ago, Miss J. A. Fletcher. In *Further field notes on the Emu-wren* (1915, p. 214), she records 'the stems of the big reeds (*Juncus pallidus*) are one of their chief sources of food supply. These reeds the birds split open, and devour the many forms of insect life which shelter therein' (later identified as spiders, ants, larvae of beetles and moths). Mallee and Rufous-crowned Emu-wrens live in habitats dominated by spinifex, in which they chiefly forage.

Grasswrens are even more terrestrial than the fairy-wrens, and much of their foraging is by hop-search on the ground, gleaning and probing in the plant litter under shrubs and in rock crevices, and by gleaning in shrubs. Most information on feeding lacks detail, and no species have been observed specifically to record feeding methods or sites; information on diets is generally from the stomach contents of collected specimens (Barker and Vestjens 1990). The Striated Grasswren has been recorded pulling the blossoms off a flowering Cactus Pea (*Bossiaea walkeri*), eating the pistils and discarding the petals (Izzard *et al.* 1973). Although most observations of the White-throated Grasswren in the Northern Territory lasted only a few seconds, Noske (1992) recorded them feeding mostly by gleaning from bare rock surfaces, from litter, or from mats of the plant *Micraira*. He saw birds leap several centimetres from the ground to snatch insects from leaves and stems of shrubs, and peck repeatedly at the ground under a shrub, probably gathering seeds. Similarly, our observations of the Thick-billed Grasswren are mainly of birds foraging on the ground in the litter that accumulates under scattered shrubs, and running fast between shrubs. The photographer Graeme Chapman watched a male Thick-billed Grasswren (*A. t. modestus*) foraging on its way into a nest, turning over small stones in search of food (*in litt.*, 3 October 1995).

For the New Guinea malurids, most of what is known of food and foraging comes from the detailed studies of New Guinea bird communities by Diamond (1972), Bell (1970, 1971, 1982*a,b,c*, 1984; Bell *et al.*1979), and the field observations recorded in Schodde's 1982 monograph and Coates' *Birds of Papua New Guinea* (1990). The White-shouldered Fairy-wren is similar in size, morphology, and foraging range to Australian *Malurus*, and obtains food by gleaning from vegetation and the ground, occasionally by snatching. Emperor Fairy-wrens forage mainly by gleaning, within 2 metres of the ground, in dense shrubbery along streams, patches of regrowth, forest edges, and secondary vegetation; about one-third of the foraging observed was on the ground or litter, half on leaves or palm fronds and the remainder on branches (Bell 1984). The bill of Emperor Fairy-wrens is slightly longer but similar in shape to that of Australian *Malurus*.

The other New Guinea species, the Broad-billed Fairy-wren, Wallace's Wren, and the Orange-crowned Wren show increasing adaptations to arboreal foraging, with shorter tarsi and tails, longer bills wider at the base, and long rictal bristles (see Fig. 4.6). They forage higher above the ground, at 2-10 metres, all in slightly different niches, but all moving through the canopy of the dense undergrowth of tall forest, in tangled shrubs, and vines, rattans, etc. They forage by leaf-gleaning (including hang-gleaning from the underside of leaves when hanging upside-down), probing in cracks, snatching, hawking, and hovering to pick insects off leaves.

Food

Many malurids live in the highly variable semi-arid and arid parts of Australia, but even the milder coastal areas are subject to severe, although less frequent, droughts. In such environments, the success of these small birds is due to their very general diet, their ability to exploit a glut, and their ability to make do on anything that is available. Most information on the diets of the Maluridae comes from records of stomach contents of birds collected as specimens (Harrison 1974; Barker and Vestjens 1990; Schodde 1982*b*). A few studies have used faecal analysis, although these underestimate soft-bodied insects (Calver and Wooller 1982). Many early records give only qualitative information on presence or absence, with no indication of relative numbers. Much of this information is summarized in *The Food of Australian Birds* by Barker and Vestjens (1990). Few studies have followed changes in diet through the year and in both good and poor seasons. All members of the family Maluridae are predominantly insectivorous, with all probably eating some plant material at different times of the year, most significantly the grasswrens.

In Table 4.5, we show the different kinds of arthropods (insects and spiders) eaten by various *Malurus*. Grouping insects into a few major categories in a table tends to obscure the great diversity of diets—the identified insects eaten by Superb Fairy-wrens as listed in Barker and Vestjens (1990) come from 40 families, as well as many unidentified 'bugs' and 'beetles'. For some species we can compare diets in summer and winter, and what adults eat with what they bring to their nestlings. For all species, diets are very varied; ants are clearly significant for Superb, Splendid, and Red-winged Fairy-wrens, especially in winter. Ants are a very diverse and abundant component of the Australian insect fauna, much utilized by small ground-foraging birds and they are an important 'last resort' food at times of drought or after fires. After a wildfire burnt out 99 per cent of their habitat, we watched Splendid Fairy-wrens sitting on a fence, waiting for, and then robbing, worker ants carrying a load back to their colony.

The food fed to nestlings is different from that eaten by adults. For the three species in which nestling diets have been quantified, the most frequent item in the nestling diet is not something commonly eaten by their parents.

Table 4.5 The diets of fairy-wrens (*Malurus*): the percentage of different invertebrate groups identified in the diet from quantitative studies of faecal samples or stomach contents.

Species of fairy-wren/season	*Col*	*Hem*	*Dip*	*For*	*Hym*	*Lep*	*Aran*	*Orth*	*Iso*	*Other*
Adults										
Splendid[1]	7	2	1	84	4	*	*	*	*	2
Splendid[2]	14	13	0	46	3	0(2)	3	4	13	3
Superb[3]	61	4	4	17	5	1(3)	1(2)	*	*	4
Variegated[3]	20	20	0	11	3	3(4)	5(31)	*	*	2
White-winged[3]	46	23	0	8	0	2(4)	3(7)	*	*	8
White-winged[4]	81	0	5	0	5	*	*	0	*	9
Red-winged—May[5]	28	0	4	51	7	*	4	6	*	0
Red-winged—Sept[5]	35	0	19	16	21	*	5	3	*	1
Red-winged—Summer[6]	15	11	7	26	6	1(15)	10	1	*	8
Purple-crowned[7]	46	9	2	34	6	*	3	1	*	0
Superb—Summer[8]	15	29	2	43	<1	1(1)	2	5	*	1
Superb—Winter[8]	9	7	8	70	2	<1(2)	<1	2	*	0
Nestlings										
Superb[3]	2	4	56	0	1	14(17)	2(1)	1	*	1
White-winged[3]	1	8	17	0	3	36(16)	8(3)	1	*	8
Splendid[2]	14	7	0	12	3	3(5)	11	35	1	0

In most cases, arthropods were identified only to family level, abbreviated as follows: Col: Coleoptera (beetles); Hem: Hemiptera (bugs); Dip: Diptera (flies); For: Formicoidea (ants); Hym: other Hymenoptera (wasps, bees); Lep: Lepidoptera (moths and butterflies) (larvae); Aran: Araneae (spiders) (eggs); Orth: Orthoptera (grasshoppers, etc.); Iso: Isoptera (termites).
Source: [1] Tullis *et al.* (1982); [2] Rowley *et al.* (1991); [3] Tidemann *et al.* (1989); [4] Wooller and Calver (1981*b*); [5] Calver and Wooller (1981); [6] Rowley *et al.* (1988); [7] Rowley and Russell (1993); [8] Rowley (1965).
* Category not listed by author.

In general the adults eat very small items themselves and carry larger items such as caterpillars, crickets, grasshoppers, or large flies back to their nestlings. In contrast to fairy-wrens which bring back items singly, we have seen Thick-billed Grass-wrens bring several items in a bill-full.

All species of *Malurus* for which a significant number of identified items are listed by Barker and Vestjens ingest seed. These are sometimes regarded as 'ingested coincidentally with insects' (Tidemann *et al.* 1989); however, other observations suggest this is not necessarily so. In some species they are a significant item; e.g. 8 per cent of 459 items listed for White-winged Fairy-wrens by Tidemann *et al.* (1989) were seeds. Rowley (1965) found seeds of the native chenopod shrub *Rhagodia* in 6–7 per cent of stomachs of Superb Fairy-wrens sampled, and also in one White-winged and four Variegated Fairy-wrens. 'The frequency of the occurrence of these particular seeds is beyond the realms of chance; in fact the ingestion of no less than 45 *Rhagodia* seeds by five different individuals makes it very unlikely that they were ingested in mistake for insects, and the fact that the fleshy pericarp had been effectively digested leaving only the hard seed suggests that *Rhagodia* spp. provide a minor source of food for the *Malurus* spp.' (Rowley 1965, Appendix 2, p. 296). We had evidence that White-winged Fairy-wrens in a coastal population near Perth, Western Australia, were also eating *Rhagodia* fruits when we mist-netted a pair of birds in heathland, and on removing them from the nests were alarmed to see that they appeared to be bleeding from the cloaca. Closer inspection revealed that they had been eating the ripe cherry-red fruits of *Rhagodia*.

Barker and Vestjens frequently recorded seeds in the stomachs of grasswrens, and material from several different plant families has been identified. In particular, seeds of several grasses, including spinifex, and of acacias have been recorded—not just one or two, but sometimes a crop full of seeds from one particular species. The bills of the Eyrean, Grey, and Thick-billed Grasswrens from the most arid environments are especially stubby and finch-like. Eyrean Grasswrens have the deepest, heaviest bills, and Parker *et al.* (1978) found seeds and insects in approximately equal amounts in their stomachs. Schodde (1982*b*) suggests the function of their specialized bill is to husk or crush the large seeds of the Dune Cane Grass *Zygochloa*, the most characteristic plant of their habitat.

5

Social organization and vocal communication

Introduction

Mormon wren was an early common name for fairy-wrens because people seeing a group with one full plumaged male and a number of female-plumaged brown birds assumed that this was a male with his many wives. The clarification of the plumages of the immatures and the two sexes, the timing of the moult, and the variable duration of the eclipse plumage in males, explained the apparent harems and left the most obvious interpretation of their social arrangements as monogamous breeding pairs with their philopatric offspring. With the advent of colour-banding, the true nature of these groups was recognized—a breeding male and female, with non-breeding adult helpers and immatures (Rowley 1957). This was Co-operative Breeding, the subject of Chapter 6. For just over 20 years, these family groups were regarded as a monogamous breeding pair with their non-breeding offspring that remained in the family group as helpers. David Lack's assessment that monogamy was the predominant mating system in birds was also generally accepted because it appeared to occur in 90 per cent of species (Lack 1968). Genetic studies have since shown how false Lack's assumption was, and allow us to understand aspects of malurid behaviour that were described over many years, but were previously impossible to interpret. Fairy-wrens of the genus *Malurus* may best be described as having a monogamous social system and a promiscuous mating system. What little is known of other genera in the family suggests that they may be similar.

The basic outline of behaviour in the Maluridae depends largely on the studies of Superb, Splendid, and Red-winged Fairy-wrens (*M. cyaneus*, *M. splendens*, and *M. elegans*), living in groups larger than a simple pair whose long-lasting members all help raise nestlings, but we have included information on other species where it is available.

Groups and territories

Group size and composition

In those species that have been studied, adult malurids are found to be resident all year round, living in pairs or groups that defend the boundaries of a territory by song and by chasing intruders. Not only can one rely on finding the same individuals in the same place month after month, but these territories also persist in much the same places year after year (Fig. 5.1). Group and territory sizes vary with species, habitat, and the recent past conditions that will have influenced productivity and survival; mean sizes of groups and territories of well-studied species are shown in Table 5.1. From many reports in the literature it appears that other lesser known fairy-wrens and members of the other four genera are also territorial and group-living at least some of the time, but without colour-banded individuals and dedicated long-term study we cannot present comparable data on group size and

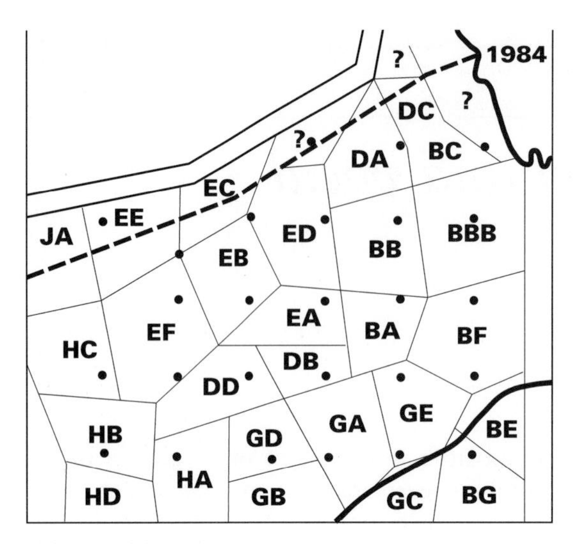

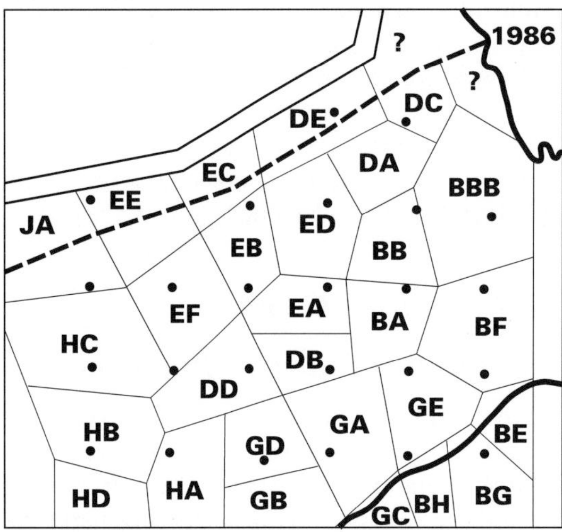

5.1 Territories of the Red-winged Fairy-wren (*Malurus elegans*) in Smith's Brook Reserve during 1984 and 1986. Territory boundaries are approximate, with all the area occupied. Dots are *c.* 100 m apart. Boundaries remained very similar over the 2 year interval, but one new territory (BH) was established when a new pair occupied a previously underutilized area (From Rowley *et al.* 1988.)

composition, although we are confident that the same patterns and processes probably apply across all the family.

Groups arise because the surviving progeny from a nesting attempt remain in the family group after they reach independence and many, especially in fairy-wrens, stay on for at least a year after they reach sexual maturity. The size of any particular group and the proportion of groups with helpers in any population wax and wane depending on how successful they were during the previous breeding season, the survival of juveniles and adults, and the availability of breeding opportunities, at home or in neighbouring territories. In the species shown in Table 5.1, the most frequent group is the simple male and female pair, but for the species that we know best, Splendid Fairy-wrens and Red-winged Fairy-wrens, most birds live in groups; most of the possible combinations of males and females have occurred at some time (Table 5.2), with one female and two males being the commonest. Larger groups are found in the Red-winged Fairy-wren (mean 4.1) where, although annual productivity is low, annual survival is very high and the habitat is more or less fully occupied by territories so that vacancies are few for dispersing males or females to occupy. In Splendid Fairy-wrens, with a mean group size of 3.1, the annual survival of breeding males and females is slightly lower and more helpers are able to disperse to breeding vacancies at an earlier age (Table 5.1). In the Australian National Botanic Gardens in Canberra, larger groups of Superb Fairy-wrens also outnumber pairs.

Groups with female helpers have been recorded in all species in Table 5.1. In Canberra, Rowley recorded only two groups of Superb Fairy-wrens where a second female was present at the start of the breeding season, and Mulder found none; in all other cases, young females dispersed from their natal territory within 12 months of fledging. Some dispersed soon after fledging, and the rest shortly before the following breeding season, generally in August, induced by persistent aggression from mothers towards their daughters (Rowley 1965; Mulder 1995). However, at Armidale, 560 kilometres further north, Nias found that, in some years, female helpers were present in almost all groups (Nias 1987). Some were still in their natal groups, but others had joined the group before the breeding season. Apart from this difference between Superb and other

Table 5.1 Aspects of social organization in fairy-wrens: duration of study, mean group and territory size, the percentage of groups that were simple pairs, and mean sex ratio with range of yearly values where available.

Species of fairy-wren	*Years (group-years)*	*Mean group size (range)*	*Territory size (ha)*	*% pairs*	*Sex ratio M/F*	*Source*
Splendid	18 (374)	3.1 (2–8)	4.4	36.1	1.3 (1.1–1.8)	Russell and Rowley (1993*b*) Rowley and Brooker (1987)
Red-winged	15 (398)	4.1 (2–9)	*c.* 1	16.8	1.2 (1.0–1.4)	Rowley and Russell, unpublished data
Blue-breasted	4 (61)	2.5 (2–4)	1–2	67.2	1.0	Rowley and Russell, unpublished data
Variegated	4 (18)	2.2 (2–3)	—	83.3	1.0	Tidemann (1983)
Purple-crowned						
Drysdale R. WA	4 (101)	2.3 (2–7)	200–300 m*	80.2	1.2	Rowley and Russell (1993)
Gregory R. Q	3 (73)	2.2 (2–7)	—	86.3	1.1	Rowley and Russell (1993)
Superb						
Canberra†	5 (47)	2.3 (2–5)	—	61.7	1.4	Rowley (1965)
Canberra‡	5 (170)	2.9 (2–5)	0.6	42.9	1.9	Mulder (1992)
Armidale	4 (51)	2.5 (2–6)	1.25	68.6	1.0–1.7	Nias (1987)
Booligal	4 (44)	2.0 (2)	1–2	100	1.0	Tidemann (1983)

From studies in which birds were individually marked; some species were studied at more than one place.

* length of linear territory.
† Gungahlin.
‡ Canberra Botanic Gardens.

fairy-wrens, the immediate cause of many differences in group size and composition appears to be the demographic environment—if there is nowhere for young birds to disperse to, they remain in their natal territory.

Groups of the White-winged Fairy-wren (*M. leucopterus*) are large and complex, with more than one breeding female and her attendant male or males. Although these females and their consorts nest in different parts of the clan area, after breeding the sub-groups coalesce and make use of the whole clan area. These complex plural breeding groups in White-winged Fairy-wrens may have derived from the occasional plural breeding groups found in other fairy-wrens (see below, Plural breeding).

For those malurids not mentioned in Table 5.1, there are many reports of birds seen in family groups, but unless the birds were individually marked, such sightings during or soon after the breeding season are not evidence that the group is more than parents with recently produced juveniles. Only groups of more than two birds seen at the start of the breeding season before young are produced suggest that a species may live in groups all year round. Confirmation of this, and of the occurrence of co-operative breeding, depends on a study of individually marked birds, and for most species this has not yet been done. Reports of group size from the literature are quoted in the accounts of individual species and we shall not repeat them here. Individually they are not very significant, but together they are enough to suggest that group-living, and probably co-operative breeding, occur throughout the family, including at least some of the little known, more arboreal, New Guinea species.

Table 5.2 Group composition of Splendid Fairy-wrens (*Malurus splendens*) (18 years, 365 group-years), and Red-winged Fairy-wrens (*Malurus elegans*) (15 years, 398 group-years).

Number of females	Number of males 1	2	3	4	5	6
Malurus splendens						
1	36.1	20.9	8.8	1.9	0.5	—
2	8.8	8.0	6.1	0.8	0	—
3	2.9	1.6	2.1	0.5	0.3	—
4	0	0.5	0	0	0	—
Malurus elegans						
1	16.8	14.3	7.8	3.8	0.5	0.3
2	9.0	11.6	9.8	3.0	0.3	0.3
3	3.0	5.5	3.5	3.8	0.5	0
4	2.0	1.3	1.5	0.5	0.8	0
5	0	0.3	0	0	0	0

Percentage groups with the number of males and females as indicated. (Russell and Rowley, unpublished data.)

For most grasswrens, there is sufficient evidence to suggest that in all species, groups of more than a simple pair may occur from time to time. The generally arid habitat where they live suggests that the incidence of groups will be low and variable, unlike species such as the Red-winged Fairy-wren that live in relatively benign environments. Some evidence of co-operative breeding is provided by records of more than two adults feeding fledglings in the Black, Thick-billed, and White-throated Grasswrens (*A. housei*, Freeman 1970; *A. textilis*, Brooker 1988; *A. woodwardi*, Noske 1992).

Changes in the composition of groups are gradual. Losses are from the death of group members or their dispersal, while additions result from the production of surviving offspring and occasionally the dispersal into the group of a replacement after the death of a senior male or breeding female. Occasionally an extra female from another group joins as a helper. In all fairy-wrens that have been studied, females tend to disperse from their natal territories, while males tend to remain where they were hatched—to be philopatric.

New groups

In our studies of Splendid and Red-winged Fairy-wrens, a few new territories were established, or territories known to have been vacant for a while were reoccupied. In most cases the new group consisted of a helper male from a territory nearby who had acquired a female partner. Sometimes a territory was divided and part of it occupied by one of the group's helper males with a female partner from outside. Some experiments with Superb Fairy-wrens relocated males and females and watched what happened in the vacancies thus created (Pruett-Jones and Lewis 1990). These experiments suggested that males did not disperse into an empty territory unless a female was available, either a helper in a nearby territory or a passing dispersing female. On the other hand, if a vacancy was created experimentally by the relocation of either a male or female, leaving the partner in sole possession of the territory, vacancies were filled very rapidly, often within hours. This may explain why territories of Splendid Fairy-wrens that had become vacant sometimes remained so for one or more breeding seasons; surplus male helpers were always available in our study area, but, at times, females were in short supply (Russell and Rowley 1993*a*).

Territory size

Groups generally maintain large, all-purpose territories that are probably bigger than they need to be when conditions are very good or group size is low, but that are large enough to support a group in a poor year or after the group is enlarged following a good breeding season. In continuous habitat, territories form a mosaic, with each surrounded by from four to eight neighbours and the boundaries changing little from year to year (Fig. 5.1). The size of a territory varies with species, condition of the habitat, population density, group size, and time of year (Table 5.1). Accurate estimates depend on individually marked birds and intensive study, and are available for only a few species.

Several studies of Superb Fairy-wrens have been carried out in different parts of Australia and serve to illustrate how the ecology of one species can vary. In the Canberra Botanic Gardens they occupied small contiguous territories of mean area 0.6 hectares; 560 kilometres further north, near Armidale, not all the habitat was suitable, but territories adjoined in suitable areas, with a mean area of 1.25 hectares; 400 kilometres to the west, near Booligal, at the edge of the species' range only groups of two occurred with territory areas of 1–2 hectares. The area used during nesting is usually smaller than the area used during the rest of the year or that defended before breeding starts. In the non-breeding season, the defence of territory boundaries in Superb Fairy-wrens is at a much lower level, resulting in extensive overlap of home ranges which are enlarged beyond territorial boundaries to include nearby 'neutral areas'. Large foraging aggregations may occur here, sharing much of the area previously occupied by several groups. However, Nias (1987) recorded that individual groups returned to their own territory at night to roost. All these studies occurred in habitat modified either for urban gardens or pastoral activities, where not all available space was suitable for breeding territories. It is not known whether similar behaviour occurs in unmodified natural habitat.

We have not seen any signs of groups coalescing into foraging flocks outside their own territories in Splendid, Red-winged, or Blue-breasted Fairy-wrens (*M. pulcherrimus*), although in Splendid Fairy-wrens with very large territories, groups wander over the whole of their areas in the non-breeding season, and even into adjacent territories when the owners are far away. Red-backed Fairy-wrens (*M. melanocephalus*), a species that particularly likes tall grass, appear to leave their territories and to wander in nomadic flocks after breeding, especially in northern Australia. As the long grass of the northern savannah woodlands dries out, collapses, and is frequently burnt, the birds abandon this ephemeral habitat and congregate in the thick cover of riverside and other refuge areas.

In contrast to the territory mosaics found in forest or heathland habitats, the territories of the Purple-crowned Fairy-wren (*M. coronatus*) in northern Australia are strung out in linear succession along the rivers they frequent. Each territory usually has only two neighbours and is 200–300 metres long, including both banks of the river, but where two rivers join, or a river has islands and a number of channels, territories may form a more complex mosaic. The areas occupied by clan groups of White-winged Fairy-wrens are considerably larger than the 1–2 hectare territories of most fairy-wrens, 2–6 hectares at Booligal (Tidemann 1990), and 4–6 hectares in coastal heath at Pipidinny, Western Australia (Rowley and Russell 1995).

For the White-throated Grasswren, the largest of the grasswrens (34 grams, adult male), Noske (1992) estimated territory areas of *c.* 10 hectares from the estimated number of groups in an area that he surveyed. Estimates of density and territory size for other grasswrens, emu-wrens, and New Guinea malurids are less exact.

Territory defence

In fairy-wrens, most territory defence takes the form of song battles, and occupancy is regularly advertised by singing (Fig. 5.2). The typical songs of fairy-wrens are described on p. 63–74. Most singing occurs early in the morning and in the evening, but intermittent song is heard throughout the day. Territory boundaries are defended most vigorously just before the breeding season, when new pairs may attempt to establish themselves in a vacant territory or in a poorly defended area. Song near a boundary receives an immediate loud reply from the neighbouring owners. If neighbours or strangers trespass, the intruders are vigorously chased back across the boundary. If chasing fails, the defender may threaten

5.2 Male Superb Fairy-wren (*Malurus cyaneus*) singing. Occupancy of a territory is regularly advertised by song, especially early in the morning and in the evening. All adults in a group sing in defence of their territory.

5.3 Threat posture of Splendid Fairy-wren (*Malurus splendens*). Feathers on the crown, nape, and face are erected, the wings are held out from the sides and depressed, and the head and tail may be lowered, while a loud threat call is uttered. Similar postures are seen in other species, directed generally to conspecific intruders or cuckoos. (From photographs by R. B. Payne.)

the intruder with body feathers part erected, wings held slightly out from the sides, and both head and tail lowered (Fig. 5.3). Group members recognize the songs of other members of their social group, and in play-back experiments responded aggressively to songs of birds from other social groups (Payne *et al.* 1988, 1991). In Superb Fairy-wrens, play-back experiments found that songs of birds in neighbouring territories elicited less response from females than songs of complete strangers (Cooney and Cockburn 1995).

Actual fights with intruders are rare, although we have seen them in both Splendid and Red-winged Fairy-wrens, and they were described in Superb Fairy-wrens by the Bradleys (1958). Physical attack is more likely to be directed at a cuckoo, as described by Payne *et al.* (1988) for Splendid Fairy-wrens presented with a stuffed cuckoo near a nest, when group members attacked the dummy vigorously with beak and claws. An intruding lone female is chased and threatened by the breeding female of the territory. Most territory defence is intraspecific, but where species overlap, interspecific song battles and chases may occur, especially in the breeding season.

An 'aggressive display' similar to the threat display of fairy-wrens was described for the Striated Grasswren (*Amytornis striatus*) by Hutton (1991), which may be followed by a very rapid 'attack flight' for 1–2 metres towards the object of the aggression.

When the intruder takes the form of a potential predator the aggressive display may be replaced by a distraction display; this occurs particularly when a nest containing young birds is threatened, and commonly takes the form of a 'Rodent-run' directed away from the predator and the nest (Fig. 5.4). The head and neck are extended and lowered, while the tail is depressed but not fanned. The wings are held slightly out from the body, and the feathers on the top of the head, throat, and mid-back are raised, giving the bird a hunched-up, fluffed-out appearance. The legs are more bent than in normal locomotion and the bird travels at a

5.4 Rodent-run distraction display in the Superb Fairy-wren (*Malurus cyaneus*). Instead of its normal locomotion, hopping, the bird runs rapidly with a body posture similar to that in threat (Fig. 5.3), with the tail depressed and often held to one side. The display is given by both sexes in most Maluridae, especially when nestlings are threatened. (From photographs by Ian Rowley.)

rapid run, moving the legs independently, quite different from the usual hop, uttering an alarm call all the while (Rowley 1962). In all aspects except the actual running and alarm call, this display is similar to the threat display.

Rodent-run behaviour has been described in all Australian fairy-wrens and the Emperor Fairy-wren (*M. cyanocephalus*) from New Guinea, in most grasswrens and emu-wrens, and it probably occurs throughout the family.

Daily routine

All the Maluridae are predominantly insectivorous and while such succulent food largely emancipates them from dependence on free water, it means that they must spend most of their waking hours searching for dispersed small prey. In winter, when the days are short and most insects inactive, dormant, or well-hidden, the group travels the length and breadth of their territory, with little pause for rest, from the time that they awake shortly before sunrise until they roost after the sun has set.

In summer when insects are most abundant and varied and the days much longer, bursts of foraging are interspersed with rests when the members of a group may huddle together and preen each other. When it is very hot, most birds seek shade around midday, and malurids are no exception. When shade is scarce, even holes in the ground such as rabbit burrows are used (Ambrose 1984). 'In a small gully vegetated with medium to sparse density of low saltbush.... were three, possibly four, grasswrens. From about 20 metres we noticed a (Grass)wren pass behind a small clump of bush, but fail to reappear. One other bird also seemed to have disappeared. We watched an adult female dart up the gully.....and it also disappeared from our view. ...There was a rabbit burrow near the bush where the first (grass)wren had disappeared. Crouching perfectly still, inside the burrow, was a Thick-billed Grasswren' (Coate 1994, p. 278).

During the spring and summer, malurids greet the dawn with a sustained barrage of song that advertises occupancy of their territory. They then start busily foraging but remain vocal at intervals throughout the day and this helps an observer to locate the birds. In late summer and autumn, most of these species enter a major wing and body moult and become very hard to find. It has been embarrassing to take a visitor to the study area where we knew there are more than 120 fairy-wrens and to be very relieved after two hours to stumble at last across one group.

The feathers and other important parts of a bird need regular attention to ensure they remain functional and durable. This is achieved by preening the feathers with the bill after it has been anointed with oil from the preen gland at the dorsal base of the tail. Preening not only applies the oil as a water-proofing to the feathers, but by mandibulating disarranged feathers, the bird reconnects those barbules that have become separated—a process rather like closing a zip fastener. Loose scales from the legs and feet as well as pieces of sheath from growing feathers are dislodged and discarded during preening. In this way, the bill, legs, and claws are kept oiled and in good condition.

In most species of birds, adults do all this for themselves and the process is called autopreening, but young nestlings with rapidly growing feathers need parental help to keep their plumage tidy. In many species, pairs preen each other during courtship; in a few species adult members of social groups preen each other—this is called allopreening. All those members of the Maluridae that have been closely studied have been found to allopreen. A group of malurids taking a rest during the day's foraging or preparing to roost will line a branch, sitting side by side in contact, and preening their neighbours and receiving preening from them in their turn (Fig. 5.5). It is amusing to watch a group busy allopreening,

5.5 Allopreening in the White-winged Fairy-wren (*Malurus leucopterus*). During a rest from foraging or when preparing to roost, members of a social group perch side by side in contact, preening their neighbours and receiving preening from them. Allopreening is an important social interaction in pairs and groups and occurs throughout the family.

because there seems to be considerable competition for the inner positions and the birds keep changing places, 'leap-frogging' over their neighbours to achieve an inside berth. There does not appear to be any clear hierarchy as to which bird preens which—males preen males and females senior or junior to them in status, as do the females. The amicable environment and obvious enjoyment of the participants must surely help to maintain the social cohesion of these groups throughout the year.

Close to sunset, the whole group moves to a favourite dense shrub with a convenient horizontal branch on which they line up, facing the same way and touching each other, and where they spend the night (Fig. 5.6)

Vocal communication

Vocal communication is an important part of the social system in the family Maluridae. It has at least two major roles: in establishing and defending a territory and in communicating between members of a social group. To what extent vocalizations play a role in the complex mating system, in advertisement by males or females, and in individual recognition is little known. Songs of individual species are illustrated by sonograms in the individual species accounts. Here we make comparisons between species and between genera, and try to characterize the vocalizations of the family.

The songs of all species of fairy-wrens and emu-wrens are structurally very similar and are clearly recognizable. Songs of grasswrens are much more varied and complex. Other vocalizations which we will call Contact and Alarm calls are more similar in all these three genera. Some vocalizations of the tree-wren (*Sipodotus*) and the russet-wren (*Clytomyias*) have been recorded but their context is unknown and they are not discussed further here.

5.6 A roosting group of Superb Fairy-wrens (*Malurus cyaneus*). Following the evening chorus of song, the whole group moves towards its customary roosting site in a dense shrub and birds take up their positions side by side on a horizontal branch.

Vocalizations of fairy-wrens

No detailed account of the complete vocal repertoire of any fairy-wren has yet been published, and an analysis of different types of song, their context, and individual variation in song has been done only for the Superb Fairy-wren.

Song of fairy-wrens

The basic song type of a fairy-wren is immediately recognizable. With experience, the songs of different species can be distinguished, and since no more than three or four species are present in any one area, it is generally possible to identify species from their song. The descriptions of songs in various field guides are not very helpful, however, and their comparisons of pitch are at variance with sonograms.

The song, which may last for 1–4 seconds, comprises a large number of short elements, where an element is the basic unit of a song, defined as a discrete, uninterrupted tracing on the sonogram. The elements are organized into phrases, sequences of several elements which always occur together and appear very similar each time they are sung (Fig. 5.7) After a number of variable, brief introductory elements, the reel continues as a series of repeated phrases, usually identical. The rate of repetition of the elements in the main part of the song (10–20 per second) is frequently such that the elements and phrases cannot be separated by the human ear. Elements are characterized by a high degree of frequency modulation (see Glossary), often very rapid, sometimes a vibrato syllable spanning a constant range of frequencies, sometimes ascending or descending (e.g. Fig. 5.7(a,b)). In fairy-wren songs, the elements themselves, the way in which the phrases are combined, and the length and complexity of the song, may all vary.

The basic elements of songs differ between individuals (Fig. 5.7(a,b)), but in most species, these differences are not detectable by the human ear. In three species of fairy-wren, play-back experiments have shown that birds can clearly distinguish between different individuals on the basis of song, as the differences apparent in sonograms would suggest. In Splendid and Red-winged Fairy-wrens, birds responded more strongly to songs of strangers than to songs of familiar group members, but their responses to songs of birds from outside their own group did not differ between kin and unrelated birds (Payne *et al.* 1988, 1991). Female Superb Fairy-wrens, also, discriminate between group members, neighbours, and strangers (Cooney and Cockburn 1995).

The songs of all species span a large range of frequencies, and because the range within individual elements is so great, it is hard to compare the pitch of songs from different species (Table 5.3). However, it is possible to estimate the frequency range that includes the majority of most elements and to estimate whether the song is higher or lower in pitch than that of another species. A significant part of the elements in the song of the Purple-crowned Fairy-wren lie below 4 kHz, and its song sounds lower than the songs of other species, and is audible to most people. At the other extreme, very little of

5.7 Sonogram of the typical territorial song 'reel' of several species of fairy-wrens. After a number of introductory elements, the song continues as a series of repeated phrases that differ between individuals and between species. (a) Splendid Fairy-wren (*Malurus splendens*) from Balingup, WA, recorded by J. Hutchinson. (b) a different individual of the Splendid Fairy-wren, southwestern WA, recorded by R. K. Templeton. (c) Red-winged Fairy-wren (*M. elegans*), Manjimup, WA recorded by R. B. Payne. (d) Purple-crowned Fairy-wren (*M. coronatus*), Drysdale R., WA, recorded by E. M. Russell. (e) White-winged Fairy-wren (*M. leucopterus*), inland SA, recorded by H. Crouch.. The slower repetition rate in the Purple-crowned Fairy-wren allows the different phrases to be distinguished, but in most species, separate phrases cannot be heard. Song is simpler in Red-winged and White-winged Fairy-wrens. The prolonged introductory elements in the Red-winged Fairy-wren are characteristic of that species.

the song of the Blue-breasted Fairy-wren lies below 5 kHz, and it is frequently inaudible to older ears, since it is also not very loud. Most other songs are intermediate in frequencies between these two extremes.

Estimates of relative loudness are somewhat subjective, and the same song may be delivered softly or with great force depending on context. Nevertheless, the loudest calls of the blue group of fairy-wrens, Superb, Splendid, and

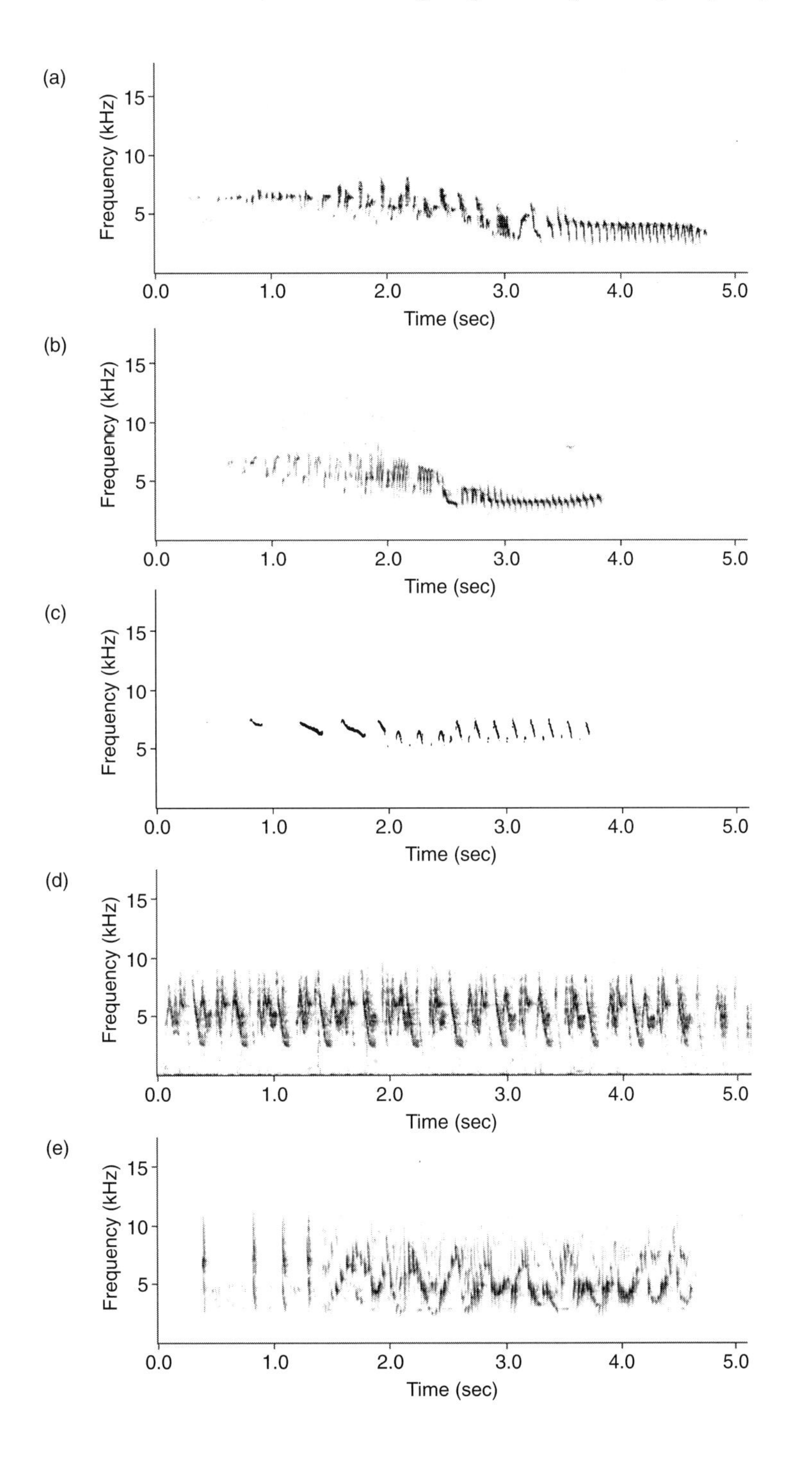

Table 5.3 Characteristics of songs of fairy-wrens and emu-wrens.

Species	*Song type*	*Frequency range (kHz)*	*Loudest frequencies*	*Maximum volume*	*Introductory notes*
Superb Fairy-wren	I	3–9	5–7	Loud	Short
	II	3–9	3–5	Loud	Short
Splendid Fairy-wren	I	2–8	4–7	Loud	Short
	?II	2–8	3–4	Loud	Short
Red-winged Fairy-wren		2–8	5–7	Loud	Long
Blue-breasted Fairy-wren		2–9	5–7	Medium	Short
Variegated Fairy-wren		2–8	5–7	Medium	Short
Purple-crowned Fairy-wren		2–8	4–7	Loud	Short
Lovely Fairy-wren		2–7	?	?	Short
White-winged Fairy-wren		2–8	3–6	Medium	Short
White-shouldered Fairy-wren		3–7	3–6	?	Short
Red-backed Fairy-wren		3–8	4–6	Medium	Long
Rufous-crowned Emu-wren		6–9	7–8	Soft	Short
Southern Emu-wren		6–9	7–8	Soft	Short

Purple-crowned, are much louder than those of the bi-coloured and chestnut-shouldered fairy-wrens, especially the Red-backed and Blue-breasted Fairy-wrens; the Variegated Fairy-wren (*M. lamberti*) is the softest of all. The Superb, Splendid, and Purple-crowned Fairy-wrens have songs with a very complex element structure and sound much harsher than Red-winged, Blue-breasted, and White-winged Fairy-wrens, where the basic elements have a simpler structure (Fig. 5.7). The strong territorial song of the Purple-crowned Fairy-wren carries clearly across rivers and from end to end of its elongated territories.

Songs are generally referred to as reels, which relates to the regular repetition of phrases. In the Purple-crowned Fairy-wren, the repetition rate is slow, the interval between elements is long, and the repetitive undulating character of the song is easily heard, as 'cheepa-cheepa-cheepa' or 'chicory-chicory-chicory', depending on the structure of the phrases (Fig. 5.7(d)). In species like the Splendid Fairy-wren, with complex elements repeated at short intervals, it is harder to distinguish the repeated phrases (Fig. 5.7(a,b)). The song of the White-winged Fairy-wren (*M. leucopterus*) (Fig. 5.7(e)) is structurally distinct and appears the least variable of the fairy-wren songs; the frequency of its very brief elements rises and falls in a regular undulating pattern like a sine wave, which appears to be common to all individuals across Australia, including the subspecies *M. l. edouardi* on Barrow Island (Baker 1995). Although the elements are brief and close together, the overall undulating changes in frequency are slow enough to be heard, and this species is easily recognized by its song. Red-winged Fairy-wrens are also easily recognizable; the introductory elements of their song are long-drawn out tones, longer than in any other species (Fig. 5.7(c)). The Emperor Fairy-wren of New Guinea is unusual in having a second warbling song in addition to the usual fairy-wren reel.

Songs are given by males and females, and by all group members, not just by the breeding male. Singing is most frequent during the breeding season and the month or so before, and is much less frequent during the moult in autumn and during winter for birds in southern Australia. Song is most frequent in the early morning, before and soon after dawn (Rowley 1965; Payne *et al.* 1988; Keast 1994). It is important in maintaining territories, and a male singing on one territory is generally answered from nearby territories. In this context, the song is loud, and an absence of responding song is probably the first indication that a territory has become vacant. In most species, individuals sing alone, and there is no obvious duetting. One or more individuals may respond with song to play-back, but any overlap appears accidental and insignificant. In the Purple-crowned Fairy-wren duetting by males and females of a breeding pair is an obvious reply to a neighbour's song or to play-back (Fig. 5.8). Softer reels are used in communication between group members foraging out of sight of each other in dense cover, when bringing food to a nest, by an incubating female who has just left her nest, or by a male calling to an incubating female. Song may also be important in mate acquisition, in the event that a breeding bird has lost its partner.

Fairy-wrens are unusual in the extent to which females sing in defence of their territories, independent of their male partner. This may be related to the remarkable mating system in which the mating activities of males and females in pairs are largely independent of each other. Females cannot rely on their male partner to defend the territory at all times, since he may be absent seeking extra-pair copulations, so they also sing. Play-back experiments by Cooney and Cockburn (1995) in Superb Fairy-wrens demonstrated that females sang most frequently in response to songs of male or female neighbours, and more intensely in reply to a stranger than to a neighbour.

In Superb Fairy-wrens, the species that has been studied in most detail, two quite different types of song have been identified, but it is not yet known whether the same two types of song are present in other fairy-wrens (Langmore and Mulder 1992). Type I song is the typical reel already described. Type II song is a relatively invariant, stereotyped song, sung only by males, both helpers and paired territory holders (Fig. 5.9). The most notable feature is a long, low frequency introduction of repeated elements, followed by a short series of elements similar to those of Type I songs. Each male seems to have his own version of this song, similar to those of the rest of his group, and birds in adjacent

5.8 A pair of the Purple-crowned Fairy-wren (*M. coronatus*) duetting. Territorial song is more co-ordinated between the pair in this species than in most other species. Male and female fly to a conspicuous perch and sing together a loud duet in response to a neighbour's song.

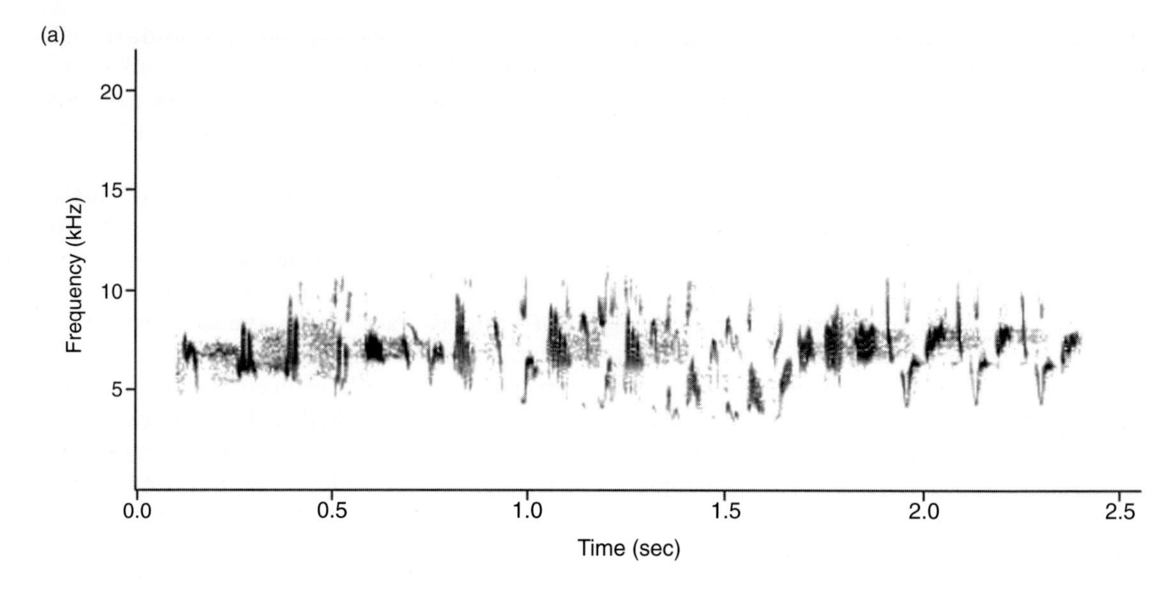

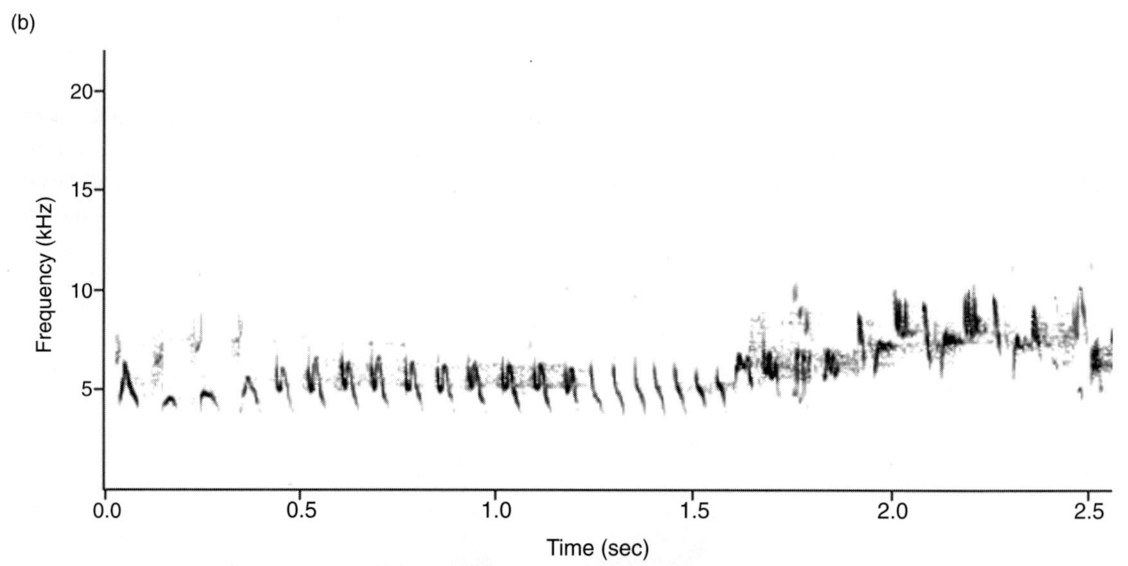

5.9 Type I and Type II songs of the Superb Fairy-wren (*M. cyaneus*). (a) Type I song is the typical *Malurus* reel, (b) Type II song is a relatively invariant stereotyped song sung only by males and triggered generally by calls of a predator (in this example, the call of an Australian Raven (*Corvus coronoides*), seen as the low frequency notes at the start of the reel). (Recorded by Naomi Langmore in the Australia National Botanic Gardens, Canberra.)

territories are more likely to have elements in common than birds separated by several territories.

The context of Type II song is unusual. It is triggered most frequently by the calls of predators or potential predators, but does not elicit any observable response from other conspecifics nearby. Langmore and Mulder suggest Type II song is a specific signal to nearby females advertising male quality. If this is so, then we would expect other species of Fairy-wrens to have similar calls.

Other vocalizations of fairy-wrens

Contact calls of all species are soft and brief (*c.* 50 ms), spanning a range of frequencies, variously rendered 'chet', 'zit', or 'prip' (Fig. 5.10). These may be given singly or in a series, and, although characterisitic of fairy-wrens, because they are so simple they are very similar to the contact calls of some sympatric finches for example, Superb Fairy-wrens and Red-browed Finch (*Neochmia temporalis*), Purple-crowned Fairy-wrens and Crimson Finch (*Neochmia phaeton*). In some fairy-wrens, the contact call is distinctive; in others, it is not so easily recognized. Thus the Purple-crowned Fairy-wren gives a strong 'chet', 1.5–8 kHz (Fig. 5.10(b)); an excited bubbling series of contact calls is also sometimes heard. This is very distinct from the higher, softer contact calls of the other species of fairy-wren that may occur in the same area, Red-backed and Variegated Fairy-wrens. Blue-breasted Fairy-wrens foraging out of sight of each other in vegetation give a single, short, weak, very high tone 'tee', not at all harsh, The equivalent contact call in the Red-winged Fairy-wren is a series, generally three, descending, 'ch-ch-ch', with the stress on the first.

Alarm calls are broad-band, loud, and noisy (Fig. 5.10(c,d)); in most species, they are a loud 'chit' of 150–200 ms, initially given singly, but repeatedly emitted if the threat continues. Although most species share a similar structure, the frequency range of the call differs between species—thus the Red-winged Fairy-wren is shrill, from 3 to 8 kHz, with a dominant frequency at *c.* 6 kHz, the Superb Fairy-wren has a slightly lower frequency, *c.* 3–5 kHz. In Blue-breasted and Variegated Fairy-wrens (Fig. 5.10(d)), the alarm call is a distinctive, loud, broad-band harsh chatter of repeated elements of less than 50 milliseconds (Jurisevic and Sanderson 1994).

Brooding purr. The female emits a very soft and almost continuous call while she is incubating or brooding that Rowley (1965) described as a 'purr'. This call is usually given after the female has been on the nest for some time, and it is quite possible that it denotes that the female is ready to leave the nest. It evokes no obvious response from the other group members, and is only audible very close to the nest.

Threat call. Rowley (1965) described a low churring threat call that he heard only during experiments with model intruders (fairy-wren). After an initial challenging song battle, the male assumed a stationary threat posture during which a low churring call was uttered.

Whining call. When presented with a mounted Horsfield's Bronze-Cuckoo (*Chrysococcyx basalis*), nesting Splendid Fairy-wrens responded giving a loud, pulsed, whining call, then flying to the cuckoo and hitting and pecking it (Fig. 1 in Payne *et al.* 1985).

Feeding-young call. This is a louder version of the brooding purr, given by any member of the group who is near the nest, and carrying food for the young. It is usually answered by the chicks if they are more than 6 days old. It is a brief reel of song without any introductory notes; the begging call of the young sounds the same. Similar calls are heard in all species with which we are familiar (Rowley 1965; Hutton 1991; Rowley and Russell, unpublished data).

Mating call. Hutton (1991) observed mating displays of Splendid Fairy-wrens in captivity at close range. She heard vocalizations in this context that she heard at no other time, despite her great familiarity with the species. These calls appeared to be forms of a 'harsh rattling chatter......zzzatt, zzzatt, zzzatt' heard at various stages during courtship when the male was displaying with Face Fan or Sea Horse Flight (see p. 74).

Vocalizations of emu-wrens

The only description of the vocal repertoire of any emu-wren is that of Hutton (1991), who observed birds in captivity. Various field guides and Schodde (1982*b*) also describe some vocalizations. Hutton describes the song and calls of the Southern Emu-wren (*S. malachurus*) as similar to those of the Splendid Fairy-wren but much softer. She identified a song reel, contact calls, alarm calls, and a nest communication call. Sonograms show that the songs of the Rufous-crowned Emu-wren (*S. ruficeps*) and the Mallee Emu-wren (*S. mallee*) are also similar in structure to those of fairy-wrens but much higher. The frequency range (6–>8 kHz) is less than in fairy-wrens and the song less harsh; the elements are more spaced and less complex. The song

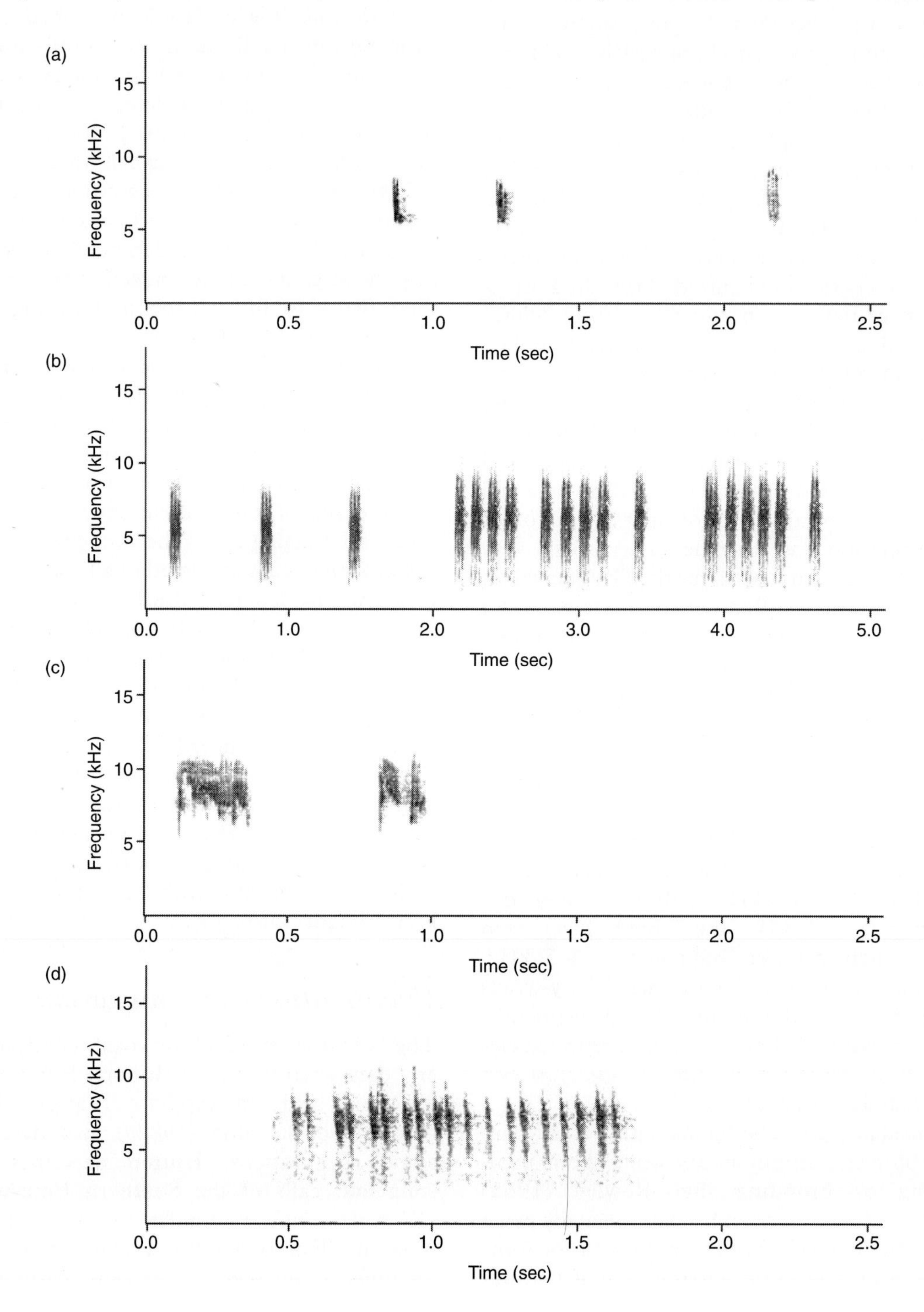

of the Southern Emu-wren is generally audible if the listener's ear can detect the higher frequencies, but the songs of the two smaller species, the Mallee Emu-wren and the Rufous-crowned Emu-wren, are very soft and high, and the contact calls so like the trills of grasshoppers and crickets that is is often hard to believe that birds are their source.

Vocalizations of grasswrens

The vocal repertoires of grasswrens are very different from those of fairy-wrens and emu-wrens. The songs are more complex, varied, and melodious, and differ considerably between species. We are indebted to Graeme Chapman for making his recordings available to us, so that we can illustrate the nature of their songs, and show their variety, but the extent of the repertoire of any species is not yet known.

Song of Grasswrens

The simplest song is that of the Grey Grasswren (*Amytornis barbatus*) (Fig. 5.11(a)). A series of high frequency (6–8.5 kHz) down-slurred notes is all that has been heard by the many observers who have visited the habitat of this, the most recently discovered, species. Even in the breeding season, and from birds described as displaying, there is no record of any song resembling that of other grasswrens. Schodde (1982*b*) identifies three evolutionary lines in the grasswrens: the white-throated *striatus* group, the dark, stripe-throated *textilis* group, and the Grey Grasswren. The distinctness of Grey Grasswren song is consistent with this suggestion. In the *textilis* group, the most complex songs are those of the Dusky Grasswren (*A. purnelli*) and the Eyrean Grasswren (*A. goyderi*) (Fig. 5.11(b)), while in the *striatus* group, the least complex songs occur in the Striated Grasswren (*A. striatus*), and the most complex in the White-throated Grasswren (*A. woodwardi*) and the Carpentarian Grasswren (*A. dorotheae*).

Schodde (1982*b*) characterizes the Thick-billed Grasswren (*A. textilis*) as the most silent of the grasswrens. There are three widely separated subspecies. The song of the nominate subspecies *A. t. textilis* recorded at Shark Bay in Western Australia (at the western edge of its range) is broadly similar to that of *A. t. modestus* from the Flinders Ranges in South Australia: short, unslurred notes, or down-slurred whistles and a short trill (see p. 239). However, the number of songs sampled was small.

Of the *textilis* group, the song of the Dusky Grasswren is the most varied. Some components of the song are similar to parts of the reels of fairy-wrens, but the variety of elements and phrases distinguish it. The sonogram lasts 3 seconds, with trills and staccato notes interspersed. The *A. p. purnelli* song recorded from Alice Springs is very different from that of *A. p. ballarae* from Mt. Isa, with its very thin, high frequency twittering song (see p. 244).

The songs of species in the *striatus* group include a considerable variety of elements and phrases—repeated staccato notes, trills, buzzes, clicks, and up- and down-slurred notes. Each species shows considerable variety in the way these phrases are assembled. While the songs of the different species in the *striatus* group illustrate the variety of grasswren songs,

---

5.10 Other vocalizations of fairy-wrens.: (a) Splendid Fairy-wren (*Malurus splendens*) contact call. (Karragullen, WA, recorded by Ian Rooke.) (b) Purple-crowned Fairy-wren (*M. coronatus*). Three single 'chet' contact calls, followed by repeated, more excited calls in a series (Drysdale R., WA, recorded by E. M. Russell.) (c) Superb Fairy-wren (*M. cyaneus*), single loud alarm call (Australian National Botanic Gardens, Canberra, recorded by Naomi Langmore.) (d). Blue-breasted Fairy-wren (*M. pulcherrimus*), a distinctive harsh alarm chatter of repeated elements (Dryandra Forest, recorded by E. M. Russell.) Alarm calls are broad-band, a loud 'chit' or 'zit' sound, initially given singly, but repeated if the threat continues.

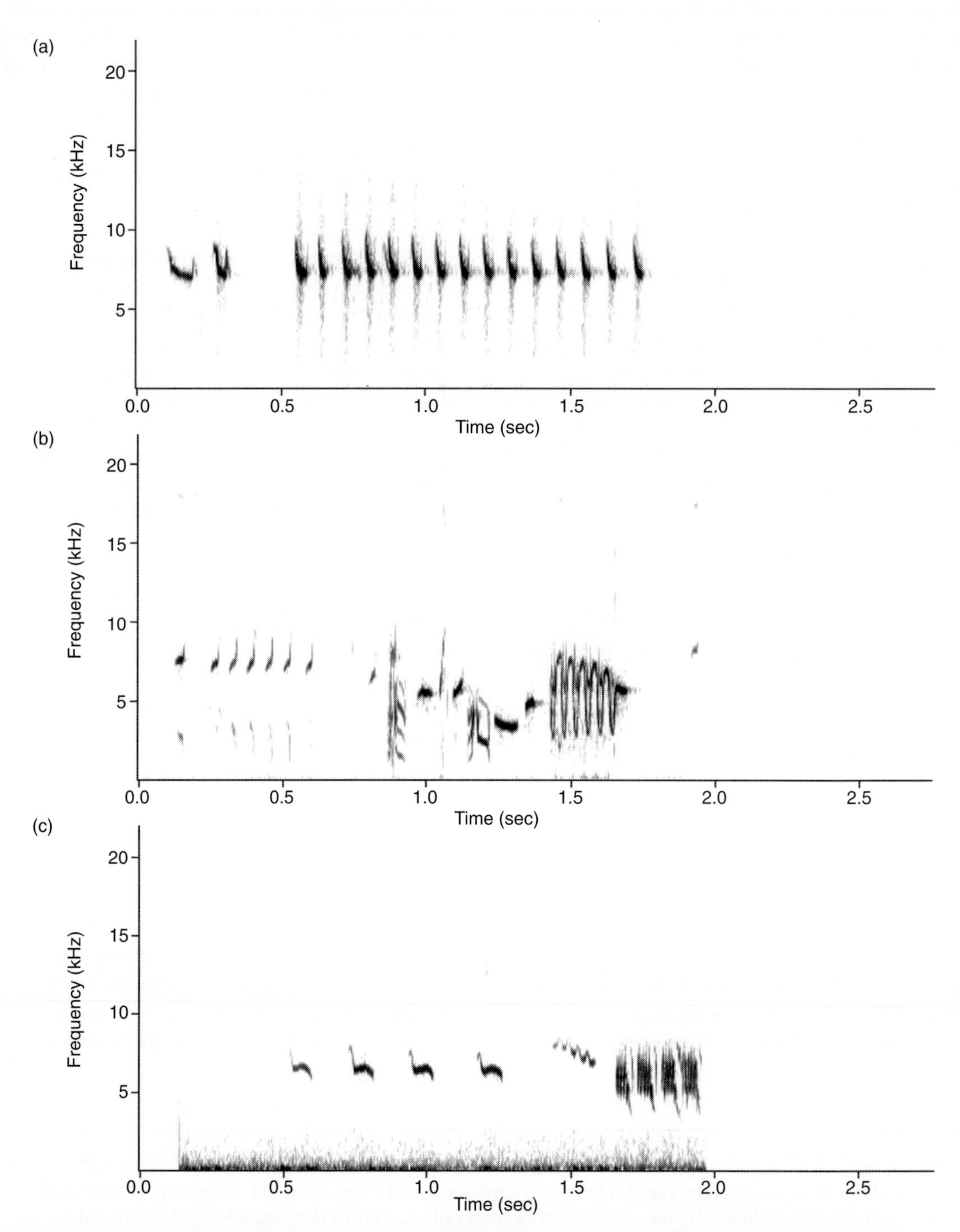

5.11 Songs of grasswrens. (a) Grey Grasswren (*Amytornis barbatus*), Teurika, NSW, (b) Eyrean Grasswren (*A. goyderi*), from Moomba, SA, (c) Striated Grasswren (*A. striatus whitei*) from the Pilbara, WA. The repeated simple elements in the Grey Grasswren contrast with the more complex whistles, trills, and buzzes of other species. (All recorded by G. S. Chapman.)

that of the Striated Grasswren shows remarkable similarity between the songs of subspecies from opposite sides of the continent: *A. s. striatus* at Yathong, New South Wales, and *A. s. whitei* from the Pilbara, Western Australia (Fig. 5.11(c) and p. 230).

Other vocalizations

The contact calls of grasswrens are a variety of narrow-band 'seeet' calls and soft twittering trills (Fig. 5.12). The Thick-billed Grasswren at Shark Bay gives a short sharp 'tik' call in alarm. Schodde (1982*b*) described alarm calls

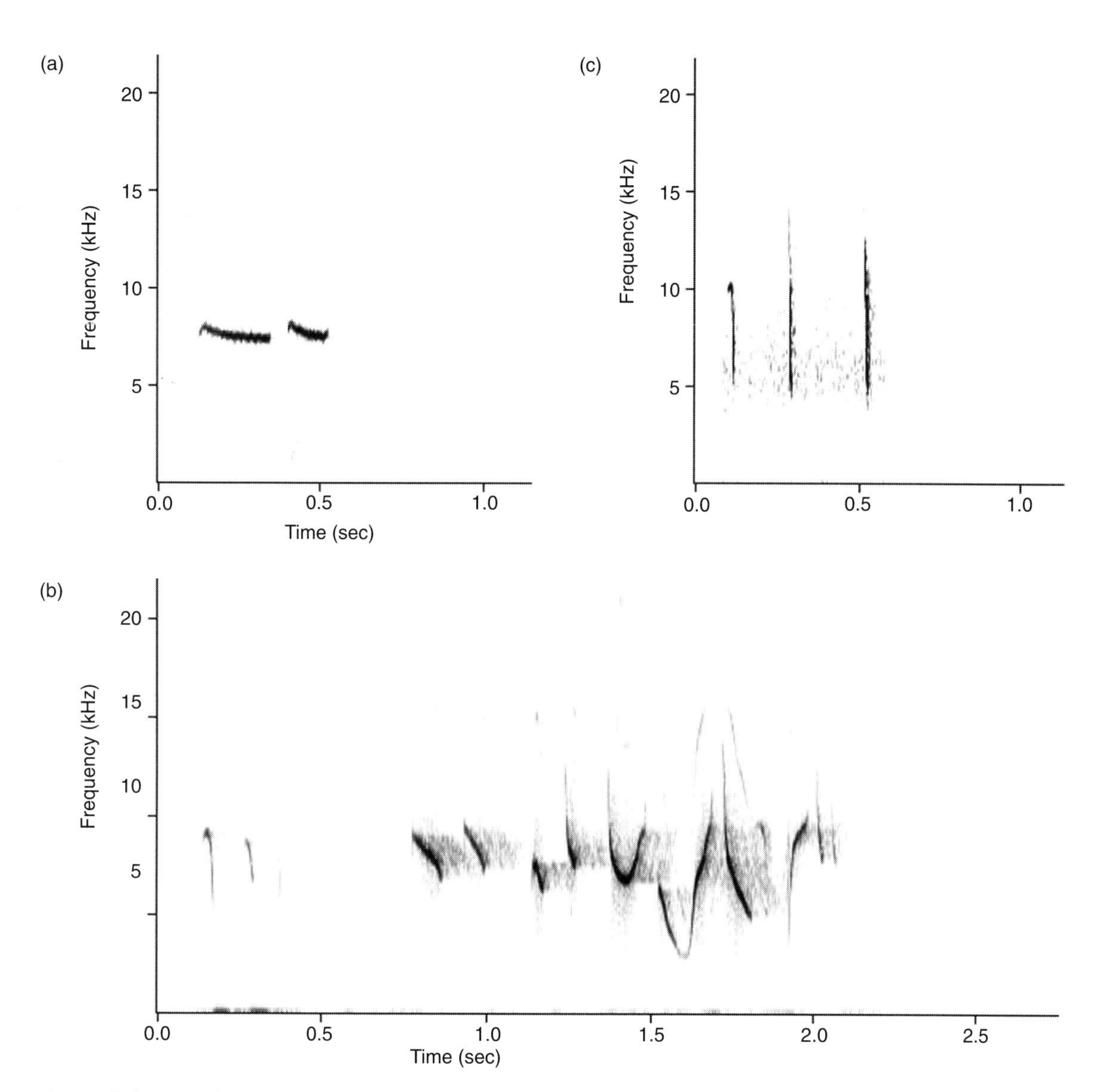

5.12 Other vocalizations of grasswrens. (a) Simple 'seet' contact call of the Eyrean Grasswren (*Amytornis goyderi*), (b) twittering trill contact call of the Thick-billed Grasswren (*A. textilis*), (c) Alarm 'tik' of Thick-billed Grasswren. ((a) Recorded by G. S. Chapman at Moomba, SA. (b) and (c) recorded by E. M. Russell at Monkey Mia, Shark Bay, WA.)

as high, drawn out 'zeeet' or 'eeep' in Grey and Eyrean Grasswrens, churring in Striated Grasswrens, and 'tchk-tchk-tchk' in Dusky Grasswrens. For the western form of the Striated Grasswren *A. s. whitei* in captivity, Hutton (1991) described an 'extremely loud, shrill, metallic call, sounding like a repeated scream followed by a machine-gun like rattle, eet eet eet rat tat tat tat'.

Narrow-band, high frequency calls of short duration, such as the contact calls of fairy-wrens and grasswrens are not long-range calls, and are likely to be audible only over short distances, and are thus less likely to betray a bird's location to a searching predator. Noisy broad-band calls with an abrupt start and finish are easier to locate, and some of the alarm calls are of this type, but there may also be narrow-band alarm calls, which alert conspecifics, without giving away their location (Jurisevic and Sanderson 1994).

Comparison of song in fairy-wrens and grasswrens

Our survey shows major qualitiative differences in song structure between fairy-wrens, emu-wrens, and grasswrens. The first two have relatively invariant songs, stereotyped sequences, and a small repertoire of phrases. Song is heard during most of the year, though more often in the breeding season, and birds in other territories reply with countersinging. Both males and females sing songs of this type, which Catchpole (1982) classified as having territorial defence and maintenance as their main function. He suggested that songs with a large syllable repertoire, variable sequencing, and thus constant variation, function more in advertisement and female attraction than in territory defence. In the North American wrens, there is a correlation between the presence of polygynous mating systems and the evolution of complicated songs (Kroodsma 1977). There is also quantitative evidence that song or other acoustic display influences female responses to males in many birds (Andersson 1994). In fairy-wrens, males are competing for extra-pair copulations and the basis for female choice appears to be brilliant plumage rather than song (Mulder and Magrath 1994). In grasswrens, sexual dichromatism is slight, and, in comparison with fairy-wrens, we might expect less competition between males vying to be chosen by females. The differences in song structure suggest that song has become the means by which male grasswrens compete, while the need for camouflage in the open desert habitat may have dictated drabber patterns of plumage. The social organization of grasswrens is little known, and mating systems not at all. They will be fascinating to study.

Mating systems

Courtship displays

A number of displays have been described, given by males in the presence of females before and during the breeding season. They occur infrequently and are very brief, seen in mainly chance observations of fairy-wrens that do not allow a definitive account. Mating is usually preceded by a sequence of displays interspersed with chases, the male chasing the female in a series of energetic flights. Aviary observations by Hutton (1991) fill in a little more of the detail, but we do not know what constraints were imposed by aviary conditions. Displays involving plumage vary with different species; for those species where the males have elongated ear tufts of contrasting colour, these are incorporated into displays, as are the erectile contrasting feathers of the wings and back in male chestnut-shouldered, White-winged, and Red-backed Fairy-wrens.

Sea Horse Flight

Most descriptions of displays by males in the presence of females during the breeding season include a display involving some form of modified flight. Rowley (1965) described an 'impeded flight' in Superb Fairy-wrens and an account of displays in Splendid Fairy-wrens by W. H. Loaring described a male 'suddenly arresting the normal forward motion of a flight, at the same time tilting the body up from a horizontal to a vertical plane. By some manipulation of the rapidly beating wings, he managed to maintain this grotesque upright position, and while making some slight forward progress, to descend to the ground with dramatic slowness..... sometimes, on reaching the ground, the birds instantly bound aloft again with astonishing agility.' (1948, p. 164). Joe Leone, one of the group working with us on Splendid Fairy-wrens, thought the displaying male looked like a sea-horse swimming, and this comparison is so apt that we have called it the Sea Horse Flight ever since. In flight, the male's neck is extended upwards and the feathers on the top of the head are erected. At close quarters in her aviaries, Rosemary Hutton could hear the louder-than-usual fluttering of the wings and a 'buzzing' vocalization by the male. In our experience, the Sea Horse Flight is usually undulating; it has been described in Splendid, Superb, and Red-backed Fairy-wrens.

Petal-carrying

In eight of the nine Australian fairy-wrens, males have been observed to pick and carry a flower petal or fruit, of a colour contrasting with their plumage (Table 5.4). Petal-carrying frequently forms part of courtship, either by an intruding male visiting a female nearby or by a male in his own territory. However, it is also performed outside the breeding season, generally by an intruding male, presumably bent on getting to know and impress a female in a territory nearby. Many accounts describe the male deliberately picking a petal, then carrying it as he flew towards a female and displayed further (Fig. 5.13). Sometimes, the male retains the petal as he displays to the female or encounters the resident male or males; sometimes he presents the petal to the female, who may or may not accept it, and may even eat it (Hutton 1991). For Splendid Fairy-wrens, 84 per cent

Table 5.4 The colour of petals carried by various fairy-wrens.

Species of fairy-wren	*Colour*
Superb	Yellow (11)
Splendid	Pink (33), Purple (20), Blue (7), Yellow (2), White (1)
Variegated	Yellow (2)
Red-winged	Yellow (7), White (1)
Blue-breasted	Orange (2)
Lovely	Yellow (1)
Red-backed	Orange (1), Red (6), Yellow (1)
White-winged	Purple (3), Blue (2), Yellow (1), White (2)

(Based on Rowley 1991, Rowley and Russell, unpublished data, and H. B Gill in Schodde 1982*b* for the Lovely Fairy-wren); number of records for each colour in parentheses.

5.13 Petal-carrying in the Superb Fairy-wren (*Malurus cyaneus*). Petal-carrying frequently forms part of courtship in fairy-wrens, chiefly by an intruding male but sometimes by a male in his own teritory. The male deliberately picks a petal, usually yellow in Superb Fairy-wrens, then carries it as he flies towards the female and displays further to her.

of 63 records were of pink and purple petals, although an array of other colours, particularly yellow, was available.

Face Fan

In those species where the males have contrasting ear tufts, these are erected to project on each side of the head (Fig. 5.14). This is a common element of many display sequences, and probably signifies in part the increased level of excitement associated with such situations. It occurs as part of aggressive and sexual encounters.

Lizard display

Hutton (1991) describes a component of the display sequence leading to mating in Splendid Fairy-wrens in which the male adopts a strange 'lizard-like' appearance, that we have not seen in the wild. His head and crown are fluffed out, his ear tufts fanned out, head and tail lowered, wings closed; the ruff round his neck makes him 'look like a Frill-necked lizard'. He gives a harsh, chattering call 'zzzatt, zzzatt, zzzatt'. The posture incorporates the Face Fan, seen in many other situations when the male is excited. It is in some ways similar to threat, and to aspects of the Rodent-run distraction display. Similar postures are described for the Superb Fairy-wren (Hindwood 1948) and it is probably the equivalent of the Blue-and-Black display described by the Bradleys (1958), who were unsure of its context. When the back is a brilliant or contrasting colour, as in Red-winged or Red-backed Fairy-wrens, it would be displayed by this posture.

5.14 The Face Fan is a common element of many aggressive and sexual display sequences, especially in those fairy-wrens with elongated, contrasting ear tufts. These ear tufts are erected, to project on each side of the head, as in the male Splendid Fairy-wren illustrated.

Wing-fluttering

In response to male displays, female fairy-wrens often adopt a posture with head lowered, tail down and somewhat fanned, wings outstretched and quivering, silent but with beak open. We consider this to be mating solicitation, but the same posture is also assumed by a juvenile begging food from an adult, by a helper giving priority to an older bird when approaching a nest with food, or sometimes by an immature male to a senior one.

Other displays in fairy-wrens

In fairy-wrens where the scapulars and mantle feathers are a striking colour and sometimes elongated, they are incorporated into displays to a greater extent. In Red-backed Fairy-wrens, the red feathers of the mantle and scapulars cover the lower back and the upper tail coverts, and are erectile, rendering the male a black and predominantly red 'puff-ball', especially in the Sea Horse Flight. A wing-shrinking display occurs in which the male holds his primaries close together and depresses the wing slightly, exposing his red back; at the same time, the crown feathers are erect (W. T. Cooper *in litt.*, December

1993; see photograph, p. 444, Schodde and Tidemann 1988). The brilliant white scapulars, secondary wing-coverts, and innermost flight feathers in White-winged Fariy-wrens are emphasized in display flights and in stationary displays when the back is exposed.

Displays by intruding males

Over many years of field work with Splendid Fairy-wrens, we were often surprised to encounter males in full nuptial plumage behaving conspicuously but not aggressively in territories other than their own (Rowley and Russell 1990). The distinctive behaviour of these males included one or more of the following.

1. Their flight to or from the site of the observation was direct and unusually prolonged for a malurid. They flew silently but conspicuously, well clear of the vegetation, in contrast to the usual inconspicuous surreptitious movement through dense cover in their home territory.
2. In some cases, part of this flight was in the undulating Sea Horse Flight.
3. Sometimes the male was even more conspicuous because he carried a coloured petal that contrasted with his plumage.
4. When the bird landed, generally near a female, he usually perched conspicuously and gave the Face Fan display.

An annotated extract from our field notes describes a typical occurrence:

On 26 November 1988, female 3096 was lining her second nest in Territory R, (a sign that she was within 10 days of laying, and therefore in her fertilization period). At 1130 h, male 554 (the senior male from Territory D 300 m to the west) flew to Female 3096 carrying a pink petal; he gave Face Fan, and she responded by fluttering her wings; he chased her but she kept moving away and finally flew to her nest area; 554 followed, flying half the distance in Sea Horse Flight; 554 was next seen to fly off west. Female 3096 was then escorted by male 3087 (a helper in her own group); she was approached by male 3048 from T territory abutting to the east) who gave Face Fan to her but was chased off by male 3087. Female 3096 then began to gather nest lining and was about to fly when male 3188 (from territory I abutting to the NW) flew in and gave Face Fan to her; she apparently took no notice but dropped the nesting material when he began to chase her; 3188 soon appeared to lose interest and flew away.

In Splendid Fairy-wrens, more than three-quarters of the intruders were senior territory-holding males, and the rest were older helpers, most of them coming from a territory next to the one where they were seen intruding. However, one intruding male had crossed five territories. Very similar behaviour has been found in Superb Fairy-wrens in Canberra where of 400 intrusions by identified males, 73 per cent were by immediate neighbours, 16 per cent were by helpers; about 20 per cent of displaying intruders carried a petal in their bills (Mulder 1992).

In Splendid Fairy-wrens, we recorded intrusions between June and January, increasing in frequency as the breeding season approached, with a marked peak in the first half of the breeding season (September to November). Most intrusions occurred in the 10 days before a clutch was laid, when a female is potentially fertile (Rowley and Russell 1990). Similarly in Superb Fairy-wrens, females were most likely to be visited by extra-territorial males 4–5 days before they laid a clutch, but intrusions occurred throughout the reproductive cycle, and even before the start of the breeding season, as in Splendid Fairy-wrens. In both species, females in their most attractive period just before egg-laying were frequently visited by males whose own female would have been fertilizable. Males do not seem to guard their females closely, and were frequently absent when an intruder visited; even if they were present, about one-third made no attempt to repel the visitor. The response to a group that attempts to take over all or part of a territory is quite different from the response to a male who visits and courts the female.

Copulation

Copulation itself frequently occurs in the cover of dense vegetation, and is so brief that in the course of 17 years' field work on Splendid Fairy-wrens we saw only 39 events. In both Splendid and Superb Fairy-wrens the few copulations observed have been extremely brief—less than 5 seconds—making identification very difficult. In Splendid Fairy-wrens, of the 29 cases where both individuals were known, 15 were between the expected senior male and female of the group, nine were between the senior female and a helper male in her group, and five were with intruding males. Brian Coates describes copulation in the White-shouldered Fairy-wren (*M. alboscapulatus*) at Sogeri, Papua New Guinea: 'After the nest was built mating was observed. This took place without ceremony; the group was active in a *Hibiscus* bush when the male mounted the female for a few seconds and then flew off while the female stayed and preened.' (1990, p. 94).

Species other than fairy-wrens

In the grasswrens and emu-wrens, with less elaborate plumage, what little is known about pre-mating displays suggests that they are far less elaborate. The only accounts are for captive birds (Hutton 1991). In Striated Grasswrens, the male frequently feeds the female during a week or so before mating and egg-laying. Eventually, when the female is ready to accept matings, she wing-quivers, the male hops up and down briefly in front of her, and then mates. Similar courtship feeding occurs in Thick-billed Grasswrens (Rowley and Russell, unpublished data). Courtship in Southern Emu-wrens appears to follow a similar pattern to that in grasswrens (Hutton 1991).

Who fathers nestlings?

The first intimation that the mating system of the fairy-wrens was anything unusual came from our study of Splendid Fairy-wrens, where we investigated the parentage of nestlings, using allozyme electrophoresis (Brooker *et al.* 1990). While this technique cannot prove which male fertilized the egg that produced a particular individual, it can show which individuals could not have been the genetic parents. From this parental exclusion we found that of 91 offspring, 66 were not fathered by the senior male and 59 of these were incompatible with all the males in their group. Since allozyme electrophoresis does not detect all the cases of extra-group fertilization, the actual proportion of nestlings not sired by any male in the group was even higher. Twenty-eight of 37 broods examined contained at least one nestling not fathered by any male in the group.

Subsequently, the modern molecular analyses of parentage using DNA fingerprinting have been applied to the Superb Fairy-wren population in Canberra. Males outside the social group fathered 76 per cent of 181 genetically typed young. In 40 broods, every nestling was typed, and 95 per cent of broods included at least one extra-group young, and in 48 per cent, all of the nestlings were fathered by extra-group males. Helper males fathered only four of the 181 nestlings, so extra-pair paternity was in effect extra-group paternity. Pairs had fewer extra-group offspring than multi-male groups (Mulder *et al.* 1994).

For seven territories in the centre of the study area, the actual fathers of 34 of 38 extra-group young were identified. Although immediate neighbours performed three-quarters of the courtship displays observed, they accounted for only 15 per cent of extra-group fertilizations, while males from two territories away gained 41 per cent although they performed fewer displays. Of 68 potential fathers (all the adult males from the seven core territories and a ring of two territories wide around them), three males sired 18 of the 38 extra-group offspring, with one fathering seven. One particular lineage was very successful (Fig. 5.15): male BRY and his descendants fathered 18 extra-group young in six of the seven core territories; of these paternities, 11 (including seven by the most successful individual)

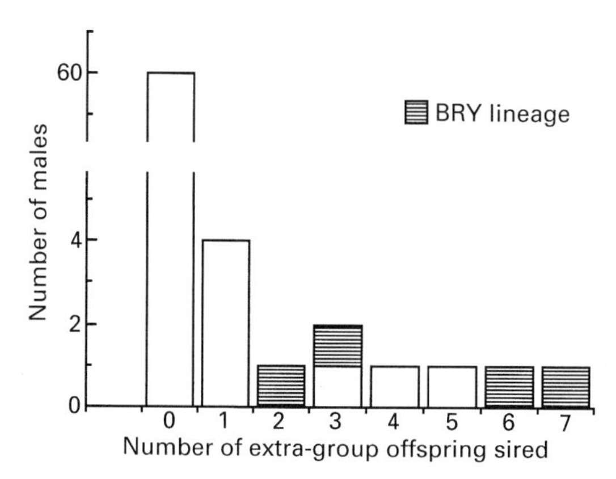

5.15 The distribution of extra-pair paternities among males in the central core study area of the Superb Fairy-wren (*Malurus cyaneus*) in the Australian National Botanic Gardens, Canberra, ACT. All successful males came from within two territories of the perimeter of the core area. Shaded bars identify males belonging to the lineage of male BRY (with six extra-pair offspring). The three most successful fathers (two of which were not from the core area) each fathered extra-group offspring in at least three territories. BRY's within-group son sired three extra-group young, his extra-group son sired seven young, and an extra-group grandson sired two young. (Redrawn from Fig. 4 of Mulder *et al.* 1994.)

were obtained while the male was a subordinate helper. Clearly females mated with males irrespective of their status, and preferred the characteristics of male BRY and his descendants.

Petal-carrying displays by intruding males have been recorded in most fairy-wrens, which implies a high incidence of extra-group fertilization is likely to occur throughout the genus. DNA fingerprinting studies have identified extra-group young in Variegated, White-winged, and Splendid Fairy-wrens occurring together at Brookfield Conservation Park in South Australia (Tuttle *et al.* 1996 and personal communication).

This pattern of extra-group fertilization may be the result of female choice of male. Females solicit copulations and can easily evade courting males; forced copulations do not occur. Many intruding males were visiting females that were clearly not fertilizable—before the start of the breeding season, long before a clutch was laid, or after the clutch had been completed. Most intrusions did not lead to copulation, even when the resident male was absent. Mulder *et al.* (1994) suggest that the primary purpose of displays by intruding males may be to advertise themselves as prospective candidates for extra-group copulation. Females may assess extra-group males through their displays, and subsequently copulate with selected individuals (Mulder *et al.* 1994).

This unrestricted mating by male and female fairy-wrens takes place within the framework of long-term, pair-bonded males and females living in stable social groups that are resident in permanent territories over many years. Because extra-pair or extra-group copulations cast doubt upon the sexual role of a senior male in a group, we need to differentiate between the social *pairing* of a male and female in a group and the sexual *matings* that result in the progeny raised in a nest tended by the group. We therefore refer to the senior male and senior female in a group as being paired, without implying that they mate together or that they are the only two birds living together. When a helper manages to fill a vacancy as a senior male, we therefore refer to him as 'pairing'.

Pair-bonds

The basic social unit in all malurids is the socially monogamous pair, a male and female that remain together as long as both survive, but which frequently mate promiscuously outside the pair-bond. The pair or group forage and defend their territory as a unit, resting and allopreening together during the day, and roosting together at night (Figs 5.5, 5.6). Both members of the pair feed the nestlings, although the female generally makes more feeding visits. Breeding adults tend to be long-lived and may stay together for several years, although the mean durations for Splendid and Red-winged Fairy-wrens were only 1.7 and 2.4 years (Table 5. 5). By counting one pair studied for a year as one pair-year,

Table 5.5 Pair-bonds in Splendid Fairy-wrens (*Malurus splendens*) and Red-winged Fairy-wrens (*M. elegans*): pair duration, mean tenure as senior males (M) or females (F), and the mean number of social partners during life.

	Splendid Fairy-wren			*Red-winged Fairy-wren*		
	Mean	*Range*	*n*	*Mean*	*Range*	*n*
Pair duration	1.7	1–7	206	2.4	1–10	156
M tenure	3.1	1–12	116	3.7	1–14+	101
F tenure	2.4	1–8	146	3.5	1–11	96
No. of partners—M	1.9	1–7	116	1.6	1–4	101
No. of partners—F	1.5	1–4	146	1.5	1–3	96

n is the sample size.

in Splendid Fairy-wrens a fifth of the pairs in 351 pair-years were together for 3 years or more; in Red-winged Fairy-wrens half of the pairs were together for 3 years or more, with 10 per cent lasting 6 years or more.

The duration of pair-bonds is generally less than the period of tenure by each individual (Fig. 5.16). Although the majority of birds had only one partner, some had several; as many as seven in Splendid Fairy-wrens and four in Red-winged Fairy-wrens. The comparison of tenure between these two species illustrates the effects of higher survival in Red-winged Fairy-wrens, where nearly a third of 101 senior males held that position for 5 years or more, including five individuals that lasted more than 10 years, one of whom was still there in the 1994 breeding season, after 14 years as senior male (Fig. 5.16). Breeding lifespans as long as these would not be remarkable in larger birds such as the Florida Scrub Jay (*Aphelocoma coerulescens*) or Acorn Woodpecker (*Melanerpes formicivorus*) (Woolfenden and Fitzpatrick 1984; Koenig and Mumme 1987), but in 10-gram passerines, they are long relative to the breeding life of New and Old World warblers of similar size.

The history of group HB of Red-winged Fairy-wrens illustrates how this works in practice (Fig. 5.17). Between 1981 and 1994, the group had only three senior males, M 617, M 672, and M 6579. Over the 10 years illustrated, the group had three breeding

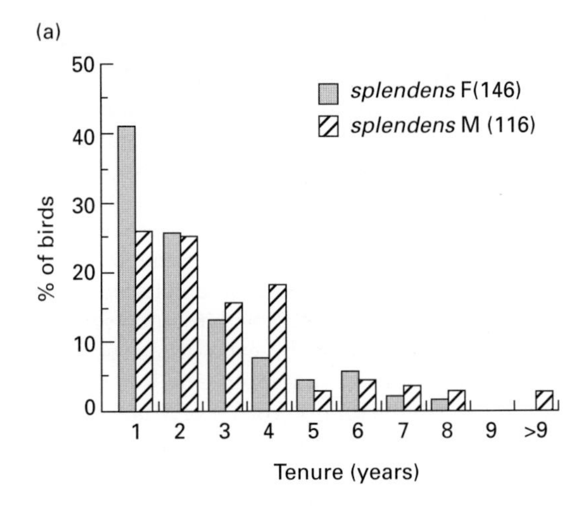

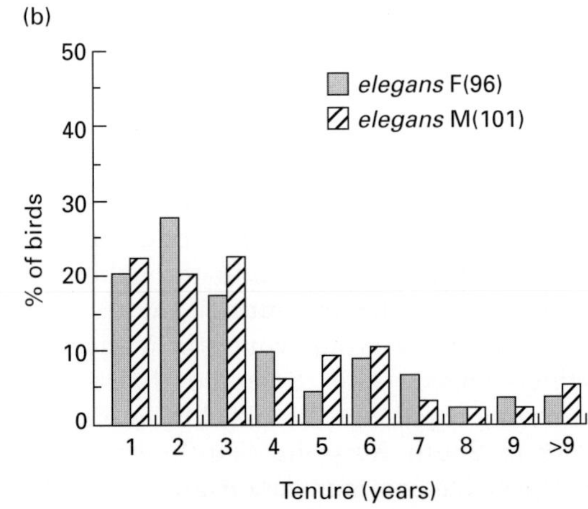

5.16 Breeding tenure in (a) Splendid Fairy-wrens (*Malurus splendens*), (b) Red-winged Fairy-wrens (*M. elegans*). The percentage of males (M) and females (F) that held the position of senior male or breeding female in a group for one or more years. Number of males and females in parentheses.

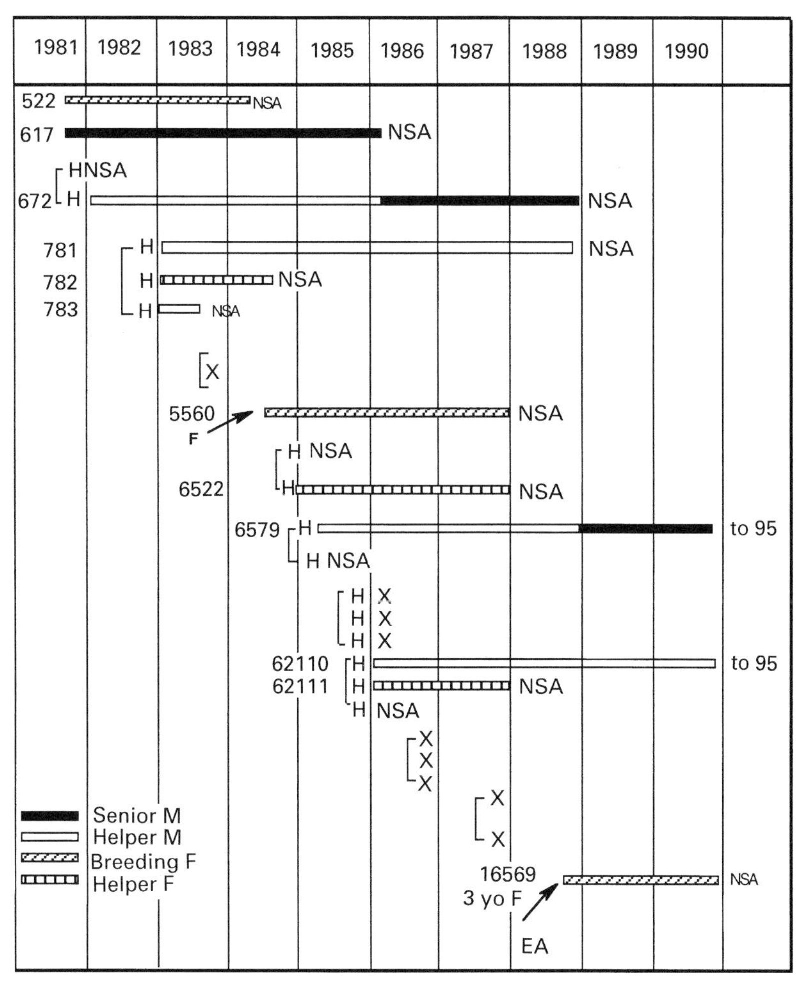

5.17 Red-winged Fairy-wren (*Malurus elegans*) in Smith's Brook Reserve, near Manjimup, WA: the history over 10 years of territory HB. The breeding female F 522 was replaced by an incoming F 5560. After the death of senior male M 617, the oldest male helper became senior male, and new partner to F 5560. She in turn was replaced by F 16569, a 3-year-old female from EA, two territories distant, who had helped for 2 years before moving to HB. Her first male partner was M 672 for 1 year, and after his death, M 6579, who took over as senior male. Broods are bracketed together; H: hatched; X: nest failed, NSA: not seen again. Arrows indicate immigration.

females, F 522, F 5560, and F 16569, the two latter known to have come in from outside. During these 10 years, there were five different pairs lasting three, two, two, one, and two breeding seasons, as females and males were replaced in turn. Because individuals remained with the group longer than the period of any one pair, the group retained its identity.

Pairs that have been together for several years are more productive than newly formed pairs. However, this improvement in productivity is due to the increasing age and experience of the female and not to the duration of the pair-bond—an older female in the first year of a pair-bond with a young male is just as productive as a female of the same age paired with the same male for several years.

Table 5.6 The filling of vacancies in territories of Splendid Fairy-wrens (*Malurus splendens*) and Red-winged Fairy-wrens (*M. elegans*), based on all occasions where a breeding male or female was known to be replaced.

Vacancy filled by	*Splendid Fairy-wren*		*Red-winged Fairy-wren*	
	Male (n = 80)	Female (n = 126)	Male (n = 79)	Female (n = 70)
Helper (natal)	71.3	26.2	77.2	22.9
Helper (immigrant)	0.0	0.0	0.0	7.1
Disperser	28.2	73.8	22.8	70.0

Helpers (natal) were birds in the territory in which they hatched. Helpers (immigrant) were females that dispersed into a territory and remained as helper for one or more years before taking over as breeding female. Dispersers came into a territory to fill a breeding vacancy. For each species, values are the percentage of total vacancies (n) for each sex.

Older females are more likely to have more helpers and this raises productivity (Russell and Rowley 1996).

When senior males died they were replaced by a helper if one was available (Table 5.6). When more than one male helper was present, the oldest invariably took over (Fig. 5.17). Only if no other male was present did a male from outside come into the group. In both Splendid and Red-winged Fairy-wrens, at least half the vacancies for males in existing groups were filled by helpers from within the group. Breeding females of both species were replaced by female helpers if one were available and this happened in *c.* 20 per cent of vacancies. In Superb Fairy-wrens, no female vacancy was filled by a female hatched within the group, although a female that had dispersed into a group in autumn could replace one that died during the following winter. In a similar manner, some Red-winged Fairy-wren females dispersed into groups and remained there as helpers for one or more years; they were then available to replace a breeder in that group or one nearby. Paired birds dispersed rarely, only after the death of a partner or as a consequence of divorce. In Splendid Fairy-wrens, when one partner died, he or she was replaced from within the group if a helper of the appropriate sex were available, but if there was no immediate replacement, the survivor might move to a vacancy nearby.

Divorce, where both members of a pair were alive but paired with a new partner, is rare in all species studied in detail. In Splendid Fairy-wrens, we recorded only 10 cases in 206 pairs over 17 years; similarly in Red-winged Fairy-wrens, we knew of only two divorces in 156 pairs over 12 years. We could find no evidence in either species that the few cases of divorce were a consequence of reproductive failure (Russell and Rowley 1996). It is possible that one of the pair is able to move to a vacancy in a territory of better quality by breaking its pair-bond, but this aspect of mate choice is not yet understood.

Plural breeding

In most groups only one female built the nest and laid the eggs, but all members helped to feed the nestlings, a situation described as singular breeding, in contrast to plural breeding, where two females in a group maintain separate nests (Brown 1987). We occasionally came across plural breeding in both Splendid and Red-winged Fairy-wrens; the former provides the best example since we were most familiar with the conditions under which it occurred (Rowley *et al.* 1989).

For the first 8 years of our study, we had only one case of plural breeding. From 1973 to 1977 a succession of small fires kept the vegetation sparse and the wren population low. From 1978 to 1984 there were no fires, the vegetation grew taller and thicker and the population became denser as both the average group size and the number of female helpers increased. Plural breeding usually occurred where a large group contained more than one female. From 1981 to 1987, we recorded 32 cases, but following an extensive wildfire in January 1985, the population gradually declined, and after 1987, we recorded it no more.

In most cases of plural breeding the primary (X) breeding female was busy with her own nest by the time the second (Y) female began hers, and there was no obvious aggression between them. The Y female was generally a 2 year old that had helped in the group during the previous breeding season, and was always related to the breeding female, most often her daughter. Usually we had no idea which male fertilized the Y female's eggs but in the one Y nest that was included in the paternity determinations, two of three eggs were fertilized by an extra-group male (Brooker *et al.* 1990). There was no reason to suppose Y females were any different from X females in their selection of partners to mate with.

Whether any of the group helped the Y female to feed her nestlings depended on the timing of the two nests. If they hatched close together, some group members fed at each nest. With a longer interval between nests, the group began to feed at the earlier one, and after establishing that routine, the other female was often left to raise her brood alone. If one nest failed, all group members fed at the surviving nest and we once watched an X female feeding at her daughter's nest. If both nests succeeded, the two broods coalesced and were cared for by the group as a whole. Y females were not very successful, but were no worse than any other novice female with no helpers. Some bred as a Y female for a second year, others succeeded their mother as sole breeder in their natal territory, while others dispersed.

We have suggested that the complex groups of White-winged Fairy-wrens with more than one breeding female are groups in which a form of plural breeding is the normal condition (Rowley and Russell 1995).

Overview

In all fairy-wrens studied in detail, and probably in the other species as well, socially monogamous pairs or groups live in contiguous territories throughout the year. The remarkably high incidence of extra-pair copulations means that more of the nestlings are sired by extra-pair males than by their mother's social partner. Divorce is rare, and not related to reproductive failure; social pairs remain together in their territory as long as both survive. What explanation can be given for such a remarkable social system?

The social system in any particular species is the result of past evolution and present environmental conditions. Fairy-wren pairs are together in their territory not only for the 4–5 months of the breeding season, but for the whole of one or more years. The life of the pair outside the breeding season, their survival and that of other group members in the territory, assume an importance that they do not have in birds that live together for a short breeding season and live independent lives for the rest of the year. The pair-bond is part of everyday life throughout the year, unlike a partnership that is renewed each year during the breeding season, as happens in those passerines where choosing a mate is coupled with choosing a territory. Where available habitat is occupied all year round by established pairs or groups, a most important limiting resource for males and females is a vacancy in an established territory. What we know about the rapid filling of vacancies in social pairs suggests there is little opportunity for choice. The social structure in *Malurus* of permanent territories occupied by resident pairs or groups is coupled with a promiscuous mating system; choice of a social partner and choice of a male to fertilize eggs are separate. In other words, rather than choosing their social partner, fairy-wrens may choose partners for copulation.

The availability of territories and competition for them are important factors in the choice of partner and territory. Birds have the option of remaining in their natal territory, moving to a known vacancy nearby, or wandering in search of a chance opening. Most birds that become paired fill a vacancy in an established territory where a bird of the opposite sex is already present. Such vacancies are rare, can occur at any time, and are filled very rapidly; choice is very limited. We suggest that it is the importance of the permanent territory that dictates social partners and contributes to the persistence of the pair-bond.

For females, a territory is a prerequisite for breeding. In a fully occupied habitat, it is important for a female to take the first available territory vacancy, by dispersal if necessary, unless the only available territory is a very poor one, in which case it may be better to delay. Where survival of adults is high and vacancies scarce, it does not benefit a female to break a pair-bond and set out to look for another territory (and partner), changing a present certain breeding opportunity for a possible future one, with the added risks of dispersal to an unfamiliar area. A female can improve her lifetime reproductive success by having a permanent territory with helpers, since helpers can improve both her productivity and her survival. Her increased survival further ensures that female vacancies are scarce.

The costs of an enduring pair-bond depend on the quality of the partner and the territory. Mate quality can be addressed by appropriate choice of partner for extra-pair copulation. Since males survive better than females, a female could be paired with a male of inferior quality for the whole of her breeding lifespan; promiscuous mating provides a choice of males, and the chance to mate with high quality males. The territory also provides a focal point where she can be found by neighbouring males. Any loss of help by the senior male because of his activities seeking copulations elsewhere is compensated for by the helpers in the co-operatively breeding group. Thus the promiscuous mating system compensates the female for the lack of choice of male and territory.

For males, with a sex ratio generally male-biased, a territory is a prerequisite to gaining a partner. A current female partner is more certain than competing for another one when females are scarce. A new partner would most likely be young and inexperienced and therefore less productive. A persisting pair-bond ensures that a male retains the higher productivity of an older female even though he does not father all her progeny. In a strictly monogamous mating system, a male would be limited to only one female at a time, who might be of poor quality. By inseminating more than one female, the risks of failure are spread and reproductive success may be higher for a few preferred males. The territory provides a familiar area in which to forage, and from which to make excursions to surrounding territories, seeking copulations with females with which he is familiar. The production of offspring that remain as helpers has several benefits to a male, regardless of whether they are his progeny or not. They help to defend the territory and they assist the female, releasing him to pursue extra-pair copulations.

6

Co-operative breeding

Introduction

A paper by Alexander Skutch in 1935 is generally acknowledged as the first formal recognition of 'helpers-at-the-nest', but because most ornithologists lived in Europe and North America where such behaviour is uncommon, the significance of that paper or his even later compilation (1961) were not appreciated for some time. Many early accounts of *Malurus cyaneus,* the common Superb Fairy-wren of south-eastern Australia, made reference to groups of more than two birds, either as 'mormon' wrens with their harems, or groups with more than one blue male, leading to speculation about polyandry. As long ago as September 1931, the Misses Wigan and Travers at the RAOU Campout at Wyperfield National Park in Victoria watched closely at a nest of the Splendid Fairy-wren (*Malurus splendens*) and recorded that the nestlings were attended by two females, a full-plumaged male and an immature male (Wigan 1932). It was not until the advent of colour-banding, however, that helpers-at-the-nest were recognized to occur widely in Australian birds. By 1957, a preliminary report of Ian Rowley's study in Canberra described how members of a group in addition to the mated pair helped the parents feed the young, both in the nest and when fledged. Most of these helpers were at least 1 year old, but in addition, birds fledged from early broods sometimes helped to feed the young of later nesting attempts.

David Lack's inclusion of a chapter on co-operative breeding in his 1968 book *Ecological Adaptations for Breeding in Birds* gave the subject greater prominence and as more and more reports of co-operative breeding appeared, the paradox posed by such apparently altruistic behaviour created widespread interest. Since natural selection favours those genotypes that leave most offspring, how could apparently altruistic behaviour such as helping evolve, since the helpers forgo breeding and aid others to reproduce? This paradox was originally identified by Darwin in 1859. Early explanations suggested that helpers were primarily young birds acquiring experience before setting out to breed on their own, but later studies showed that helpers often delayed breeding for several years. Such birds were unlikely to produce as many offspring in their lives as those individuals that bred without delay. Under these circumstances, natural selection could not favour helping.

In his 1987 review, Jerram Brown documented 222 species of birds in which co-operative breeding had been recorded. The total is undoubtedly higher, since the recorded incidence is highest in the tropical and sub-tropical regions of Australia, Africa, Asia, and the Americas, where the biology of many species is little known. In Australia, co-operative breeding is recorded in at least 85 species, of which 74 are passerines, all from the old endemic families (based on compilations by Clarke 1995 and Cockburn 1996). In total, 28 per cent of 266 old endemic Australian passerines are co-operative breeders compared with 3 per cent of passerines world-wide as recorded by Brown (1987). This raises the possibility that co-operative breeding may be more common in some phylogenetic lineages than others (Russell 1989; Edwards and Naeem 1993; Ligon 1993; Cockburn 1996). Whether some environments provide conditions that are particularly favourable to the evolution of co-operative breeding is a question that has been debated at length, but is not yet adequately answered (Brown 1987; Ford *et al.* 1988).

At first the solution to the problem of altruism, as seen in social insects, was seen to lie in the concept of inclusive fitness developed by Hamilton (1964), who realized that since individuals share genes with relatives, aiding one's kin to produce offspring also leads to the transmission to the next generation of some genes that are identical with one's own. Hamilton proposed that what is maximized by natural selection is not individual fitness but 'inclusive fitness' such that the fitness of an individual depends both on its own survival and reproductive success and on that of its kin. Thus by helping their parents to rear more offspring than they could have done if unaided, helpers could gain an indirect fitness benefit (in the terminology of Brown 1987). However, the subsequent discovery of high levels of extra-pair copulation in many species, especially in several co-operatively breeding fairy-wrens, means that many birds are less closely related than was once thought, and any kin benefit from helping is much reduced.

In most co-operatively breeding birds and mammals, groups form because offspring old enough to disperse and breed independently delay their dispersal and remain in their natal group for months or years after they reach sexual maturity. As well as questions about phylogeny and environment, to understand the evolution of co-operative breeding, two basic questions about helpers need to be answered: why do they defer breeding and stay at home and why do they help? It has proved remarkably difficult to find answers to these questions. Many attempts have focused on the ecological factors that promote delayed dispersal. Others have discussed whether helping occurs simply because helpers respond automatically to begging young, or whether helping behaviour has evolved because helpers themselves benefit from helping. Although no attempt at a comprehensive review is made in this chapter, we shall return to these questions after considering how the co-operative breeding system operates in the family Maluridae, particularly in the fairy-wrens that we have studied intensively.

Co-operative breeding in the family Maluridae

Group-living, and probably co-operative breeding, appear to occur throughout the family Maluridae (Table 6.1). For most Australian fairy-wrens, except the Lovely Fairy-wren (*M. amabilis*), studies of colour-banded birds have confirmed co-operative breeding. Most species have been studied at only one place, but Superb Fairy-wrens have been studied at three different localities, enough to indicate that co-operative breeding does not always take the same form. It may vary within a population, between populations, and between species. Experience with well-known species suggests that not all groups will have helpers, and that in some places or seasons, no groups will have helpers.

Within one population, our long-term study of the Splendid Fairy-wren found that in a period when productivity and adult survival were high, population density increased, female dispersal to breeding vacancies was reduced, and several plural breeding groups occurred. In Superb Fairy-wrens, two studies in Canberra found almost no female helpers; young females were driven out of their natal groups by female aggression. At Armidale, 560 kilometres further north, almost as many female as male helpers were found in two out of four years, while at Booligal, in an environment less favourable for Superb Fairy-wrens, Tidemann (1983) found no helpers at all in three seasons. Differences between species include the common occurrence of female helpers in all species except the Superb Fairy-wren. Two of the bi-coloured

Table 6.1 Evidence for co-operative breeding in the family Maluridae.

Species	*Marked birds*	*>2 birds at nest*	*Seen in groups*	*Source*
Superb Fairy-wren	+			Rowley 1965, Mulder *et al.* 1994
Splendid Fairy-wren	+			Rowley 1981*a*
Purple-crowned Fairy-wren	+			Rowley and Russell 1993
Variegated Fairy-wren	+	+	+	Tidemann 1983, Rowley and Russell, unpublished data
Lovely Fairy-wren		+	+	White 1946, Gill 1970, Rowley and Russell, unpublished data
Red-winged Fairy-wren	+			Rowley *et al.* 1988
Blue-breasted Fairy-wren	+			Rowley 1981*b*, Rowley and Russell unpublished data
White-winged Fairy-wren	+			Tidemann 1986, Rowley and Russell 1995
Red-backed Fairy-wren	+		+	Rowley and Russell, unpublished data
White-shouldered Fairy-wren		+	+	Nicholson and Coates 1975
Broad-billed Fairy-wren		+	+	Diamond 1981
Emperor Fairy-wren			+	Filewood 1971 and unpublished data
Wallace's Wren		+	+	Bell *et al.* 1979
Orange-crowned Wren			+	Frith and Frith 1992, 1993
Southern Emu-wren			+	Fletcher 1915, North 1904
Mallee Emu-wren			+	Howe 1933, Garnett 1992*a*
Rufous-crowned Emu-wren			+	Hitchcock and Jarman 1944, Rowley and Russell, unpublished data
Striated Grasswren			+	Izzard *et al.* 1973, Miller 1973, Schodde 1982*b*
Eyrean Grasswren			+	Schodde 1982*b*
Dusky Grasswren		+	+	Carruthers *et al.*1970, Glass 1973
Thick-billed Grasswren	+		+	Brooker 1988, B. Brooker, unpublished data, Rowley and Russell, unpublished data
Black Grasswren		+	+	Freeman 1970
Grey Grasswren		+	+	Favaloro and McEvey 1968, Robinson 1973, Schodde1982*b*
Carpentarian Grasswren			+	Schodde 1982*b*, Whitaker 1987, Beruldsen 1992, Harris 1992
White-throated Grasswren	+			Noske 1992

Three levels of evidence are recorded. (a) Study of colour-banded birds indicates that more than a simple pair feed at the nest or care for fledglings. (b) Observations of more than two birds (unmarked) at or near a nest. (c) Records of sightings frequently report birds occurring in groups of three or more. Records of (a) are sufficient evidence of co-operative breeding. Records of (b) or (c) alone are insufficient evidence, but if both (b) and (c) are recorded, the species is probably co-operative breeding. By itself, (c) is evidence of probable delayed dispersal.

fairy-wrens, the White-winged Fariy-wren (*M. leucopterus*) and probably the Red-backed Fairy-wren (*M. melanocephalus*), appear to have complex plural breeding groups (clans), with a single full plumaged male and several breeding females, each with her own male consort usually in brown plumage.

For most of the remaining species, evidence of co-operative breeding is weaker, but for grasswrens, emu-wrens, and the five malurids in New Guinea, the occurrence of groups at all times of year suggests that at least delayed dispersal is general and for many species there are one or two records of more than two birds seen near a nest or looking after fledglings. In the White-shouldered Fairy-wren (*M. alboscapulatus*) in New Guinea, delayed dispersal and helping certainly occur (Nicholson and Coates 1975). For the White-throated and Thick-billed Grasswrens (*Amytornis woodwardi* and *A. textilis*) more than two colour-banded birds have been seen to attend the nest. Without long-term study of individually recognizable birds, it is not certain that groups are formed by delayed dispersal, but it is most likely since the species are all resident and territorial. Little is known about the almost invisible emu-wrens, but juvenile dispersal appears to be delayed until winter or early spring.

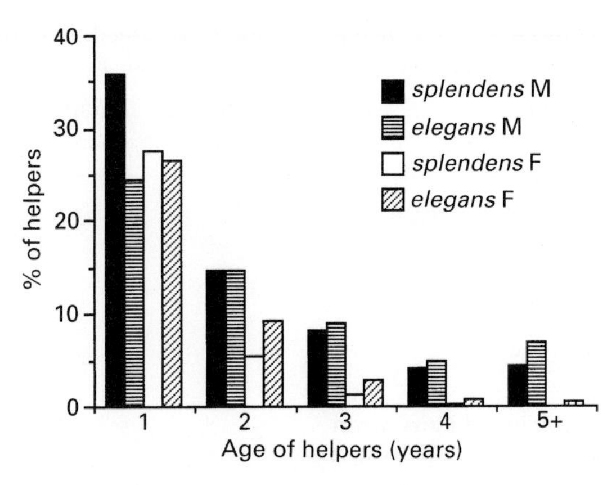

6.1 The percentage of helpers (M: male; F: female) that were aged from 1 to 5 or more years old in two studies: (a) Splendid Fairy-wren (*Malurus splendens*) (*n* = 422 helper-years from 348 group-years), and (b) Red-winged Fairy-wren (*M. elegans*) (*n* = 669 helper-years from 315 group-years). (One bird helping for 1 year is 1 helper-year. A bird that helped for 5 years contributes 5 helper-years.) (Russell and Rowley 1993*b* and unpublished data.)

Who are the helpers?

In all fairy-wrens, most helpers are surviving progeny from earlier years that remain in the family group after they reach sexual maturity. In the Splendid Fairy-wren virtually all the birds that helped were in the group where they hatched; 76 per cent of 171 males and 68 per cent of 140 females that survived to 1 year old helped for at least one breeding season. The rest were paired as breeders when 1 year old. The majority of males and females helped for only 1 year before either disappearing or pairing, while 30 per cent of males helped for 2 years or more. One male remained as a helper for 7 years before moving to a neighbouring territory as senior male, where he survived for 2 more years. A few females helped for 2 years and one for 3; all but one of these subsequently became established as breeders. Although a few birds remained as helper for a number of years, we found that the majority of helping was done by 1-year-old birds (Fig. 6.1).

In the Red-winged Fairy-wren (*M. elegans*), because of the high survival of paired adults, vacancies for dispersing birds were rare, and very few males or females paired at 1 year old. This is reflected in the different distribution of helper ages, and the higher proportion of birds that helped for 2 or more years. One-year-old helpers were more or less equally divided between males and females (Fig. 6. 1). Over 95 per cent of young that survived to 1 year old helped in their natal group for at least one breeding season and very few males or females left their natal group in their first year. The majority of males helped for 2 or more years and 20 per cent were 4 years old or more (Fig. 6.1); one remained as helper with a particularly long-lived senior male for 10 years before finally succeeding him, and as at November 1994, he has had 3 years at the top. More than a third of females helped for 2 years or more, one doing so for 6 years. Not all female helpers were in their natal group; some dispersed into other groups after their first year and remained as helpers for a year or more in their adopted group.

Where colour-marked birds have been studied in other species, helpers have similar origins and fates—the differences between species depend on productivity and on the mortality of adults and juveniles. Superb Fairy-wren populations in Canberra had no female helpers, but in Armidale, female helpers occurred in some seasons, and some females dispersed into groups as helpers, as in the Red-winged Fairy-wren. The length of time a bird remained as a helper depended on whether an opportunity arose to inherit its natal territory, to fill a vacancy in an established territory nearby, or, if a partner was available, to establish a new territory in a previously vacant area. In all the species so far studied, however, more birds help and then disappear without achieving senior status than help and subsequently become members of a social pair.

Relationships within the groups

The argument that helpers in a co-operative breeding group do not entirely sacrifice their own fitness by helping requires that they should be closely related to the nestlings they help to rear. If a helper assists its parents to raise a brood of its own siblings, and if more young are produced with help than without it, then the inclusive fitness of the helper is enhanced. In those cases where the young are not closely related or are unrelated to the helper, then no fitness benefit is gained, and inclusive fitness explanations for the evolution of co-operative breeding are not supported.

Many of the members in fairy-wren groups are related to each other, since the average group grows by the retention of progeny as helpers. Genetic relationships can be extremely complex, and, owing to the high level of extra-pair copulation, they are not always what they seem to be. All nestlings in a brood have the same mother, but one or more may be fathered by a male from another group. If two nestlings share the same father, they will be full siblings, but if they have different fathers, they will only be half-siblings.

The recent study of Superb Fairy-wrens in Canberra has used DNA fingerprinting in conjunction with long-term genealogies to analyse how helping behaviour is affected by kinship. From a knowledge of the parentage of helpers and nestlings the Canberra workers were able to determine whether helpers were full siblings or half-siblings of the nestlings they were helping to feed. The results of this study are summarized in Table 6.2. Younger helpers were more likely to be helping their mother, and about one-third were helping to feed a brother or sister, but about 20 per cent of young were unrelated to the helpers feeding them. As males became older, their mothers frequently had died or moved to a different group, so that more than half of the older helpers were unrelated to the nestlings they fed. Helpers were living with their genetic mother in 53 per cent of helper seasons, but only 31 per cent were living with their true father. One genetic father actually sired his helper son when he was living in a neighbour ing group, and then took over the territory where his son lived when the previous senior male died! All helpers helped at all nests, and did not distribute food to the nestlings according to how closely they were related—a helper feeding full siblings provided no more food than a helper that was completely unrelated to the same young.

Table 6.2 Superb Fairy-wren (*Malurus cyaneus*): relationship of helpers to the nestlings they feed, determined by DNA fingerprinting and genealogies.

Relationship of helper to nestlings	*% of helpers* 1 year old	*% of helpers* 2+ years old
Sibling	35	23
Half-sibling	46	23
Unrelated	19	54

Siblings have the same mother and father, half-siblings generally have the same mother and different father; unrelated includes all nestlings less closely related than half-sibs. Helpers are divided by age; 138 were helping for the first time at 1 year old, 133 were 2 years old or more. (From data in Dunn *et al.* 1995.)

We did not have DNA fingerprinting information to establish precise paternity in our studies of Splendid or Red-winged Fairy-wrens, but we have long-term genealogical data to show whether helpers were feeding young of their own mother or of an unrelated female. In Splendid Fairy-wrens, 66 per cent of helpers were feeding young of their own mother, and 20 per cent were feeding young of an unrelated mother; in Red-winged Fairy-wrens, the corresponding values were 63 per cent and 28 per cent. In both species, over 60 per cent of helpers were feeding young that were at least half-siblings, only some of which would have been full siblings. Since many helpers were assisting an unrelated female paired with their social father who may also have been their genetic father, probably less than 20 per cent were feeding completely unrelated young, lower than in Superb Fairy-wrens. This difference is due to the higher survival of females in both Splendid and Red-winged Fairy-wrens, and the fact that more females remain as helpers in these species than in Superb Fairy-wrens.

Overall, since most helpers are feeding half-siblings or unrelated young, indirect fitness benefits gained through helping to rear relatives are less than would be gained from helping to rear full siblings.

The role of helpers

Territory defence

In all fairy-wrens, helpers assist in the defence of territories. Every year at the start of the breeding season, boundaries are re-negotiated and reinforced. Even though groups remain in the same place year after year, a single territory increases or decreases in area according to whether there are more or fewer males (Brooker and Rowley 1995). If a senior male has one or two male helpers, the group is able to encroach on neighbouring territories and defend the extra area. Song battles joined by all group members are an important part of territory defence. If a neighbouring group intrudes, the reel song is given loudly by all group members and if that fails to deter the intruders, all members threaten and give chase; it is very rare for combatants to come to physical blows although one of us once caught three battling male Red-winged Fairy-wrens by dropping his hat on them! The helpers' role in defence is particularly important when a senior male leaves the territory to go philandering. If, at that time, the female is incubating, then the helpers are the only visible occupants of the territory.

Defence against cuckoos

During long watches at nests where a cuckoo was expected to lay, Lesley Brooker saw little effort by the Splendid Fairy-wrens to drive it away (Brooker *et al.* 1988). Later, helpers join enthusiastically in defending a nest with eggs or nestlings against cuckoos, which are thought to be nest predators as well as nest parasites. A stuffed Horsfield's Bronze-Cuckoo placed near nests of the Splendid Fairy-wren was usually discovered by the breeding female, who gave a loud whining call that attracted the other group members; all gave threat displays, flying at the dummy, pecking at it, and vigorously mobbing it (Payne *et al.* 1985). The intense mobbing by a large group may be sufficient to drive a cuckoo away.

Defence against predators

The main value of helpers in defence lies in the enhanced early warning given by many pairs of eyes. We found it much more difficult to approach a nest unnoticed when a group had helpers. If an incubating female was absent foraging, escorted by her partner, then one or more helpers would be close enough to the nest to sound the alarm if a predator approached.

When nestlings are handled for banding, all members of the group join in the agitation, with repeated loud 'zit' alarm calls, and the Rodent-run distraction display, which has been described in all Australian malurids. The displaying birds appear to take quite amazing risks on behalf of nestlings to which they may not be related. We watched a group of four trying to lead a large predatory lizard, a goanna (*Varanus* sp.), away from a nest containing three 9-day-old nestlings. Time and again, the goanna climbed half a metre up in low shrubs near the nest and launched itself towards an adult wren Rodent-running on the ground, only to miss narrowly as the bird flew away at the last moment.

Feeding nestlings

Although only the breeding female builds the nest, lays the eggs, and incubates them, all members of the group bring food to the nestlings and remove faeces. At first the female broods the nestlings more or less continuously and she intercepts most of the food brought to the nest, eats some of it, and passes the rest on to the nestlings. She can thus cover naked, very young nestlings almost continuously, especially when it is cold. Later, when more feathers have emerged, the female leaves the nest for increasing periods and the other attendants deliver their load directly to the nestlings.

Determining the contribution of individual group members to the feeding of nestlings involves scoring how many feeding visits each makes to the nest over a period of time. We have watched many nests from hides placed so that we could identify each colour-marked bird as it delivered food to the nest. No two watches were ever the same—nests differed in the number, age, and sex of the helpers, the age and number of wren or cuckoo nestlings, the weather, and the time of year (and thus the type of food available). The efforts of males in particular were very irregular, with occasional long absences followed by a period of frequent visits, so that short watches could give a very misleading picture. We have summarized the contributions of members in groups of different size in Table 6.3, and illustrate the variety of helper and breeder contributions with examples from two nests of the Splendid Fairy-wren and one of the Red-winged Fairy-wren (Table 6.4).

Table 6.3 The effect of helpers on the work load of breeding males and females.

	Group size			
% feeds given by	2 (9)	3 (10)	4 (17)	5+ (18)
Female breeder	67.2	53.9	37.4	33.2
Male breeder	32.8	11.5	16.4	17.5
Male helper		20.6	18.8	14.0
Female helper		45.6	34.6	28.0
Female breeder with MH		65.5		
Female breeder with FH		42.3		
Juvenile helper			14.4	9.0
Feeds per hour	11.1	10.9	13.6	15.3

Data from hide watches at nests of Splendid Fairy-wrens (*Malurus splendens*) with two or three nestlings 6–9 days old (not including nestling cuckoos). The figures in each column are the mean percentage of feeds delivered by the different categories of group member to nests with 2, 3, 4, and 5 or more attendants (number of groups in parentheses). Because the number of male and female helpers in larger groups varies, columns do not add up to 100%. (MH, male helper; FH, female helper). (Rowley and Russell, unpublished data; hide watches lasted for a mean of 183 min, range 80–779 min).

Helpers reduced the amount of work done by the breeding female. In simple pairs, she provided on average more than half the feeds delivered, and in many cases much more, since some old senior males contributed little, even when they had no helpers. Other males in simple pairs worked as hard as the female. Female helpers made a more significant contribution than male helpers in Splendid Fairy-wrens, providing nearly as much food to the nestlings as the breeding females, and sometimes more. This difference is clear for groups of three, in which the breeding female had either a male or a female helper; the effect of a single male helper was chiefly to reduce the work done by the senior male and probably release him to spend more time philandering. The greater contribution by a female helper reduced the work-load of the breeding female as well (Table 6.3).

Table 6.4 The role of helpers in feeding nestlings of the Splendid Fairy-wren (*Malurus splendens*) and the Red-winged Fairy-wren (*M. elegans*).

Details of nest	*Bird number*	*Sex and status*	*Age (years)*	*% of feeds*
Splendid Fairy-wren				
Nest **7705**	082	MB	5+	15.1
3 nestlings, 6 days old,	077	FB	5+	38.6
10.1 feeds/hour in 779 min	154	MH	3	16.7
	222	FH	1	29.5
Splendid Fairy-wren				
Nest **8405**	301	MB	5	0.0
3 nestlings, 9 days old,	406	FB	4	24.0
19.4 feeds/hour in 462 min	580	MH	1	25.3
	581	MH	1	22.6
	720	FH	1	28.0
Red-winged Fairy-wren				
Nest **8682**	672	MB	5	12.5
2 nestlings, 3 days old,	560	FB	4	12.5
19.5 feeds/hour in 120 min	781	MH	4	20.0
	580	MH	3	20.0
	110	MH	1	12.5
	522	FH	2	10.0
	111	FH	1	12.5

For each nest is shown the contribution by each group member (expressed as a percentage of all feeds delivered during the time the nest was under observation), the number and age of the nestlings, the feeding rate, and the length of the watch. Each nest is identified by its number, e.g. 7705. Each group member is identified by its number, sex, and status: FB: breeding female; MB: senior (breeding?) male; MH: male helper; FH: female helper. (Rowley and Russell, unpublished data.)

In larger groups, the breeding female's work-load was reduced still further, and, at the same time, the feeding rate was higher (the feeding rate increases with the age of the nestlings, but the data in Table 6.3 are from nestlings 6–9 days old). The contribution of senior males was variable in groups of all sizes and ranged from zero to more than half of all feeds in some pairs. Individually, juvenile helpers (less than 6 months old) seldom fed as much as adult helpers, but when two or three juveniles survived from an early brood, their combined efforts could be significant, although the items delivered were often small. All members of the group removed faecal sacs, which were carried 20 metres or more away from the nest before they were dropped.

At nest 7705 of a long established pair (Table 6.4), the old male and his 3-year-old male helper between them delivered about one-third of 132 feeds in the course of a whole day. However, at 8405, even with 9-day-old nestlings and a feeding rate of 19.4 feeds per hour, the old male brought no food to the nest, although he visited the nest several times in the course of the day. Despite his lack of contributions, the breeding female made only one-quarter of the feeding visits, with two 1-year-old males and a 1-year-old female all sharing equally in the rest. In our experience, the greater number of birds attending a co-operative nest did not increase the predation risk because the feeding rate remained more or less the same however many attendants there were.

A wide range of insects, spiders, and other arthropods are harvested, and different individuals appear to develop different hunting capacities and skills. A variety of hunting methods for different types of prey should reduce competition between helpers and lead to more efficient exploitation of food in the territory, but this is difficult to quantify when many of the food items brought to the nest are unrecognizable because they are so small or already partly processed. One adult male appeared to specialize in catching crickets, while at the same nest juvenile helpers, which had not yet mastered the techniques of catching and dismembering such large prey, brought only small items such as flies and ants.

Estimates of the contributions by group members to feeding nestlings in Superb Fairy-wrens have shown similar results. Males in pairs feed significantly more than males with helpers, and the presence of one adult male helper (the most frequent situation) allows the senior male to reduce his effort but has little effect on the contribution of the female breeder. Where female helpers do occur they contribute more than males (Tidemann 1986; Nias 1987; Mulder 1992; Mulder *et al.* 1994). In contrast more than half of all groups in the Red-winged Fairy-wren have two or more helpers, frequently female. The seven birds that feature in Table 6.4 delivered food at a very high rate to the 3-day-old nestlings, but no one individual provided more than 20 per cent of the feeds and the breeding female provided only 12.5 per cent. In the large groups of Red-winged Fairy-wrens, individual contributions were usually less than 25 per cent of the total.

Only in the Canberra study of Superb Fairy-wrens has it been possible to match the feeding rate of helpers with their relationship to the young they provisioned. There was no evidence that helpers fed closely related young more than unrelated young (Dunn *et al.* 1995).

Does helping really help?

Before considering how co-operative breeding may have evolved, we need to ask whether helping behaviour increases the fitness of the breeding adults of the helper's group. There are several ways in which this might occur. As described above, helpers might increase the success of individual nests by their provisioning of nestlings, caring for fledglings, increased awareness, and deterrence of predators and cuckoos. By reducing the work-load of the breeding female, they may enable her to lay more clutches per season or to survive for longer. They may allow the senior male more time to seek extra-pair copulations by reducing his work-load in feeding nestlings and caring for fledglings and in being there to defend the territory in his absence.

Increasing reproductive success

The reproductive success of a group in any year is affected by many things: the timing of rainfall, the incidence of predation and the frequency of parasitism by cuckoos, and the previous experience of the breeding female. In Splendid and Red-winged Fairy-wrens, if all groups with helpers are lumped together for analysis, groups with and without helpers produced the same numbers of independent young in a breeding season, and in Splendid Fairy-wrens, helpers made no difference to the mean production from individual nests (Table 6.5). However, in some circumstances, helpers do enhance their group's success, but this increase is by no means universal. In good conditions, female Splendid Fairy-wrens with at least 2 years' previous breeding experience and at least two helpers were more productive than females with the same experience and no helpers; they produced more than twice as many fledglings per year as females in their first year as a breeder with helpers (Table 6.6).

The main reason for this difference lies in the number of nesting attempts that a female can make in a year (Fig. 7.7). The presence of helpers reduces the time to re-nesting after one brood has fledged, and this effect is greatest for older females with more helpers. They may be able to re-nest sooner partly because other group members take over the care of the fledglings and partly because the female's lower

Table 6.5 The effect of helpers on the success of individual breeding attempts and on the annual reproductive success of their group: the number of independent young (surviving to 4 weeks) for pairs and groups (number in parentheses).

Species	*Production of independent young*	
	Pairs	*Groups*
Superb Fairy-wren (all females)		
1991/2	2.5 (18)	2.7 (21)
1992/3	3.7 (25)	3.5 (29)
Splendid Fairy-wren (experienced females)		
per nest	0.81 (190)	0.85 (456)
per group per year	2.2 (57)	2.2 (145)
Red-winged Fairy-wren (experienced females)		
per nest	0.88 (24)	1.4 (145)
per group per year	1.3 (30)	2.0 (98)

Sources: Superb Fairy-wren (*Malurus cyaneus*), Dunn *et al.* 1995; Splendid Fairy-wren (*M. splendens*) and Red-winged Fairy-wren (*M. elegans*) Rowley and Russell, unpublished data.

Table 6.6 The effects of breeding experience and group size on reproductive success in the Splendid Fairy-wren (*Malurus splendens*).

Female experience	*Group size*	*No. of females*	*Fledglings*		*% re-nest after success*	*Nests/female/year*
			/nest	*/year*		
Novice	2	41	1.1	1.7	15	1.5
	>2	32	1.3	2.1	13	1.6
Experienced	2	15	1.8	3.3	40	1.9
>2 years	>3	35	1.6	4.3	69	2.6

Data for 1973–85. (Based on data from Russell and Rowley 1988.)

work-load when raising the earlier brood enables her to build up her condition again more quickly and to lay a new clutch sooner. Larger groups are also more likely to have female helpers, which work harder in feeding nestlings and are associated with higher reproductive success (Brooker and Rowley 1995). The work-load of the breeding female is not much reduced by a single male helper, and it is only if she has at least two male helpers, or a female helper in a group of three, that the breeder's contribution is much reduced. The Canberra population of Superb Fairy-wrens had few, if any, female helpers, and only 25 per cent of groups had more than one helper, so the circumstances when we might expect helpers to increase reproductive success were infrequent.

In the Red-winged Fairy-wren, on the other hand, 1-year-old females generally remained in their natal territory as helpers; more groups included one or more female helpers, survival of females was high, and novice females few. Over all groups, the presence of helpers significantly increased annual reproductive success (Table 6.5). As in the Splendid Fairy-wren, those females with more than 2 years' experience and two helpers were the most productive, rearing a mean of 2.3 independent young per year. Fewer females produced two broods, but of the 24 that did so, 19 had more than 2 years' breeding experience and were in groups with two or more helpers.

There is now evidence that the presence of helpers leads to increased reproductive success for senior males in the Superb Fairy-wren. Their reduced contribution to feeding nestlings in groups with helpers allows senior males more time for visiting nearby females, displaying, and achieving extra-pair copulations (Green *et al.* 1995).

Increasing breeder survival

Contributions by helpers to feeding nestlings could increase a breeder's fitness by increasing its survival, thereby allowing it to breed for another year or more. A reduction in workload, particularly that of the breeding female, should mean that she has more time to feed herself and should be in better condition at the end of the breeding season. A reduction in her feeding visits to the nest should also decrease her exposure to predators. Overall, helpers had only a slight, not statistically significant, effect on the survival of breeding Splendid Fairy-wren females; 63 per cent with helpers survived to the next breeding season compared with 55 per cent with no helpers. In the Red-winged Fairy-wren, very few females were without helpers and during the 12 years 1980–91, their annual survival (76.9 per cent) was almost identical with that of the majority of females who had helpers (76.0 per cent). Helpers appear to have no effect on the survival of senior males in either species, and we have no information on the effect of helpers on survival in any other malurid.

Why do helpers stay home?

Constraints on dispersal

In the family Maluridae, co-operative breeding groups usually result from the delayed dispersal of grown young. What are the relative advantages of staying at home or dispersing early? How does the ecology of co-operative breeders favour delayed dispersal? One approach has been to look for the 'ecological constraints' that prevent offspring from dispersing early and breeding independently as soon as they reach sexual maturity (Emlen 1982, 1991). One of the most frequently invoked is habitat saturation, where all suitable breeding space is already occupied, and young birds must wait for a vacancy to occur in an established territory. The safest place to wait is a familiar home territory.

Other constraints are associated with early departure. Dispersal into unfamiliar areas carries a risk of increased mortality (see p. 126). Even if a territory is available, there may be a shortage of mates; and even if both territory and mate are available, two birds on their own may not be able to reproduce successfully. This is illustrated by another Australian co-operative breeder, the White-winged Chough (*Corcorax melanorhamphos*), in which pairs without helpers are unable to raise any young (Rowley 1978; Heinsohn 1995).

The benefits of philopatry

Another approach to delayed dispersal emphasizes the 'benefits of philopatry' (Stacey and Ligon 1987, 1991). Staying at home may lead to the inheritance of the natal territory, or it may give the young bird a competitive edge in filling a vacancy in a neighbouring territory, without the risks of dispersal. In situations where territories differ greatly in quality, an individual may do better to wait and inherit a good territory than to disperse to a vacancy in a poor quality territory. These two approaches share the same driving factor of 'strong competition for a limited number of suitable territory vacancies' (Emlen 1991, p. 313).

The exciting work of Jan Komdeur on the endangered Seychelles Warbler (*Acrocephalus sechellensis*) has shown that the expression of delayed dispersal and co-operative breeding may be influenced by both habitat saturation and territory quality (Komdeur 1992). After 23 years of a recovery programme on Cousin Island in the Seychelles (1959–82), the population had grown from 26 to 320 individuals in 115 territories. Until 1973, no co-operative breeding was recorded, but by the 1980s, it was common, and between 1986 and 1990, only 41 per cent of yearlings dispersed; the rest remained in their natal territories as helpers. In this case, co-operative breeding occurred only when the available habitat became saturated. One-year-old birds from low quality territories were more likely to disperse than those from higher quality territories. When birds from Cousin Island were translocated to a new site on Aride Island, the vacancies created on Cousin Island were almost immediately filled by previously non-breeding helpers. However, birds dispersed only to a vacancy in a territory of the same or higher quality than that from which they came, and none dispersed from a high quality territory to one of lower quality. On Aride Island, where there was plenty of vacant habitat, all the offspring produced by the translocated birds dispersed as yearlings; with no habitat saturation, co-operative breeding did not occur. Only when all the good habitat on Aride was occupied did young birds begin to remain as helpers in their natal territories. The Seychelles Warbler is a good illustration of the point made by Koenig *et al.* (1992) that no one factor by itself can provide an explanation of delayed dispersal and co-operative breeding, and this is true of *Malurus*.

The significance of sex ratio

Populations of *Malurus* frequently have an excess of males, but this is not such a distinctive characteristic as was once thought. Some long-term studies have shown that sex ratios vary over the years, from equal numbers of males and females to a large excess of males (Fig. 6.2). In the Splendid Fairy-wren, the percentage of groups with helpers remained stable over a wide range of sex ratios, providing no support for the hypothesis that a short-

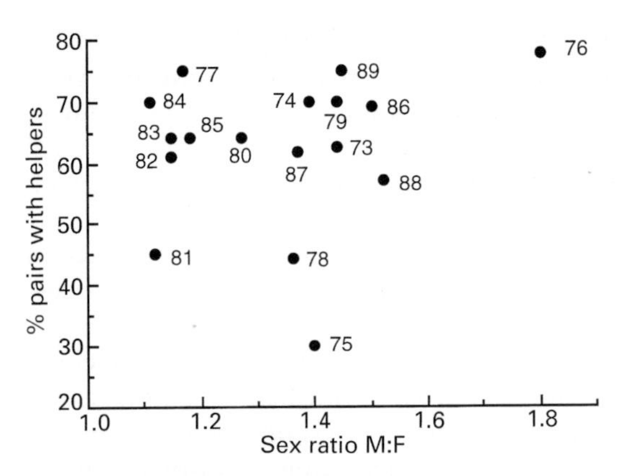

6.2 The relationship between the adult sex ratio (number of males per female) and the percentage of groups with helpers in the Splendid Fairy-wren (*Malurus splendens*) between 1973 and 1989. During 17 years, the percentage of groups with helpers has been stable over a wide range of sex ratios and there is no correlation between the two.

age of sexual partners constrains males to remain as helpers, as was suggested by Brown (1975) and Emlen (1978).

In populations where no groups, or very few, have helpers, the sex ratio is more or less 1M:1F, with a slight excess of males, as in Superb Fairy-wrens at Booligal, Blue-breasted (*M. pulcherrimus*) and Purple-crowned (*M. coronatus*) Fairy-wrens (Table 5.1). Some populations of Superb Fairy-wrens clearly have a large excess of males, and this generally occurs when most groups have only male helpers, as in the Canberra population. However, this is not always so, as in Nias' study near Armidale, where the sex ratio ranged from *c.*1:1 in a year when numbers of male and female helpers were almost equal, to 1.7:1 when only male helpers were present. Similarly, in our study of Splendid Fairy-wrens near Perth, sex ratio ranged from 1.1:1 to 1.8:1. The sex ratio approached 1:1 after a period of population increase when density was high, vacancies for breeders were few, and young females stayed in their natal groups rather than disperse. When density was low, there were surplus males, suggesting that females were able to disperse to vacancies elsewhere. As would be expected in a species where females disperse and probably incur increased mortality, at no time has there been a significant excess of females.

Delayed dispersal in fairy-wrens

Intensive studies of fairy-wrens suggest that shortage of territories, scarcity of females, and a high probability of territorial inheritance are all possible factors influencing delayed dispersal. In the Splendid Fairy-wren study, the continuous suitable habitat was fully occupied by territories in most years. Male helpers were generally more abundant than female ones, and, at times, there were not enough spare females to fill vacancies that became available (Fig. 6.3). At other times, both males and females were available in excess of vacancies, and presumably could have dispersed if space had been available. On average, six male and two female helpers were available for every vacancy that occurred, with a maximum of 53 males for four vacancies and 37 females for four vacancies over one 6-month period in 1984. In most years, unpaired females were available at the start of the breeding season, but unpaired males still did not disperse, suggesting that suitable habitat was not available to establish new territories. During the years 1978–86, the numbers of helpers fluctuated considerably, but the number of occupied territories remained relatively constant.

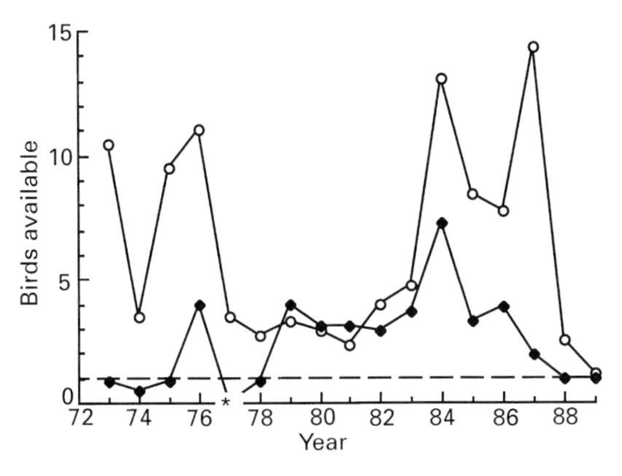

6.3 Evidence of habitat saturation in the Splendid Fairy-wren (*Malurus splendens*). Birds available to fill breeding vacancies, Gooseberry Hill, WA, 1973–89, expressed as birds per vacancy, based on vacancies occurring between March and August and the number of non-breeding birds at 1 July. Open symbols, males; filled symbols, females. The horizontal line represents one bird per vacancy. Points below that line indicate that not enough birds (female) were present to fill vacancies that occurred. In 1977 (*), two non-breeding females were present, but no vacancies occurred. (Data from Russell and Rowley 1993*a*.)

In the Superb Fairy-wren, experiments by Pruett-Jones and Lewis (1990) showed that males would not disperse into a vacant territory unless a female was present (see Chapter 5, p. 58). Mulder estimated that the mortality of late dispersing females was 50–60 per cent, leading to a shortage of females to fill vacancies that occur during the breeding season. In such circumstances, territories remain vacant even though unpaired males remain as helpers in neighbouring territories (Mulder 1995).

All these observations suggest that delayed dispersal is governed by a shortage of both territories and females. The balance between these two factors will differ within and between species from time to time depending on the variation in the environmental and demographic situations.

One benefit of philopatry is the prospect of inheriting the natal territory or being on hand to fill a vacancy next door as soon as it occurs; the only data so far available are for Splendid and Red-winged Fairy-wrens. Of 80 Splendid Fairy-wren males replaced in established groups, 71 per cent were succeeded by their oldest helper, and 29 per cent of 171 yearling males eventually became senior male in the territory where they hatched. Another 13 per cent moved to a neighbouring territory, while only eight males dispersed further afield. Female helpers, too, may inherit their natal territory; 26 per cent of breeding females were replaced by their oldest helper, and another 25 per cent by a helper from next door. Similarly in Red-winged Fairy-wrens, 75 per cent of male vacancies were filled by the oldest helper, and 33 per cent of vacancies for breeding females were filled by a helper in her natal territory.

Why do helpers help?

The benefits that helpers gain from remaining at home and not breeding by themselves do not explain why they should help to rear offspring that are not their own. Most explanations of co-operative breeding assume that young birds staying at home have the option of helping or not helping, and that they help because the benefits from doing so outweigh the costs. In the fairy-wrens studied in detail, helping (in the narrowest sense of feeding nestlings) varies widely between individuals. Many different ways in which 'helpers' may benefit from helping have been suggested (Emlen 1991; Clarke 1995), while, on the other hand, it has also been suggested that helping occurs simply because of universal selection for the behaviour of feeding young when they beg (Jamieson and Craig 1987). This suggestion has been hotly debated without resolution.

Among the reasons for helping that have been proposed, we can reject some, and for none of them is the evidence overwhelmingly strong. As we have shown above, it is unlikely that helping is maintained by significant indirect benefits from the production of non-descendant kin. Although in some species, the presence of helpers may sometimes lead to slight increases in the production of young and the survival of breeding females, the low level of relatedness between the helpers and breeders and between helpers and nestlings means that they add little to their inclusive fitness by increasing the production of relatives.

The suggestion that helpers increase their own future reproductive success has some support, particularly for females. Dunn *et al.* (1995) suggest that since senior males do not often work hard at feeding nestlings, experience gained as a helper is of no great benefit to a male. However, for female helpers, the experience gained in helping may well be of value. Female Splendid Fairy-wrens breeding for the first time were less succesful than more experienced, older females, and there is evidence that females with experience as a helper were more successful when they bred for the first time a year later (Russell and Rowley 1988). Females that helped for a year had higher lifetime reproductive success than females that bred first at 1 year old (Russell and Rowley 1993*a*).

Many helpers, especially males, inherit the senior position in their natal territory. By helping to rear young, a male helps to maintain group size, which may benefit him if he becomes senior male, because the presence of helpers will allow him to leave the territory and seek extra-pair copulations. Maintaining group size may also benefit a female helper if she becomes the breeding female, since she inherits a large group. Survivorship of helpers may be enhanced by being in a larger group, through improved vigilance and anti-predator behaviour.

Dunn *et al.* (1995) argue that their data on the Superb Fairy-wren suggest that helping is a 'payment' to the breeders for remaining on the territory, with access to its survival-enhancing benefits. They cite experiments by Mulder and Langmore (1993) in which helpers were removed from the terrritory for 24 hours; on their return, they were frequently subject to intense aggression from senior males, especially when the group was caring for nestlings or fledglings. We are not entirely convinced by this, since our experience with some of the very large groups found in Splendid and Red-winged Fairy-wrens is that one or more helpers may do very little to feed nestlings with no sign of aggression towards them, while, on the other hand, the senior male is frequently absent and unable to enforce such payment. More experimental and field evidence is needed to support this view.

Yet another suggested benefit of helping is that it may enhance the future probability of getting a territory or a mate. Splendid Fairy-wren groups with more male helpers were able to expand their territory boundaries (Brooker and Rowley 1995); this benefits the incumbent male, but there is little evidence that it enhances the future probability of breeding for a helper. We have only very few examples, mainly in Red-winged Fairy-wrens, of territories enlarged in this way budding off a new territory for a helper, as occurs in the Florida Scrub Jay (*Aphelocoma caerulescens*) (Woolfenden and Fitzpatrick 1984).

The story so far

This seems a good point to review where we have reached in our discussion of co-operative breeding in the Maluridae. We know that helpers are usually birds helping in the territory where they hatched, even if they are not the offspring of the present male and female breeders. They are involved in most of the activities of group life—defence against conspecific intruders, predators, and cuckoos, and the feeding of nestlings and fledglings. Birds that remain at home have a good chance of inheriting their natal territory or one of the territories surrounding it, and their survival is higher than that of birds that disperse (see p. 126). In their natural habitat, since groups are resident all year round and opportunities to establish new territories are rare, the best chance for a young bird is to replace a member of a pair; such vacancies may occur at any time. In all the species studied so far, helpers do not dramatically increase the production of offspring or the survival of breeders, so that, because of the high incidence of extra-pair copulation, most birds add little to their inclusive fitness by aiding close relatives. This leaves helping as the aspect of co-operative breeding that seems hardest to explain.

Until now, we have followed a well-trodden path in our discussion of co-operative breeding starting from the usual point that the fundamental social system in birds is a monogamous pair whose offspring disperse (or migrate) as soon as possible, and that anything different from this must have evolved from it. In 1968, Rowley was a lone voice suggesting 'should we perhaps regard the communal way of life as normal and investigate the different ways by which other species avoid prolonged communal relationships after successful reproduction' (p. 367). However, a process of reappraisal began with the change in views of avian phylogeny brought about by the DNA hybridization studies of Sibley and Ahlquist (1990; see Chapter 3). Recently, a phylogenetic bias in co-operative breeding has been explored by several writers (Russell 1989; Edwards and Naeem 1993; Ligon 1993; Cockburn 1996), and raises the possibility that in the ancestral state, offspring remained with their parents beyond independence, and only later did early dispersal arise. One problem that has always remained at the edge of attempts to understand co-operative breeding is its very high incidence in Australian birds. The phylogenetic approach goes a long way to explaining this (Cockburn 1996), and in the process turns much of the thinking about the evolution of co-operative breeding on its head.

Why are there so many co-operative breeders in Australia?

The prevalence of co-operative breeding in Australia has prompted many attempts to identify environmental circumstances that favour it. In 1965, in the first detailed account of co-operative breeding in an Australian species, Rowley suggested that it was an adaptation to a widely varying climate. Since then, it has been variously attributed to variability and unpredictability of climate, aridity, equable conditions, and lack of marked seasonal fluctuations. Rowley (1968, 1976) suggested that lack of a severe winter allows non-dispersal, and that in an irregular climate, delayed dispersal may favour the survival of offspring. In 1980, Dow attempted to relate the incidence of co-operative breeding to vegetation type and some 11 environmental variables; he could find no common ecological factor in its occurrence. However, his analysis suggested that co-operative breeding was least common in the monsoonal far north and the arid centre, and more frequent in the south-east and south-west, inland from the coast. Ford *et al.* (1988) demonstrated that these concentrations were in *Eucalyptus* and semi-arid woodlands, in which there is evidence that resources do not show marked seasonal fluctuations. They suggested also that the open nature of these habitats would favour non-dispersal in species vulnerable to predation. However, these areas are in the region where birds are most studied, which may bias the pattern.

The year-round availability of resources is considered important for co-operative breeding, and in his 1987 book, Jerram Brown concluded that the primary pattern in the geographical distribution of co-operative breeding species is of a higher frequency in regions where species are permanent residents. Several other attributes of the Australian environment favour co-operative breeding (Russell and Rowley 1993*b*).

1. Australian eucalypt and acacia woodlands are evergreen, although new seasonal growth is associated with a seasonal flush of leaf-eating invertebrates.
2. The whole of the Australian landmass lies between 10°S and 44°S, temperate and tropical latitudes that appear to favour co-operative breeding.
3. Few Australian birds (apart from seabirds and waders) are long-distance migrants over water; many species are resident.
4. Australia receives few immigrants from elsewhere (apart from seabirds and waders), so the seasonal peaks of available food do not have to be shared with an influx of migrants from the northern hemisphere.
5. Winters are not generally severe; although there is seasonal variation in the availability of food, the extent of areas covered for long periods by snow that precludes foraging is a minute fraction of the total area (Osborne and Green 1992). Seasonal peaks of insect abundance occur, but the peaks are not as great as in the northern spring (Ford *et al.* 1988).

However, the family Maluridae defies any attempts to find simple correlates of co-operative breeding. Species of *Malurus* range from temperate south-eastern Australia and Tasmania (*M. cyaneus*), the temperate south-west forests (*M. elegans*), through the arid centre (*M. leucopterus leuconotus*, *M. lamberti assimilis*, and *M. splendens callainus*), tropical grasslands (*M. melanocephalus* in Australia; *M. alboscapulatus* in New Guinea), tropical river fringes (*M. coronatus*), margins of tropical forests (*M. amabilis* in Australia; *M. cyanocephalus* in New Guinea), to primary tropical forests (*M. grayi*). The genus *Amytornis* ranges from southern semi-arid mallee woodlands through arid shrublands to spinifex grasslands on rugged hillsides in the arid centre and monsoon tropics. Probably the only attributes shared throughout the family are that they are resident, and have the low reproductive rates and relatively high adult survival typical of Australian passerines, not only the co-operative breeders.

Does phylogeny matter?

As we began to discover that each new species of *Malurus* that we studied was a co-operative breeder, despite living in such very different habitats, it became more and more difficult to accept that this was something that had evolved independently for each species in response to separate environmental constraints. It seemed much simpler to think of co-operative breeding as a characteristic that had perhaps evolved early in the family history. However, in other parts of the world, general opinion in the past was summarized by 'communal breeding is clearly not a trait whose phylogeny can be usefully analysed across phyletic lines...' (Brown 1987, p. 34). Brown was following the traditional classification based on morphological similarities, in which convergent morphological characters resulted in the grouping of Australo-Papuan species with groups based on European, Asian, and African types. Reinterpretation of the classification by Sibley and Ahlquist (1985) placed most Australian passerines as members of a single lineage (the Corvida) with a major radiation taking place in Australia, from which branches arose that have radiated in Africa, Eurasia, and the Americas. Members of the other major lineage of the passerines in Australia and New Guinea are relatively recent colonists that arrived after the Australian continent drifted closer to Asia, *c.* 15–20 million years ago. They include starlings, swallows, cisticola, white-eyes, reed warblers, larks, and finches. This was at about the same time, presumably, that members of the Corvida spread outwards to south-east Asia, Africa, and the Americas.

It was with great excitement that we realized that all Australian passerine co-operative breeders belong to the old endemic lineage, the Corvida, and that some of the families with a high incidence of co-operative breeding (such as Maluridae and Climacteridae, the tree-creepers) are some of the oldest passerine families. This led us to suggest looking to the distant past rather than the present for the origins of co-operative breeding (Russell 1989).

This approach was extended by Edwards and Naeem (1993) who confirmed that the taxonomic distribution of co-operative breeding is definitely non-random. Phylogenetic reconstruction indicated that co-operative breeding had arisen fewer times than would be expected had each species evolved it independently. It persisted in a group long after its origin, and sometimes was lost. Since then, the role of phylogeny in the evolution of social systems has received more attention (e.g. Koenig *et al.* 1992; Ligon 1993). Where once social systems were seen as flexible adaptations to current environmental variables, now there is recognition that many aspects may be determined by phylogeny but shaped by adaptation to current environmental circumstances.

More recently, Andrew Cockburn (1996) has shown that co-operative breeding is an even more fundamental characteristic in the Corvida than was previously realized. He examined the extent of co-operative breeding in the Corvida worldwide, restricting himself to species with definite records of the presence or absence of helping behaviour. Co-operative breeding occurs in at least 61 genera (of 140 for which reliable data are available), and the proportion of genera with co-operative species is the same outside Australia/New Guinea as within. The distribution of mating systems among the branches of the Corvida shows a pattern in which the early branches are typically groups with either a high level of co-operative breeding or complete male emancipation from parental care along with display site promiscuity, (as in lyrebirds, birds of paradise, and bower birds). Monogamous lineages are found as terminal branches or twigs, rarely major branches (Fig. 6.4). This supports the view that monogamy in the Corvida is a specialized state that has evolved from group-living. Many non-co-operative Corvida show extended natal philopatry.

The occurrence of co-operative breeding, delayed dispersal of offspring, and mating systems with male emancipation from parental care as basic to the Corvida leaves monogamy as a derived state and suggests that co-operative breeding or at least delayed dispersal was present in ancestral Corvida. What was the ancestral state in birds is at present only speculation (Wesolowski 1994). It is not known in what sort of habitats the primary diversification of the Corvida took place; possibly they favoured co-operative breeding. Cockburn (1991) pointed out that since evidence from palaeobotany suggested that the genus *Eucalyptus* is of relatively recent origin, the association of co-operative breeding with *Eucalyptus* woodlands is likely to be secondary.

Overview

One problem with many of the attempts to understand co-operative breeding is that to date they have been made by people who

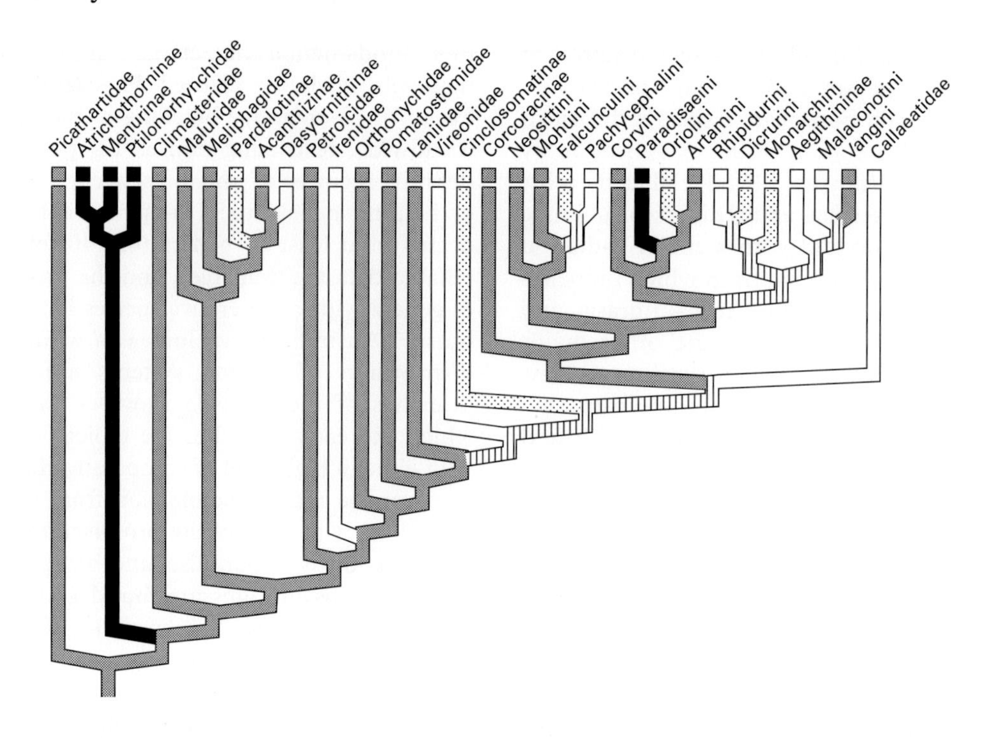

6.4 The distribution of co-operative breeding in the Corvida. The figure distinguishes where co-operative breeding occurs in most genera (dark stipple), where co-operative breeding occurs but is not common (light stipple), where co-operative breeding is not recorded (unshaded), and where the predominant mating system is one where males take no part in care of young (black bars). Bars with horizontal lines indicate where there is no evidence for the direction of evolution. (From Cockburn 1996, Fig. 3.)

regard it as the exception that has to be explained and for whom the normal situation is the monogamous pair, either migrant or, if resident, not necessarily remaining as a pair through the winter. When one lives in an environment with few migrants, where family groups resident all year round are a common occurrence, one realizes that the term co-operative breeding and the use of 'helpers-at-the-nest' are misleading. It is an all year round phenomenon, with helping or co-operative breeding only a small part of that year. The important resource for all the group is the territory that supports them through the year. The group in its territory is the important social unit. In the case of the Maluridae, for example, the group moves round the territory together, feeds together, roosts together, works together to repel intruders or to detect predators. Although males may seek copulations outside the group, most of their time is spent as active group members. One gets the subjective impression that feeding nestlings and looking after fledglings is just another group activity, and part of the social currency. This makes it easier to accept Jamieson and Craig's (1987) suggestion that the origin of helping is that all birds are programmed to feed nestlings, even if it has been modified or lost due to adaptation in some species.

As for delayed dispersal, it should be seen in the context of the relatively high adult survival and low reproductive rates of many Australian passerines, and the absence of severe winters and long-distance migrations. In resident species, dispersal into the unknown when vacancies for breeders are rare is risky, which increases the advantages for young birds of staying at home and waiting for a vacancy. In a long life, forgoing one breeding season is a small price to pay for this. For adults, the costs of retaining the young in the natal territory are minimized by the low reproductive rate, so that there are rarely more than two or three additions each year. Where vacancies for breeders are few, if allowing the young to remain increases their chance of survival and of ultimately attaining a breeding territory, then delayed dispersal of young can increase the fitness of parents and can be seen as a form of extended parental investment.

7

Life histories

Introduction

Most long-term studies of avian life histories have been made in the temperate habitats of northern Europe and North America, where the majority of ornithologists live. Such studies are very much rarer in the tropics and the temperate regions of the southern continents. In Australia, where the passerine avifauna has a long history of evolution in isolation, it appears likely that life histories have developed along a different course. At present we cannot be dogmatic about this because there have been relatively few long-term Australian studies. Fortunately, several studies of fairy-wrens provide adequate long-term demographic and behavioural data from which we can construct a generalized life history for the family, even though there is no comparable information for emu-wrens or grasswrens, or for any of the New Guinea malurids.

Breeding biology

Breeding seasons

The length of the year available for birds to breed in varies with latitude, being shorter at high latitudes. In a review of the length of avian breeding seasons Wyndham (1986) found that in Australia and Africa they lasted two months longer than on other continents at equivalent latitudes. Because the Australian continent stretches from latitudes 10 to 42°S, the timing and quantity of rainfall vary considerably and, in consequence, the breeding seasons of birds and their likely controlling factors differ from place to place. A review by Henry Nix (1976) related avian breeding cycles to seasonal changes in plant growth, as measured by indices of temperature, light, and moisture. Cold winters or high summer temperatures may reduce plant growth in some areas, but moisture is the major limiting factor, in summer in the south and in winter in the north. Nix's predicted best times of year for plant growth are spring in the south and summer in the tropics and northern half of the east coast with adjacent ranges. Inland, moisture is deficient most of the time, and plant growth may follow rain in any season, especially in summer when tropical depressions occasionally penetrate to the interior. Nix found broad agreement between the breeding seasons of birds and the periods of highest plant growth, with some anomalies in areas where the climate was less seasonal.

Members of the family Maluridae illustrate the diversity of breeding cycles across Australia and New Guinea, despite a scarcity of systematically collected long-term data from inland areas. The breeding of a few species has been studied in detail, but generally only at a single site. The most extensive data available come from the Nest Record Scheme (NRS) of the Royal Australasian Ornithologists' Union (RAOU) whose records for Superb (*Malurus cyaneus*), Variegated (*M. lamberti*), and White-winged (*M. leucopterus*) Fairy-wrens between latitudes 30 and 40°S were analysed by

Tidemann and Marples (1987). We have drawn on that analysis here, as well as assembling more recent data from the NRS and published records for these and other species. The NRS records are concentrated on species breeding in the coastal fringe of south-eastern Australia. In most places, late winter to early summer appears to be the main breeding period, but in the north and inland, breeding may occur at other times (Fig. 7.1).

Temperate

Fairy-wrens in southern Australia breed in spring and early summer, with few clutches laid after December (Fig. 7.1(a–f)). In more northern populations with summer rainfall (Variegated Fairy-wren *M. l. lamberti*, Fig. 7.1(f)) breeding may start as early as July or August; with winter rainfall in south-western Australia, the Red-winged Fairy-wren lays few clutches before October (Fig.7.1(d)). The timing of breeding may differ for different populations; Blue-breasted Fairy-wrens (*M. pulcherrimus*) at Dryandra generally begin to breed about a month later than at Kellerberrin, 150 kilometres north-east (M. Brooker, personal communication). Similarly, near Perth, during 17 years, we found no nests of Splendid Fairy-wrens (*M. splendens*) before 20 August, but on the Murchison River, 500 kilometres further north, the Brookers found them nesting in July.

Arid

The extent of variation within a species is best shown by the White-winged Fairy-wren which, in coastal habitat in south-western Australia, exhibits strict seasonal spring breeding, but in more arid areas has been recorded breeding in all months except June (Fig. 7.1(e)). At Ivanhoe in inland New South Wales (average rainfall less than 300 millimetres per year), during two years of above average rainfall in 1973 and 1974, breeding was recorded continuously from July 1973 to the end of May 1974. Even though the same individuals did not breed continuously, clearly the environmental conditions remained favourable (Fig. 7.2). At Barrow Island off the north-west coast (20°S, annual rainfall 324 millimetres), White-winged Fairy-wrens (*M. l. edouardi*) bred in May following 300 millimetres of rain between February and April 1992 (Ambrose and Murphy 1994).

The inland form of the Variegated Fairy-wren *M. l. assimilis* also shows a concentration of breeding records between July and January (Fig. 7.1(f)), but in an arid region of Western Australia (22–23°S), following heavy summer rain from tropical cyclones, they bred in March and April (Carnaby 1954; Robinson 1955), as did White-winged and Splendid Fairy-wrens. Further south (30–34°S), widespread, sometimes successful, autumn reproduction may take place in years when there are abnormal southerly penetrations of summer cyclones from the north (Serventy and Marshall 1957).

Tropical

Even in species found predominantly in tropical summer rainfall areas, a peak of breeding occurs in spring–early summer, although there is some breeding in most months, especially in autumn after summer rain. The main breeding season therefore precedes the peak growing season, possibly because heavy rain at the height of the wet season leads to flooding, makes feeding difficult, or saturates nests. On the east coast, Red-backed Fairy-wrens (*M. melanocephalus*) breed predominantly from August to December (Fig. 7.1(g)), but across northern Australia, they are more likely to breed from January to April (Frith and Davies 1961; Lavery *et al.* 1968). The few nest records for the Lovely Fairy-wren (*M. amabilis*) are for April, June, July, August, December, and January (White 1946; Lavery *et al.* 1968; Gill 1970; Rowley and Russell, unpublished data). Breeding records for the Purple-crowned Fairy-wren (*M. coronatus*) are concentrated in spring and autumn (Fig. 7.1(h)). The timing and extent of the wet season in northern Australia is variable, and our experience with Purple-crowned Fairy-wrens suggested that a good wet season was

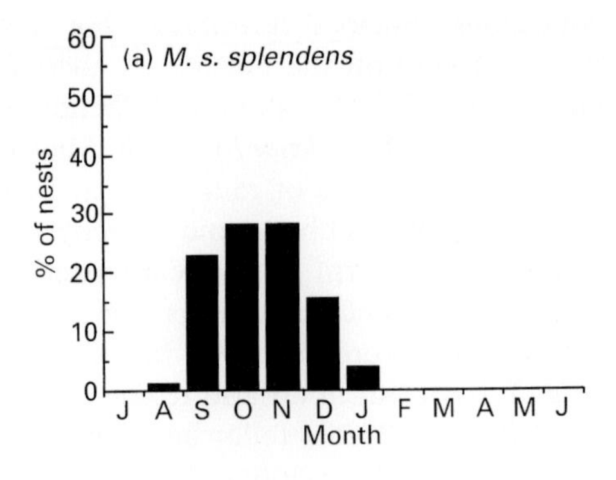
(a) M. s. splendens
% of nests
0
10
20
30
40
50
60
J A S O N D J F M A M J
Month

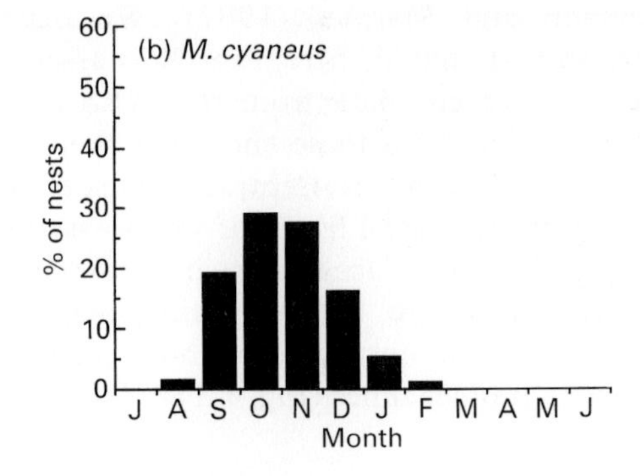
(b) M. cyaneus
% of nests
0
10
20
30
40
50
60
J A S O N D J F M A M J
Month

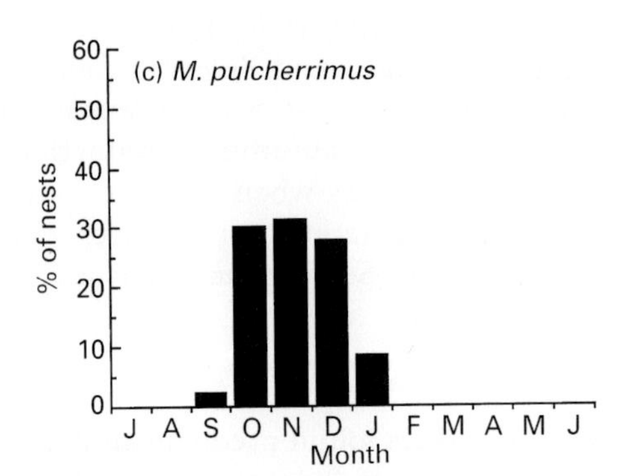
(c) M. pulcherrimus
% of nests
0
10
20
30
40
50
60
J A S O N D J F M A M J
Month

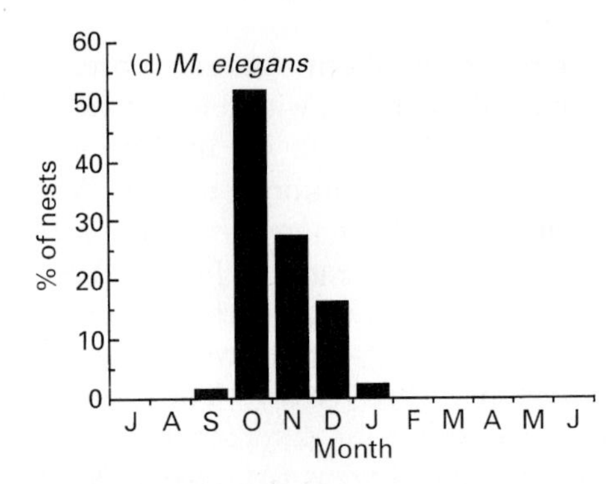
(d) M. elegans
% of nests
0
10
20
30
40
50
60
J A S O N D J F M A M J
Month

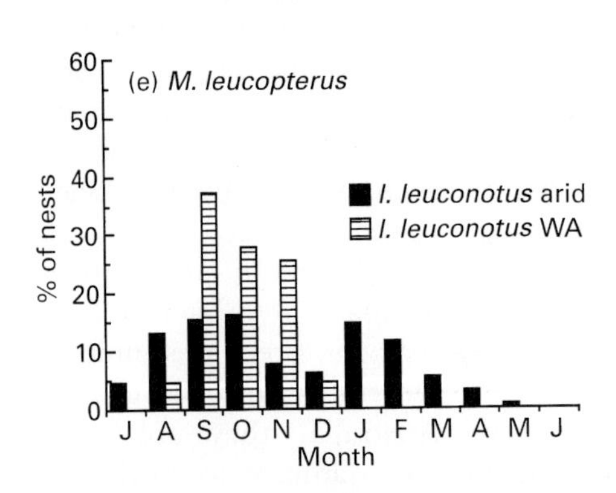
(e) M. leucopterus
l. leuconotus arid
l. leuconotus WA
% of nests
0
10
20
30
40
50
60
J A S O N D J F M A M J
Month

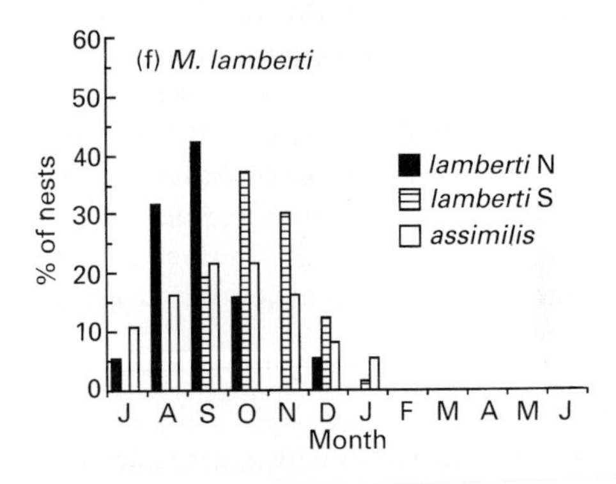
(f) M. lamberti
lamberti N
lamberti S
assimilis
% of nests
0
10
20
30
40
50
60
J A S O N D J F M A M J
Month

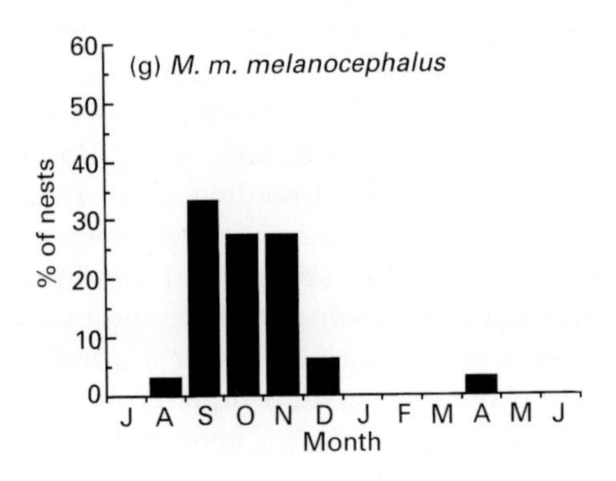
(g) M. m. melanocephalus
% of nests
0
10
20
30
40
50
60
J A S O N D J F M A M J
Month

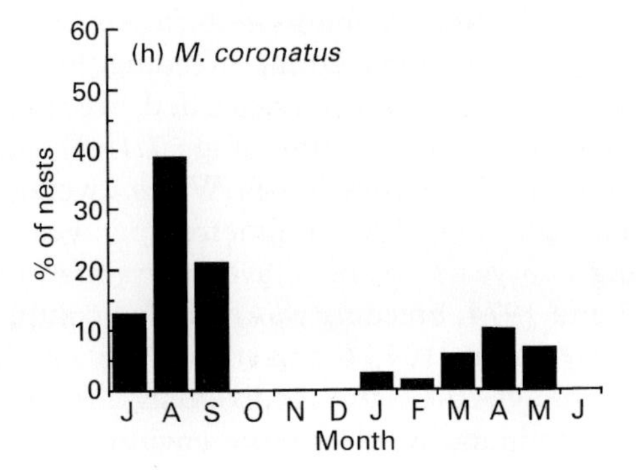
(h) M. coronatus
% of nests
0
10
20
30
40
50
60
J A S O N D J F M A M J
Month

◀ **7.1** The breeding seasons for different species and populations of fairy-wrens, shown by the months when eggs were laid. (Data from literature and RAOU Nest Record Scheme (NRS).
(a) Splendid Fairy-wren (*M. splendens*): at 32°S, Perth, WA; 741 clutches; Rowley *et al.* 1991.
(b) Superb Fairy-wren (*M. cyaneus*): SE Australia; 633 clutches; Tidemann and Marples (1987).
(c) Blue-breasted Fairy-wren (*M. pulcherrimus*): at 32°S, Dryandra, WA; 93 clutches; Rowley and Russell (unpublished data).
(d) Red-winged Fairy-wren (*M. elegans*): at 34°S, Manjimup, WA; 246 clutches; Rowley *et al.* 1988.
(e) White-winged Fairy-wren (*M. leucopterus leuconotus*):
1. arid (<300 mm rain); 128 clutches; Tidemann and Marples (1987);
2. at 32°S, Pipidinny, WA; 43 clutches; Rowley and Russell (1995).
(f) Variegated Fairy-wren (*M. lamberti*):
1. *M. l. lamberti*: N of 29°S; 19 clutches; NRS;
2. *M. l. lamberti*: S of 29°S; 73 clutches; NRS;
3. *M.l. assimilis*, inland; 37 clutches; Tidemann and Marples (1987).
(g) Red-backed Fairy-wren (*M. melanocephalus melanocephalus*): east coast; 33 clutches; NRS.
(h) Purple-crowned Fairy-wren (*M. coronatus*): North Australia: 70 clutches; Rowley and Russell (1993).

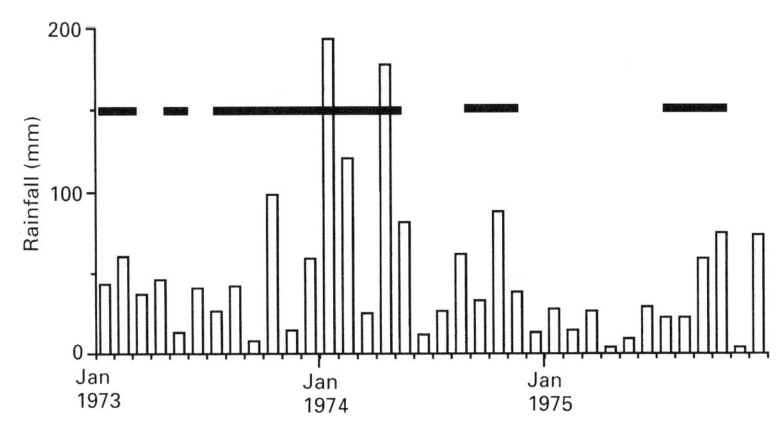

7.2 Breeding of White-winged Fairy-wren (*Malurus leucopterus*) in response to rainfall at Ivanhoe, NSW, during three years. (Based on data from Tidemann and Marples (1987) and from John Hobb's records in the Nest Record Scheme). The open bars represent monthly rainfall, and the dark bars the incidence of breeding.

followed by breeding early and late in the following dry season, with little if any from mid-May to mid-July. After a very poor wet season, we found no signs of any breeding in either autumn or spring (Rowley and Russell 1993).

For most New Guinea malurids, breeding information is sparse. Studies of other insectivorous species from New Guinea indicate that the height of the rainy season is not a favoured time for breeding, with most breeding recorded before or after the peak of the wet season, as in tropical Australia. Analysis of breeding records from the 1938–39 Archbold Expeditions to New Guinea suggests that in coastal West Irian, insectivorous birds showed a peak of breeding in autumn at the close of the wettest period, but in the highlands, the breeding peak was in spring (Nix 1976). In rainforest and savannah woodland near Port Moresby, most nesting activity of insectivorous birds occurred between August and December, although some breeding was recorded in all months (Bell 1982*a*,*b*).

Variation

A breeding season expressed as nests initiated per month for a particular site or region gives only an average representation of when breeding can occur. At any one place, the time when breeding begins and ends varies from year to year, as shown in Fig. 7.3, where the initiation of first clutches for the season spanned more than 5 weeks, from 20 August to 29 September (Rowley *et al.* 1991). The most obvious factor delaying the start of breeding was heavy rain late in August. The date when the last clutch of the season was laid varied by an even greater amount, since, in one year, a summer fire terminated breeding prematurely. However, even in years without fire, the last clutch of the season was laid between mid-December and the end of January. At other intensively studied sites in southern Australia, breeding can occur over a period of 5–6 months, but in any one year, breeding over 3–4 months is most likely. If this amount of variation in the timing of breeding is found in the more predictable southern climates, even greater variation may be expected in those parts of Australia with less predictable climates, although at present long-term data from individual sites are lacking.

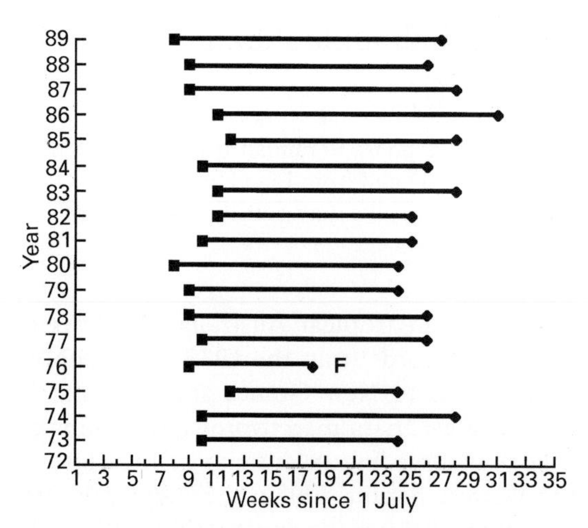

7.3 The length of the breeding season varies from year to year. Data for Splendid Fairy-wrens (*Malurus splendens*) on Gooseberry Hill, WA, showing the weeks when eggs were laid in 17 years 1973–89. F indicates fire terminated breeding in that year. (Data from Rowley *et al.* 1991.)

Breeding in emu-wrens and grasswrens is little known with few nest records and no long-term field studies of any species. The Southern Emu-wren (*Stipiturus malachurus*) and the Mallee Emu-wren (*S. mallee*) in southern Australia appear to have similar breeding seasons to fairy-wrens in the same areas, and the Rufous-crowned Emu-wren (*S. ruficeps*) of inland Australia appears similar to White-winged and Variegated Fairy-wrens in similar habitats. Even less is known about breeding in the grasswrens. The few records suggest that northern species are influenced by the wet season in much the same way as Purple-crowned Fairy-wrens, with breeding before, during or after, depending on the amount and timing of rainfall. Records for grasswrens in inland Australia suggest spring breeding if conditions allow, and at other times in response to good rains. In the Thick-billed Grasswren (*Amytornis textilis*) at Shark Bay (25° 47′S), with annual rainfall of about 200 millimetres concentrated between May and August, most eggs are laid in July and August (M. G. Brooker 1988; B. Brooker, unpublished data).

Nests

Size and shape

Most malurids whose nests have been found build a domed structure which is taller than it is wide with a slightly hooded side entrance near the top (Fig. 7.4). Size ranges from the large rough bulky nests of the Purple-crowned Fairy-wren, resembling piles of flood debris, to the small neatly woven nests of White-winged Fairy-wrens (Table 7.1). Grasswren nests are much more variable; some are fully domed (Striated *A. striatus* and White-throated *A. woodwardi* Grasswrens), while others are intermediate, and those of the Thick-billed Grasswren hardly domed at all, just slightly higher at the back.

7.4 The domed nest of the Superb Fairy-wren (*Malurus cyaneus*), showing nestlings and adult male in attendance with food.

Table 7.1 Nest dimensions and height above ground of nests built by six species of fairy-wren (*Malurus*).

Species of fairy-wren	*Height above ground (mm)*	*Height (mm)*	*Diameter (mm)*	*Volume (cc)*
Purple-crowned[1]	880 (25)	152 (25)	100 (23)	935
	250–2900	100–190	65–140	
Red-winged[2]	215 (184)	130 (154)	77 (165)	478
	40–1200	90–200	70–90	
Superb[3]	720 (55)	120 (55)	80 (55)	467
	200–1300	90–150	70–90	
Splendid[4]	782 (115)	125 (115)	64 (113)	333
	300–2500	90–190	45–90	
Blue-breasted[5]	585 (71)	111 (21)	66 (21)	326
	100–1530	80–160	60–70	
White-winged[6]	237 (15)	105 (15)	63 (15)	261
	120–370	90–140	60–70	
Variegated[5]	193 (11)	97 (11)	65 (11)	251
	0–450	90–110	50–90	

The mean, number of measurements (n), and range are given for height above ground, nest height, and diameter. Volume is calculated from the mean measurements.
[1] Rowley and Russell (1993); [2] Rowley *et al.* (1988); [3] Tidemann (1983); [4] Rowley *et al.* (1991); [5] Rowley (unpublished data); [6] Rowley and Russell (1995).

Building

The female alone collects nest material and builds the nest, accompanied by her male partner as she does so. She starts with a framework of spider's web fastened at the site, into which she weaves grass and strips of bark to form the base and walls. After the framework and walls are established, she lines the nest with finer grass and fibre, and an inner layer of fur, feathers, or vegetable down. In the Splendid Fairy-wren, building females often appeared to have a ginger moustache as they carried a bill full of red fibrous bast from the cycad *Zamia* to line their nests. The entrance is a hole 20–40 millimetres in diameter, about two-thirds of the way up one side, generally with the roof projecting slightly beyond the entrance. The nests of some tropical species from high rainfall areas incorporate moss into the roof (Beruldsen 1980; Lovely Fairy-wren, Rowley and Russell, unpublished data; Emperor Fairy-wren *M. cyanocephalus*, Ripley 1964). The only nest of the Broad-billed Fairy-wren (*M. grayi*) that has been found was placed within thick moss on a tree trunk (Diamond 1981). It appears as if moss may have some water-repellent characteristics; certainly, the interior of the nest and the incubating Lovely Fairy-wren female we watched remained remarkably dry even during several days of very heavy tropical rain. Nest-building females are more conspicuous than at other times, as they move about busily, often in relatively open areas, collecting material, and then flying directly to the nest-site with their load; construction takes 3–7 days. Although they are accompanied by their male partner, they are often visited by intruding, displaying, males from other groups since the week before egg-laying starts is the female's most fertile period.

Placement of nests

Most nests are placed near the ground in a low shrub, grass tussock, or spinifex hummock that provides dense cover. Few nests of any species are more than 1 metre above the ground and most are up to 0.5 metres from the outside of the bush, not at the outer edge. At any one site, there is generally a preferred direction towards which the nest opening faces—this may be away from prevailing wind and rain, or from the hot afternoon sun (Rowley *et al.* 1991). In some species, early nests are closer to the ground than later ones. Higher nests are more exposed to cooling breezes in summer, and are probably less accessible to reptilian predators, which are most active in summer. Apart from such obvious changes in nest-site preference, it is hard to generalize about preferred nest-sites. Characteristics such as a shrub's height and density and its branching structure are probably important, but these can vary with season and the fire history of the area, as well as the vegetation available. At any one site, two or three species of shrub provided the favoured nest-site for the local malurids—thus for Splendid Fairy-wrens in heathland near Perth, *Acacia pulchella*, *Hakea*, and *Dryandra* were preferred. In the tropics, Purple-crowned Fairy-wrens preferred *Pandanus aquaticus*, placing nests in the angle between leaf blades and the main stem (Rowley and Russell 1993). Superb Fairy-wrens willingly accept a wide range of exotic plants, among them weeds such as brambles (*Rubus* sp.) and ornamentals such as *Crataegus*. Old mature hummocks of spinifex (*Triodia* and *Plectrachne*) which provide nest-sites for the Rufous-crowned Emu-wren and several grasswrens are easily destroyed by fire, and too frequent fires can eliminate suitable nest-sites and render the habitat unsuitable. For a relict population of Thick-billed Grasswrens in Western Australia, cover and nesting substrate proved to be a very important feature of the habitat. The tangle of grasses, herbs, climbers, and rubbish within which the grasswrens built their nests were shielded from grazing, by recumbent shrubs battered by old age or wind action (Brooker 1988).

Eggs

Laying

Females do not always lay as soon as they have finished building the nest. Some early nests of Splendid Fairy-wrens near Perth, Western Australia remained empty for more than a week after they were fully lined and we had assessed them ready for eggs; 28 days was the longest such interval. After mid-October, as days became warmer and the chance of rain decreased, most nests were built and laid in within a week (Rowley *et al.* 1991). We have recorded similar delays before laying in early nests of Red-winged (*M. elegans*), Blue-breasted, and White-winged Fairy-wrens in south-western Australia, and in Canberra, Rowley (1965) found delays of up to a month before laying by Superb Fairy-wrens.

For those fairy-wrens that have been studied in detail, eggs are laid early in the morning on successive days. Schodde (1982*b*) quotes unpublished information from G. B. Ragless indicating that Striated and Thick-billed Grasswrens laid on successive days. However, Thick-billed Grasswrens at Shark Bay, Western Australia, laid on alternate days, early in the morning (B. Brooker, unpublished data), as reported by Marchant (1986) for Australian passerines in the family Pardalotidae, including probably most species of *Gerygone* and *Acanthiza*. Marchant suggested that the Southern Emu-wren laid at 48-hour intervals, but the careful observations of Fletcher (1913, 1915) reported eggs laid at intervals of 24 hours.

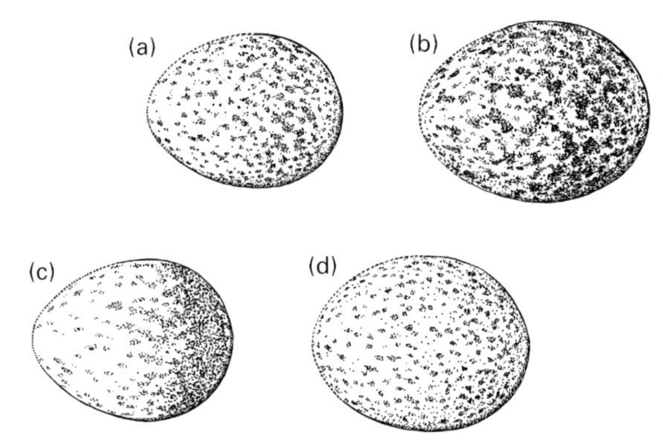

7.5 Eggs of the family Maluridae. (a) Southern Emu-wren (*Stipiturus malachurus*). Eggs relatively large for the size of the bird; creamy-white, reddish-brown blotches denser at larger end. (b) Thick-billed Grasswren (*Amytornis textilis*). Eggs creamy-white, thickly freckled with red-brown, often in a zone at larger end; among the most heavily marked malurid eggs. (c) Superb Fairywren (*Malurus cyaneus*). Eggs pinkish-white with red-brown blotches often concentrated in a zone at larger end. (d) Striated Grasswren (*Amytornis striatus*). Pinkish-white eggs, sparsely freckled with brown or red, often concentrated at larger end.

Size and shape

All malurids lay very similar eggs: a dull white, tapered oval, sparsely to moderately spotted and blotched with red-brown, particularly towards and over the larger end, where the markings sometimes form a band or cap (Fig. 7.5). The most densely spotted eggs are found in the Thick-billed and Dusky Grasswrens of the *textilis* group. All known eggs are well illustrated in Schodde's 1982 monograph. As yet, eggs have not been described for the Broad-billed Fairy-wren or Orange-crowned Wren (*Clytomyias insignis*) from New Guinea; although Schodde illustrated the eggs of the Emperor Fairy-wren, he doubted their authenticity.

As in all birds, egg size increases with body size (Fig. 7.6), but smaller malurids lay larger eggs relative to their body mass. Egg mass as a percentage of adult body mass in the Maluridae is comparable with estimates for

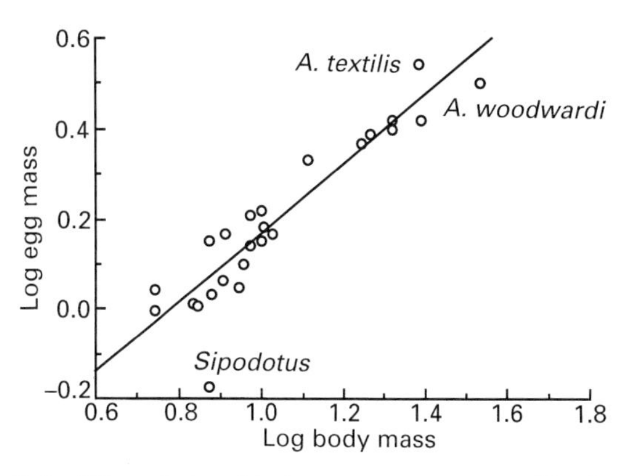

7.6 The relationship between egg mass and body mass in the family Maluridae. Egg mass was estimated from egg dimensions according to the formula Egg mass = 0.548 × Length × Breadth^2 (Hoyt 1979). The scale for both axes is logarithmic, and the regression line plotted has the equation $\log_{10}$ Egg mass = 0.77$\log_{10}$ Body mass − 0.61 (r^2 = 0.85, SE of slope = 0.097, n = 25 species). The slope of 0.77 indicates that smaller species or subspecies have relatively larger eggs.

other passerine families, ranging from 20 per cent in the Mallee Emu-wren to 9.3 per cent in the White-throated Grasswren, and the slope of the regression line (0.77) lies within the range for passerine families calculated by Rahn *et al.* (1975). In Fig. 7.6, most points lie close to the regression line, but two species (Wallace's Wren *Sipodotus wallacii* and White-throated Grasswren) appear to have an estimated egg mass lower than might be expected from their body size; since only one or two eggs have been measured, and few adult weights are available, more information is needed to confirm that these differences are real. More measurements are available for the western subspecies of the Thick-billed Grasswren, suggesting that their larger than predicted eggs are a valid difference.

Clutch Size

Clutches are small, with a narrow range of variation, as in most other Australian passerines. For most fairy-wrens, clutch size ranges from two to four (Table 7.2), with three eggs most frequent. For Purple-crowned, Blue-breasted, and Red-winged Fairy-wrens, no clutches of four have been recorded, and for Red-winged Fairy-wrens, clutches of two are only slightly more frequent than clutches of three. In the emu-wrens, the tiny Rufous-crowned Emu-wren of arid areas lays clutches of two to three eggs, with two most common, while the Southern Emu-wren in coastal habitats lays clutches of two to four, with three most common. For all the grasswrens, detailed information on breeding is sparse, but data from egg collections suggest that clutch size is two or three eggs, with two most common. Little is known about the breeding of any New Guinea malurids, but what information there is suggests that clutches of two are most usual.

Clutch size within species of birds often varies markedly with latitude in the northern hemisphere, with the largest clutches at high latitudes. With small clutches such as are found in Australian passerines, one egg more or less constitutes a very large relative change and analysis of NRS records for the widespread species Eastern Yellow Robin (*Eopsaltria*

Table 7.2 Breeding in fairy-wrens (*Malurus*): clutch size, number of clutches, brood parasitism, and occurrence of multiple broods for the five most intensively studied species.

Species of fairy-wren	*n*	*Mean*	*Clutch Mode*	*Range*	*Nests/ female*	*Cu %*	*% females Re-nest after young fledge*	*% females Raise 2 broods*
Superb[1]	67	3.4	3	3–4	n.a.	3.4	40.6	34.4
Superb[2]	64	3.1	3	2–4	2.5	17.6	+	n.a.
Superb[3]	107	3.4	3 = 4	1–4	2.6	13.0	+	n.a.
Superb[4]	343	3.2	3	1–4	3.5	3.0	+	n.a.
White-winged[2]	27	3.1	3	2–4	2.0	35.7	+	n.a.
White-winged[5]	44	3.4	3	2–4	n.a.	3.0	n.a.	n.a.
White-winged[6]	22	3.5	3 = 4	3–4	1.3	0	26.3	26.3
Red-winged[7]	117	2.4	2	1–3	1.5	2.8	16.6	14.7
Splendid[8]	308	2.9	3	2–4	2.0	20.1	36.3	22.2
Blue-breasted[9]	35	2.8	3	2–3	2.3	11.4	13.3	13.3

n: number of clutches from which mean clutch size is calculated; Cu%: % of nests parasitized by cuckoo; +: occurs, but incidence not measured; n.a. no figures available.
[1] Rowley (1965) (Canberra, ACT); [2] Tidemann (1983) (Booligal, NSW); [3] Nias (1987) (Armidale, NSW); [4] Mulder (1992), Dunn *et al.* (1995) (Canberra, ACT); [5] NRS (J. Hobbs, Ivanhoe, NSW); [6] Rowley and Russell (1995) (Pipidinny, WA); [7] Rowley *et al.* (1988) (Manjimup, WA); [8] Rowley *et al.* (1991 and unpublished data) (Perth, WA); [9] Rowley and Russell (unpublished data) (Dryandra, WA).

australis), Willie Wagtail (*Rhipidura leucophrys*), and Welcome Swallow (*Hirundo neoxena*) have failed to show any variation in clutch size with latitude (Marchant 1974, 1984; Marchant and Fullagar 1983). For the wide-ranging Variegated, White-winged, and Red-backed Fairy-wrens, records from tropical Australia are too few to allow comparisons of clutch size at different latitudes.

Within the breeding season, clutch size varies little. For Splendid Fairy-wrens, more than 90 per cent of all clutches were of three eggs, and this did not change with successive nests during the breeding season (Table 7.3). In Red-winged Fairy-wrens, we knew the clutches laid by each of 43 females in two successive nests in the same season; 23 of the second clutches were the same size, six were larger and 14 smaller (Rowley *et al.* 1988). Where studies have continued over several years, no evidence of between-year variation in clutch size has been found (Rowley *et al.* 1988, 1991).

Most species for which we have good data have been studied at only one site, so that comparison of clutch size in different habitats is possible for only a few wide-ranging species (Table 7.2). The Superb Fairy-wren has been studied at three locations; clutch size was 3.4 at Canberra and Armidale, and 3.1 at Booligal in the semi-arid region of New South Wales, the western extreme of its range. From NRS records of Superb, Variegated, and White-winged Fairy-wrens, clutch size was higher in areas of higher rainfall (Tidemann 1983). However, the clutch size of 3.4 that we found in the diminutive White-winged Fairy-wren in coastal heath north of Perth (annual rainfall *c.* 800 millimetres) is the same as that at Ivanhoe, New South Wales, at the same latitude, during a period of above average rainfall (annual rainfall less than 300). The clutch size of grasswrens, almost all living in arid habitats, is most frequently two, but it is not possible at this stage to separate the effect of low rainfall from that of larger body size.

Table 7.3 Splendid Fairy-wren (*Malurus splendens*), Gooseberry Hill, Western Australia: clutch size for first and subsequent nests (n = 296 nests, 1978–88).

Clutch size	*Order of clutch in season*			
	1	2	3	4
2	15	10	3	3
3	135	72	32	16
4	1	2	—	—
mean	2.91	2.91	2.91	2.84
SD	0.31	0.37	0.28	0.38
n	151	84	35	19

Mean clutch size did not differ significantly between successive nests (one-way Analysis of Variance, $F = 0.77$, df = 4, $P>0.05$).

Small clutches which vary little with latitude are not just an Australian phenomenon. It has long been recognized that clutches in the tropics are smaller than at higher latitudes in the northern hemisphere (Lack 1968; Skutch 1985), but it has now been shown that in the three southern continents, clutch size is smaller than at equivalent northern latitudes. From 298 species of Australian passerines, the mean clutch size of 2.69 is significantly smaller than 4.26 from 119 species at equivalent latitudes in North Africa (Yom-Tov 1987). The mean clutch size of passerines in 'old endemic' families of insectivores and nectarivores (including Meliphagidae, Maluridae, Pardalotidae, and Pachycephalidae) is smaller than that of more recently arrived species, especially the seed-eating finches. Mean clutch size for South African passerines, from 353 species, is 2.8, while that of passerines in South America (south of 22°S) is 2.98, both appreciably smaller than the clutch size of passerines at similar latitudes in the northern hemisphere, with no evidence for a latitudinal gradient (Rowley and Russell 1991; Yom-Tov 1994; Yom-Tov *et al.* 1994). As in Australia, the clutch size of more recent arrivals in South America, the oscines, is larger than that of the suboscines (see Glossary), which are thought to have evolved there. Smaller clutches do not necessarily mean that birds in southern continents lay fewer eggs in a season, however, because southern breeding seasons are longer, with individuals making several attempts and maybe raising more than one brood per season (Rowley and Russell 1991).

Number of broods

As in many Australian passerines, the reproductive effort of most malurids varies by means of the number of small broods attempted, rather than through the size of a single large clutch. Most birds will lay a second clutch if the first one is lost through predation or bad weather. In multi-brooded species, as well as replacing failed attempts, females may re-nest after fledging one brood, and they may continue re-nesting throughout the breeding season. To study breeding in a multi-brooded species, one has to follow individually recognizable females throughout the breeding season, monitoring the fate of one brood while looking for the start of the next. Information of this quality is available for some fairy-wrens but for no emu-wrens, grasswrens, or New Guinea malurids. The five species listed in Table 7.2 are known from studies of colour-banded birds to be multi-brooded. Our limited studies of Purple-crowned Fairy-wrens in northern Australia showed four cases of two, and one of three, broods being raised in a season (Rowley and Russell 1993). Nicholson and Coates (1975) described an instance of two broods raised by White-shouldered Fairy-wrens (*M. alboscapulatus*) in New Guinea.

Immediately after the young of the first brood fledge, the female joins her male partner and helpers (if any) in feeding the young and generally shepherding them. Some time before the young become completely independent of parental feeding, the female begins to build a new nest, and is soon incubating the next clutch, while her male partner (and helpers) look after the first brood. When the second brood hatches, the nestlings are fed by the female, male, and helpers. The juveniles of the first brood still follow the adults about, begging for food from them, and may also bring food to the new nestlings.

An ideal cycle of overlapping broods and the division of parental responsibilities is shown in Fig. 7.7. In the real world, such a sequence seldom happens because it is rare for three clutches to be reared in succession without mishap. More often, one or more clutches are taken by predators, parasitized by cuckoos, or chilled and deserted during a period of bad weather. In that case, the female builds a new nest and lays another clutch, continuing to do so until either the nest succeeds or the breeding season ends. In one season, a female Superb Fairy-wren in Canberra built six nests and laid four clutches (16 eggs), none of which hatched (Rowley 1965). In the Splendid Fairy-wren (Table 7.4), from 1973 to 1989,

Table 7.4 Re-nesting by Splendid Fairy-wrens (*Malurus splendens*) after success or failure: females (n = 314) with various combinations of successful (S) or failed (F) nests, on Gooseberry Hill, Western Australia, 1973–89.

Number of clutches (%)	*Success or failure*	*Number*	*%*
1 (24.2)	S	50	15.9
	F	26	8.3
2 (45.9)	SS	46	14.6
	FF	20	6.4
	SF	78	24.8
3 (22.3)	SSS	2	0.6
	FFF	20	6.4
	SFF	29	9.2
	SSF	19	6.1
4 (6.1)	SSFF	3	1.0
	SFFF	11	3.5
	FFFF	5	1.6
5 (1.6)	SFFFF	3	1.0
	FFFFF	2	0.6
At least one success		241	76.8
Re-nest after success		116	36.9
No success		73	21.6
>1 Success		70	22.3
Succeed in re-nest after success		70	60.3
Re-nest after fledge Cuckoo		36	11.5

A successful nest is one where at least one young was raised to independence; fledging a cuckoo is scored as a failure. Not all sequences are shown individually: FFFS includes all possible combinations FFFS, FSFF, SFFF, etc. Females present for only part of the breeding season are not included.

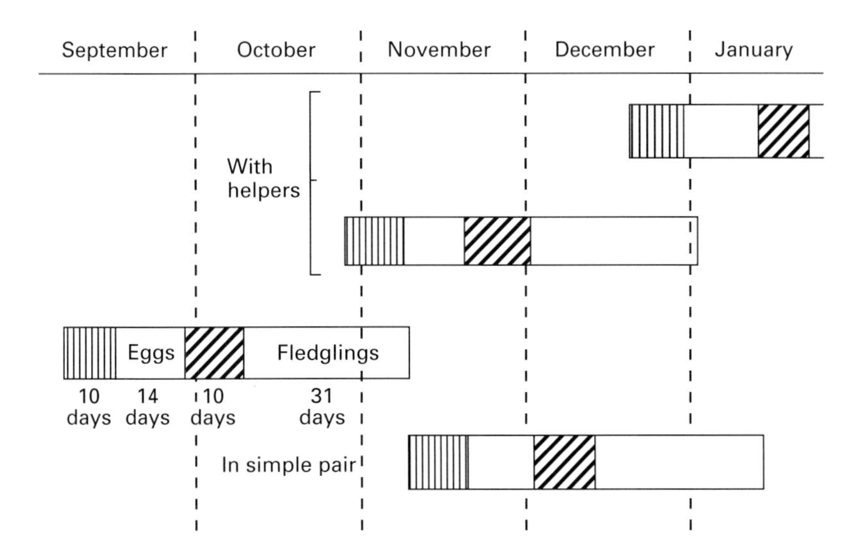

7.7 Idealized overlap of successive broods in a group of fairy-wrens with and without helpers, over 5 months of the breeding season. After the young fledge from the first nest, a simple pair (below) cares for the young until they are independent, and only then does the female re-nest. When helpers are present, they may take over care of the young after fledging, freeing the female to re-nest sooner (horizontal hatching-building). If seasonal conditions are good, a female with helpers may have time to fit in three nests during the season, or at least two successful broods and a failure or a cuckoo.

five females each laid five clutches—three succeeded in rearing one brood only, the other two failed completely.

The time between nests is significantly shorter if there are helpers that take over feeding and shepherding the fledglings, freeing the female to re-nest earlier than if she and the male alone look after them. For an experienced female Splendid Fairy-wren with no helpers, the mean time between clutches was 65 days (52–87, n = 19); an experienced female with one or more helpers took 57 days (33–82, n = 57), and for an elite group, experienced females with two or more helpers, re-nesting was even quicker, with only 53 days (33–77, n = 28) between clutches. Allowing (17 + 12) days for laying, incubation, and feeding nestlings of the first clutch, these females laid the first egg of their next clutches *c.* 24 days after the first broods fledged, so they must have started nest-building about 14 days after the nestlings fledged, or even sooner.

Over 17 years, 37 per cent of female Splendid Fairy-wrens re-nested after successfully raising one brood to independence and 22 per cent reared more than one brood (Table 7.4). Only two females produced three broods of wrens and another eight fledged a cuckoo as well as two broods of wrens. The normal breeding season from September to December is barely long enough to allow three clutches to be raised, and the incidence of predation and brood parasitism means that the chances of this happening are low (Fig. 7.7). Females that re-nested after success were usually those that laid their first clutch on average a month earlier than those that did not re-nest. Females breeding for the first time were likely to lay their first clutch later in the season and to lay fewer clutches than experienced females.

In the Red-winged Fairy-wren, the other species for which there are long-term data on the incidence of multiple broods, the breeding season started later and was shorter. Fewer females attempted to raise a second brood (16.6 per cent), 14.7 per cent succeeded, and 57.6 per cent of females had only one nest. No female laid more than three clutches, and only one female raised three broods. Again, females laying more than one clutch were those that laid their first clutch early in the season, and the mean interval between successive clutches was 51 days (33–75, n = 35; Rowley *et al.* 1988).

Incubation

In fairy-wrens, only the female incubates. She starts when she lays the last egg in the clutch, and bouts of incubation last for about an hour, after which the male usually calls her off the nest. He accompanies her, remaining vigilant nearby while she forages urgently, without interruption, for 15–30 minutes; later in the breeding season when days are warmer, she may leave the eggs for longer. An incubating female returning to her nest characteristically makes an obvious, relatively long, direct flight. In the confined space of the domed nest, she sits facing the nest entrance and since her long tail has to be stowed somewhere it generally develops a characteristic bend—a useful indicator that she is nesting (Fig. 7.8).

Incubation in fairy-wrens is typically 13–15 days. In Splendid Fairy-wrens, we found a median incubation period of 14–15 days (i.e. the eggs hatched on the 15th day of incubation), with a range of 12–13 to 16–17 days; incubation was about a day shorter in December–January. Similar incubation periods have been recorded for Superb and Red-winged Fairy-wrens, with later clutches requiring a day less (Rowley 1965; Rowley *et al.* 1988). A few reports for other fairy-wrens indicate similar incubation periods (Marchant 1980; Tidemann 1983, NRS).

In the Southern Emu-wren and the Thick-billed Grasswren, there are reports of the male playing some part in incubation. Rosemary Hutton (1991) observed at close quarters a pair of Southern Emu-wrens nesting in captivity; although the female performed the majority of incubation, she was relieved for short periods by the male, whose tail did not develop the marked bend of the female's.

7.8 Female Superb Fairy-wren (*Malurus cyaneus*) during the incubation period, showing the characteristic bent tail that develops in the confined space of the enclosed nest.

Lawson Whitlock collected the first known nest and eggs of the Thick-billed Grasswren (*A. textilis*) (= *A. gigantura*) near Wiluna, Western Australia. He found a nest and heard grasswrens nearby, but since he was not certain whose nest it was, he '...resolved ......to steal silently up to the bush, and if possible, surprise the sitting bird, and procure it before it could slip away. I carried out this plan, and very cautiously approaching the bush, I peeped into the nest. There was a veritable *Amytornis gigantura* sitting on the eggs! It flushed with a cry of alarm, but I secured it. It was a male.' (Whitlock 1910, p. 203). Brooker (1988) flushed a male from each of two nests of the Thick-billed Grasswren, one with eggs and one with nestlings. In an aviary, a male Striated Grasswren occasionally took over incubation for short periods when the female left to feed (Hutton 1991).

Nestlings

Hatching is usually synchronous, with all nestlings appearing within 24 hours of each other. From nests where at least one nestling hatched we estimated fertility in Splendid Fairy-wrens at 97 per cent and in Red-winged Fairy-wrens at 94 per cent. Only the female broods the newly hatched young and then she is often given food by other group members; sometimes she passes this to the hatchlings, sometimes she eats it herself. Later all group members feed the nestlings; in Splendid and Superb Fairy-wrens, young of the year from early broods sometimes also feed nestlings, but we did not record this for Red-winged Fairy-wrens. The division of labour in feeding nestlings is discussed in Chapter 6.

The stages in development of nestlings are similar in all fairy-wrens. When Red-winged Fairy-wrens hatch, they are raw-red, naked and blind, but within a day they darken, becoming blue-grey as the feathers develop below the skin. On the third day, the sheathed primary feathers begin to protrude through the skin. By the fifth day, the eyes start to slit, and on the sixth day, they are usually open, the primaries start to burst from their protective sheaths, and the central rectrices start to emerge. Only then do the body feathers start to protrude, but the young are well covered by the time they leave the nest, even though their tails are mere stubs, less than a quarter of their final length (reached *c.* day 40–45). Fledglings can fly only weakly when they leave the nest; their wings continue to grow for another 10 days, which corresponds to the time when they usually remain well hidden in the undergrowth—the cryptic stage. Thus, their tails, which would be hard to accommodate within the cramped confines of a domed nest, grow most of their length outside the nest, whereas the wings are nearly fully grown by the time they leave the nest (Fig. 7.9). Unlike fairy-wren nestlings which are naked at hatching, nestling grasswrens are covered with charcoal-grey down (Brooker and Brooker 1987).

The length of time nestlings remain in a nest is variable, although the whole brood leaves the nest at the same time. For both Splendid and Red-winged Fairy-wrens, the median nestling period is 11–12 days, with nestlings fledging on

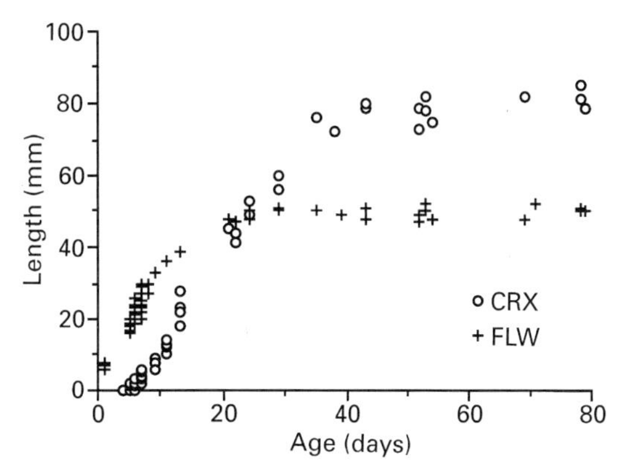

7.9 Growth of folded, flattened left wing (FLW) and central rectrix (CRX) of Red-winged Fairy-wren (*Malurus elegans*) nestlings and juveniles whose age was known. Wings and tail continue to grow after the nestling fledges. Measurements of folded left wing, seventh primary and central rectrix from nestlings whose hatch date was accurately known were used to calculate the regression of length on age for periods of linear growth, and we used these regression equations to estimate the age of juveniles whose date of hatching was unknown (see Rowley *et al.*1988, 1991 for regression equations to estimate the age of nestlings of *M. elegans* and *M. splendens*, respectively).

the 12th day after hatching. We suggest that the period is influenced by weather, brood size, and external disturbances, including predators; there is no clear trend for nestlings to fledge at a younger age later in the season. A brood of Splendid Fairy-wren nestlings left the nest on the 10th day while we were watching, when a goanna (*Varanus* sp.) approached close to the nest. After much agitated calling by the parents, the nestlings scrambled out and were led away safely.

Reproductive success

Nest failure

The progression from nest-building and laying through incubation to nestling development and growth is not without hazard. The main causes of nest failure are brood parasitism, predation, unfavourable weather, and food shortages, the incidence of which varies with species, habitat, nest location, and season. We have summarized the information available in Table 7.5. For other fairy-wrens, all emu-wrens, grasswrens, and New Guinea malurids, few records of individual nests are available.

Parasitism

Brood parasitism, by several species of cuckoos, is a significant cause of nest failure in the Maluridae. Seven of the 10 species of cuckoo that occur in Australia are recorded as parasitizing at least one malurid (Table 7.6). In their review of cuckoo hosts in Australia, Brooker and Brooker (1989*a*) distinguish between regularly parasitized successful host species (biological hosts) and 'accidental' or unsuitable hosts.

For emu-wrens and grasswrens, records are too few to identify a species as a biological host, but there are sufficient records for most fairy-wrens to identify them as biological hosts for Horsfield's Bronze-Cuckoo (*Chrysococcyx basalis*) (Table 7.6). This is probably true for other genera as well, since Horsfield's Bronze-cuckoo clearly prefers to parasitize species with enclosed, domed nests, and 'throughout Australia, the local malurid species is undoubtedly the major host' (Brooker and Brooker 1989*a*, p. 28). Hosts of cuckoos breeding in New Guinea are little known.

Superb and Splendid Fairy-wrens are also identified as biological hosts for the Shining Bronze-Cuckoo (*Chrysococcyx lucidus*). The northern subspecies of the Brush Cuckoo (*Cacomantis variolosus dumetorum*), which also prefers enclosed nests, parasitizes the Purple-crowned Fairy-wren in Australia and the White-shouldered Fairy-wren in New Guinea (Coates 1985; see Fig. 7.10). The Fan-tailed Cuckoo (*Cacomantis pyrrhophanus*) (mean female weight 44 grams) is larger than the Brush Cuckoo (32 grams) and the two Bronze-

Table 7.5 Nest failure in fairy-wrens (*Malurus*).

Species of fairy-wren	*No. of nests laid in*	*% nests fail*	*Cause of failure (% nests)*			*% nests succeed*
			Desertion	*Predation*	*Cuckoo*	
Red-winged[1]	206	32.5	2.4	27.2 (27.7)	2.9	67.5
Splendid[2]						
1973–88	656	52.0	10.4	21.7 (27.5)	19.7	49.0
1978–84	271	40.6	5.5	10.4 (11.3)	19.6	59.4
White-winged						
Pipidinny[3]	34	11.8	2.9	8.8	0.0	88.0
Booligal[4]	28	71.4	17.9	17.9 (41.4)	35.7	28.6
Ivanhoe[5]	66	42.0	6.1	54.5(0)	3.0	36.4
Blue-breasted[6]	88	65.9	9.1	45.5 (50.0)	11.4	34.1
Superb						
Booligal[4]	74	47.3	8.1	24.3 (32.4)	17.6	52.7
Armidale[7]	92	51.1	—	35 (?)	13.0	48.9
Canberra[8]	343	61.2	4.4	55.1 (?)	1.7	38.3

Number of nests laid in includes those found with eggs and nestlings. Nests parasitized by cuckoos are scored as failing. Nests taken by predators gives two values; the first is the percentage of nests failing due to predation (not including those parasitized), the second is the percentage of nests taken by predators, including those containing cuckoos.

[1] Rowley *et al.* (1988); [2] Rowley *et al.* (1991); [3] Rowley and Russell (1995); [4] Tidemann (1983); [5] NRS (J. N. Hobbs, Ivanhoe); [6] Rowley and Russell (unpublished data); [7] Nias (1987); [8] Mulder (1992).

Table 7.6 Parasitism in the family Maluridae, based on literature records, museum and private egg collections, NRS, and various unpublished records

Host species	*Cuckoo*						
	Cuculus pallidus *Pallid*	Cacomantis variolosus *Brush*	Cac. flabelliformis *Fan-tailed*	Chrysococcyx osculans *Black-eared*	Ch. basalis *Horsfield's*	Ch. lucidus *Shining*	Ch. minutillus –
Fairy-wrens							
Purple-crowned	—	7	—	—	—	—	—
Superb	2	7	26	2	236	33	—
Splendid	—	—	1	—	183	9	—
Variegated	—	8	8	2	67	3	1
Blue-breasted	—	—	1	—	12	—	—
Red-winged	—	—	3	—	3	—	—
White-winged	—	—	—	3	88	2	—
Red-backed	2	5	2	—	20	5	3
White-shouldered	—	1	?	—	—	—	—
Emu-wrens							
Rufous-crowned	—	—	—	—	2	—	—
Southern	1	1	7	—	8	2	—
Grasswrens							
Carpentarian	—	1	—	—	—	—	—
Striated	—	—	1	1	7	—	—
Thick-billed	—	—	—	—	4	—	—

Sources: Brooker and Brooker 1989*a*; White-shouldered Fairy-wren from Coates 1985; 10 records for Blue-breasted Fairy-wren from Dryandra State Forest, WA, Rowley and Russell, unpublished data.

Cuckoos (both 23 grams), but nevertheless it has been identified as a parasite of biological hosts Superb, Red-winged, and Variegated Fairy-wrens, and all its identified biological hosts build enclosed nests; it is probably a parasite of the White-shouldered Fairy-wren in New Guinea (Coates 1985).

Most information on the relationships between cuckoos and their malurid hosts comes from the work of the Brookers on Splendid Fairy-wrens parasitized by Horsfield's Bronze-Cuckoo (Brooker and Brooker 1989*b*). Eggs of the cuckoo closely match those of fairy-wrens for size and pattern. The Brookers could find no case in which the failure of a parasitized clutch could be attributed to discrimination by Splendid Fairy-wrens; they also found that Splendid Fairy-wrens accepted dissimilar model eggs, so there is no evidence that they can recognize cuckoo eggs. No cuckoo eggs were removed or punctured and parasitized nests were not selectively deserted. Some nests laid in by cuckoos still managed to produce host nestlings; if the cuckoo laid too early, the egg became buried in the nest lining and did not hatch; if she laid too late, more than 4 days after the host started incubating, then the cuckoo nestling did not survive.

Brooker *et al.* (1988) observed cuckoos laying directly into nests of Splendid Fairy-wrens, early in the morning within 2–3 hours of sunrise, after the host female had laid. The cuckoo entered the nest head first, leaving tail and wing tips exposed; she laid in about 6 seconds, and after laying, left the nest backwards, carrying a host egg in her bill. Laying was usually well synchronized with that of the host, most cuckoo eggs hatching before or with those of the host. Nestling cuckoos evicted host eggs and nestlings, usually by the

7.10 Adult male Purple-crowned Fairy-wren (*Malurus coronatus*) feeding a nestling Brush Cuckoo (*Cacomantis variolosus*). The nest was placed in *Pandanus aquaticus*, in the angle between one of the spiny fronds and the main stem. (From a photograph by G. S. Chapman.)

second day after hatching; cuckoos remained in nests of Splendid Fairy-wrens for 15–18 days, up to 6 days longer than the host brood. Normal host broods always weighed more than a nestling cuckoo of equivalent age. A brood of three Splendid Fairy-wrens weighed 25–30 grams at fledging; a cuckoo nestling reached *c.* 20 grams by 9 days after hatching and maintained that weight until it left the nest; they were fed by their foster parents for about 19 days after fledging.

The rate of parasitism by Horsfield's Bronze-Cuckoo in nests of Splendid Fairy-wrens between 1973 and 1989 varied considerably from year to year, with a mean of 19.7 per cent, range 0–52 per cent. In only 11 of 305 group-years was the reproductive effort of a group entirely disrupted due to parasitism: one group had all three nests parasitized, four had two cuckoos, and in six cases, the group's only nest contained a cuckoo. While parasitism had a significant effect on breeding success in Splendid Fairy-wrens at Gooseberry Hill, we do not know if they are parasitized to the same extent in other areas. Some other fairy-wren populations are heavily parasitized, others much less so (Table 7.5).

Predation

Although most malurids build enclosed, domed nests that are supposed to suffer less predation than open cup nests, predators are a significant cause of nest failure, ranging from less than 10 per cent in one population of White-winged Fairy-wren to 45.5 per cent in Blue-breasted Fairy-wrens at Dryandra (Table 7.5). Higher rates of predation have been found in some open nesting Australian passerines (Robinson 1990*a*; Major 1991*a*; Bridges 1994*a*). Predation does not generally reach the high levels found in the tropics, where frequently only

20–30 per cent of nests succeed (Skutch 1985; Brosset 1990). Predators of malurids are rarely caught in the act, but a wide range of known nest predators are present in all areas. The many avian predators include the Australian Magpie (*Gymnorhina tibicen*), butcherbirds (*Cracticus* spp.), Laughing and Blue-winged Kookaburras (*Dacelo gigas* and *D. leachii*), crows and ravens (*Corvus* spp.), currawongs (*Strepera* spp.), and shrike-thrushes (*Colluricincla* spp.) ; typically they make a hole in the top or the back of the nest. There is some evidence that cuckoos themselves are significant nest predators, removing eggs and thus inducing females to lay again.

Mammalian predators are predominantly introduced, including Red Foxes (*Vulpes vulpes*), Cats (*Felis catus*) (feral or domestic, depending on proximity to houses) and rats, especially *Rattus rattus*. Foxes and cats tend to tear the nest out of the bush or tear the side or bottom of the nest, while rats rarely damage the nest. Although few authors have considered rats as a major nest predator of Australian birds, studies by Major (1991*b*) and Laurance and Grant (1994) which photographed all nest predators have shown that rats are significant predators, even at apparently undisturbed sites. Marsupial predators such as possums (*Trichosurus* spp.) and gliders (*Petaurus* spp.) have also been implicated (H. A. Ford, personal communication). Reptilian predators also rarely damage the nest. They are particularly active in hot weather, and snakes and varanid lizards (goannas) have been seen to take eggs or nestlings (Fig. 7.11). Birds and reptiles are able to reach nests in the outer parts of shrub canopies that are probably inaccessible to mammals.

Blue-breasted Fairy-wrens illustrate the variation in causes of nest failure at different times of the breeding season. Early nests are generally built near the ground, in low vegetation or dead brush, and almost all are taken by predators. Most later nests are higher, built in live vegetation, generally isolated shrubs, where many are parasitized by Horsfield's Bronze-Cuckoo, which, being migrants, do not arrive until after the first clutches are laid. For Splendid Fairy-wrens, Table 7.5 gives the rate of nest failure over two periods of time; that over the whole period (1973–88) was 52 per cent, including periods early and late in the study that were much affected by wildfire. After fire, nests were significantly more exposed as the vegetation grew again, and levels of both predation and parasitism were high. The period 1978–84 was without fires, vegetation became progressively thicker, and probably more typical of an undisturbed population; nest failure was then only 41 per cent.

7.11 A typical nest of the Superb Fairy-wren (*Malurus cyaneus*) being robbed by a Brown Snake (*Demansia textilis*). (Photograph by John Warham.)

Desertion

Desertion is due to a number of different causes, all of which include eggs or dead nestlings remaining in a nest that is no longer attended; it is a relatively minor cause of failure (Table 7.5). Sometimes nests fill with water during heavy rain, or are deserted during particularly cold weather, perhaps if a foraging female has to spend so long off the nest that her eggs become chilled. In the few cases where the female disappeared at the same time as her nest was deserted, with eggs remaining in an undamaged nest, we presume that she was killed. Some complete clutches of infertile eggs failed to hatch and incubation was abandoned. Deserted, apparently undamaged, nestlings may have died from disease, starvation, or attack by ants. A late brood of Splendid Fairy-wrens that hatched in January died when the air temperature exceeded 40°C for 4 hours around mid-day; the observer in a hide nearby nearly did the same.

Breeding success

Obviously not all nests are successful. Breeding success varies both within and between species; we have measured it both as the proportion of eggs giving rise to fledglings and the proportion of nests in which eggs are laid that fledge young fairy-wrens (Table 7.7). A study of even three or four years will not include the full range of variation in conditions, and may be unduly influenced by very good or very poor years. Over 17 years, breeding success (*a*, see Table 7.7) of Splendid Fairy-wrens ranged from 25 per cent to 76 per cent, with a mean of 52 per cent. Years of low success coincided with high brood parasitism (1979, 1985, 1986), or followed fire (1974, 1985). Breeding success remained below 40 per cent for four years after a major fire in January 1985, when vegetation cover was sparse and both predation and brood parasitism were high (Fig. 7.12).

For four of the five years that the Superb Fairy-wren was studied by Ray Nias at Armidale, rainfall in the breeding season was below average; in one year in particular, the area was in the grip of severe drought, and less than half the usual number of nesting attempts were made (Nias and Ford 1992). The same year (1982) had similar effects on the breeding of species studied by Tidemann at Booligal.

Table 7.7 Breeding success and productivity in fairy-wrens (*Malurus*): breeding success, brood parasitism, and productivity for the six most intensively studied species.

Species of fairy-wren	*n*	*Breeding success*		*Cu %*	*Fledglings per*	
		a	*b*		*Nest*	*Group per year*
Superb[1]	67	53.8	—	3.4	1.7	3.9*
Superb[2]	64	46.0	52.7	17.6	—	2.6
Superb[3]	107	38.3	48.9	13.0	1.0	2.6
Superb[4]	343	32.6	38.8	1.7	1.0	3.2
White-winged[2]	27	26.0	28.6	35.7	—	1.2
White-winged[5]	66	35.9	36.4	3.0	1.2	—
White-winged[6]	34	80.6	88.0	0.0	2.8	3.3
Red-winged[7]	190	59.7	67.9	2.9	1.3	2.5*
Splendid[8]	709	52.3	58.0	20.1	1.5	3.2
Blue-breasted[9]	35	33.3	39.2	11.4	0.8	1.9*

n: number of nests on which breeding success is calculated; breeding success, *a*: percentage of eggs that produced fledglings; breeding success, *b*: percentage of nests laid in that fledged young wrens; Cu: cuckoo; * fledglings per female per year.
[1] Rowley (1965) (Canberra, ACT); [2] Tidemann (1983) (Booligal, NSW); [3] Nias (1987) (Armidale, NSW); [4] Mulder (1992), Dunn *et al.* (1995) (Canberra, ACT); [5] NRS (J. Hobbs, Ivanhoe, NSW); [6] Rowley and Russell (1995) (Pipidinny, WA); [7] Rowley *et al.* (1988) (Manjimup, WA); [8] Rowley *et al.* (1991 and unpublished data) (Perth, WA); [9] Rowley and Russell (unpublished data) (Dryandra, WA).

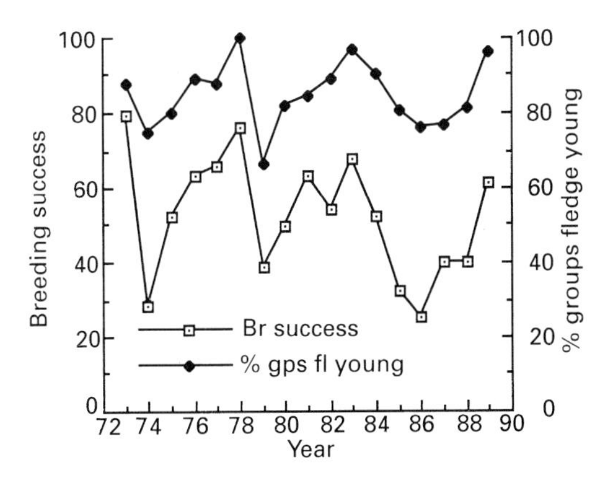

7.12 Yearly variation in breeding success (% eggs producing fledglings) and percentage of groups that fledged young in the Splendid Fairy-wren (*Malurus splendens*). Re-nesting after failure, after success, or after rearing a cuckoo means that although the mean breeding success (per nest) was only 52%, a mean of 85% of groups fledged young each year. In 1979, when only 67% of groups fledged young, 52% of clutches laid were parasitized, and for three out of 12 groups, both nests contained cuckoos. A widespread fire in 1974 during the breeding season reduced breeding success in that year; an extensive fire in January 1985 did not affect breeding success in that season, but reduced it in the seasons that followed.

The data for White-winged Fairy-wrens were collected by John Hobbs at Ivanhoe during two years of exceptionally good rainfall in that area, while that at Pipidinny reflected low predation and no cuckoo parasitism. The high level of breeding success recorded for Red-winged Fairy-wrens in the Karri forests in south-western Australia also reflects a low level of brood parasitism. Over six years, 67.9 per cent of nests produced fledglings (range 56.8–84.2 per cent).

It is not yet possible to say with any confidence how the breeding success of Australian passerines compares with that of tropical and northern hemisphere temperate passerines. For fairy-wrens, the success of individual domed nests is not as low as that of tropical passerines in Central and South America or Africa. In Gabon, 75 per cent of nests were destroyed by predators (Brosset 1990), and in lowland tropical America, where two-thirds of nests fail, Skutch (1985) estimated that predators, particularly snakes, are largely responsible. Breeding success is generally comparable with that of a range of shrub- or low-foliage nesting species in North America discussed by Martin (1995), with mean breeding success of 45 per cent in forest habitats and 40 per cent in shrub/grassland habitats. However, present-day breeding success in Australia is probably influenced by the changes wrought by European colonization—the clearing of vegetation, the increased number of avian nest predators such as currawongs, and the introduction of cats, rats, and foxes, all leading to decreased success.

Population biology

Productivity

Because the breeding season is long, and birds are able to re-nest after success or to re-lay repeatedly after failure, the success of an individual nesting attempt tells only part of the story. It is necessary to know not only production per nest but also production per pair or group for the whole season (Table 7.7). Even if a species is multi-brooded, not every female produces two broods of fledglings; if breeding success is 50 per cent and all females that succeed in rearing their first brood re-nest, on average only half of those (25 per cent of all females) will rear two broods.

The significance of longer breeding seasons and repeated re-nesting after success or failure becomes apparent when the proportion of nests producing young (= breeding success) is compared with the proportion of groups fledging young in any year. This is illustrated for the Splendid Fairy-wren over 17 years in Fig. 7.12, which shows that even in the years when breeding success was below 40 per cent, most groups were able to produce fledglings—more than 75 per cent of groups in all years except 1979, when over 50 per cent of nests were parasitized by cuckoos. Thus, while groups produced 1.5 fledglings per nest, their annual production was on average 3.2 fledglings. The production of fledglings per group ranged from a low of 2.1 in 1974, when fires in December destroyed nests and cut short the breeding season, to 5.8 in 1978, when 82 per cent of females re-nested after raising one brood, and 64 per cent reared two broods.

This strategy of producing a number of small broods allows a flexible response to seasonal conditions and allows birds to make the most of good conditions for as long as they last, with a high probability of at least some success. Nevertheless, despite their ability to produce two broods, overall productivity is low, with females producing a mean of three fledglings per year or fewer. This is below the production from one clutch by some Old World passerines, and emphasizes the low reproductive rates of Australian passerines.

The fate of fledglings

Fledging is followed by a cryptic stage lasting 7–10 days, during which the young birds remain in dense cover and are fed there by the group (Fig. 7.13). Their wings are not fully grown and their tails only 10–20 millimetres long. Nevertheless, they are able to fly if danger threatens, though fluttering slowly and rarely in the right direction. The parents do move their brood around the territory during this cryptic phase, but always through thick cover. During this week the wings complete their growth and the tail grows to about half its final length; the young are then able to fly quite well and venture into more open areas, following and begging from the adults as they forage through shrubs. The young birds gradually learn to forage for themselves, and about a month after leaving the nest, they

7.13 Fledglings of the Superb Fairy-wren (*Malurus cyaneus*) *c.* 17 days old. The first 7–10 days after leaving the nest is a cryptic phase for the fledglings which remain together in dense cover and are fed by other group members. Their wings and tail are not fully grown, but they can fly short distances and move rapidly if threatened.

become independent of adult provisioning, although they remain in the family territory, travelling and foraging with the group. This pattern is probably followed in all the malurids, even in those species that may wander from their territories after the breeding season.

Life is hazardous for the young birds, both before and after independence, and only about 50–70 per cent of fledglings survive to reach independence (Table 7.8), even though they are being fed and looked after by the adult members of the group. Because juveniles from a first brood remain with the family group after they become independent, they may be present when the adults are feeding nestlings in a later nest, and, sometimes, they help feed the nestlings. Young of the year are not always very efficient—they may bring food to the nest and not know what to do with it, or they may beg food from adults bringing food to nestlings. But at some nests, they make a significant contribution (see Chapter 6). Individuals vary in their response; even siblings from the same brood may differ. Whatever else they achieve, they do make the nest more obvious.

Dispersal

Young malurids are not expelled from their natal territories as soon as they are independent, and there is no synchronized pulse of post-fledging dispersal. Juveniles remain with

Table 7.8 Survival of juvenile and adult fairy-wrens (*Malurus*).

Species of fairy-wren	*Fledging to independence*	*Fledging to 1 year*	*Annual survival Breeding male*	*Annual survival Breeding female*
Superb				
Canberra (4)[1]	53.5	22.4	75.0	53.0
Canberra (3)[2]	57.1	28.1	67.0	70.0
Booligal (4)[3]	—	11.3	25.0	44.0
Armidale (5)[4]	82.0	25.0	44.0	44.0
White-winged				
WA (4)[5]	50.6	10.6	54.0	45.0
Booligal (5)[3]	—	4.2	64.0	48.0
Red-winged[6] (6,12)	76.1	37.9	78.3	79.3
Splendid[7] (17)	68.4	31.2	69.5	59.3
Blue-breasted[8] (4)	63.4	30.1	63.0	60.9
Purple-crowned				
WA (5)[9]	—	—	70.0	75.0
Qld (4)[9]	—	—	78.6	66.0
Variegated (3)[3]	—	10.0	46.7	31.3

Survival is expressed as the percentage of animals known to be alive after a specified interval, from recapture and sightings of marked birds in intensively studied populations. The number of years on which survival is based is shown in parentheses under the species name; for the Red-winged Fairy-wren, the first refers to fledglings and the second to adults.

[1] Rowley (1965); [2] Mulder (1992); [3] Tidemann (1983); [4] Nias (1987); [5] Rowley and Russell (1995); [6] Rowley *et al.* (1988); [7] Rowley *et al.* (1991); [8] Rowley and Russell (unpublished data); [9] Rowley and Russell (1993).

their family group and many birds of both sexes delay dispersal and help for a year or more before dispersing to breed. Some juveniles do disperse and wander, but most dispersal appears to occur in response to a vacancy for a male or female in a nearby group. Some birds made brief excursions from their natal territory and returned after a few days. This was difficult to observe directly, since the malurids were too small to track with currently available radio transmitters.

Most recorded dispersal is to nearby territories, and few birds of either sex move more than one or two territories from their natal territory (Fig. 7.14); dispersal tends to be female biased, but some males also disperse. Species appear to differ in the extent to which females remain (or are allowed to remain) with their family group. Only in Superb, Splendid, and Red-winged Fairy-wrens has the dispersal of large numbers of marked juveniles been followed, and two quite different patterns have emerged. In Superb Fairy-wrens, female helpers are rare, and females that remain with a group during the winter are generally expelled by the start of the breeding season. In most other fairy-wrens, female helpers are recorded, and are common in Red-winged Fairy-wrens (Rowley *et al.* 1988).

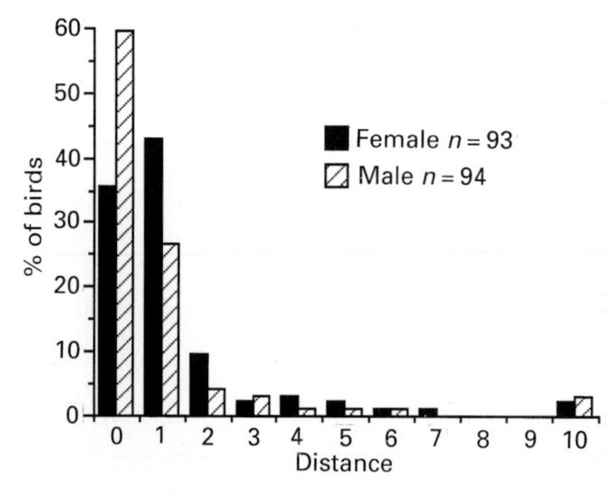

7.14 Distance dispersed by male and female Splendid Fairy-wrens (*Malurus splendens*), measured in territory widths; breeding in the natal territory scored 0, in an adjacent territory 1, and in a territory with one intervening territory, 2 (After Russell and Rowley 1993*a*.)

In the Superb Fairy-wren population studied in Canberra by Mulder (1995), every young female emigrated from her natal territory within a year of fledging. Females that fledged early in the breeding season tended to disperse within 4 months of leaving the nest; females that fledged later tended to remain with their natal group through the winter and to disperse shortly before the next breeding season. Late dispersing females were clearly expelled by aggression from the breeding female, but no such aggression was apparent towards the early dispersers. Some early dispersing females settled in a group that already had an adult female, but before the start of the next breeding season they had either inherited the breeding position in that group or in a group nearby, or else been expelled from the group. During the breeding season, no floating, unpaired females existed; they either gained a breeding position somewhere or they died. Mulder estimated that late dispersing females probably experienced 65–75 per cent mortality and that they had a mean dispersal distance of 11.8 territory widths (range 1–36). Most of Mulder's males were philopatric; some remained unpaired in their natal territory for up to 5 years, ultimately disappearing (probably dead), without achieving a territory or a partner. More than half the males that eventually became paired did so in their natal territory; overall, males dispersed a mean of 0.2 territories distant (0–5).

In some years at Armidale, Superb Fairy-wren female helpers were present in almost all groups at the start of the breeding season. Some dispersed in the course of the breeding season, but others remained. Experimentally removed breeding females were replaced rapidly from this pool (Ligon *et al.* 1991). Female aggression, if it occurred, had clearly not succeeded in expelling them before the breeding season (Nias 1987). The reason for this difference is not clear. The Armidale population was isolated by surrounding unsuitable habitat, and the year when most female helpers occurred was a year of severe drought when normally marginal habitat was quite unsuitable. In contrast, the artificially watered habitat of Mulder's Canberra study area may have provided sufficient opportunities of space and partners to encourage females to disperse.

Splendid and Red-winged Fairy-wrens in south-western Australia present a very different picture to Superb Fairy-wrens; females are not forced to disperse, so that groups frequently contain female helpers, and these may inherit a breeding position in their natal territory. In Red-winged Fairy-wrens, a dispersing female may join a group in which there is no breeding vacancy and may remain there as a helper for one or more breeding seasons; she may then fill a breeding vacancy either in her adopted group or in one nearby. We ascribe this behaviour to the low level of vacancies for female breeders when female survival is nearly 80 per cent. Of 90 Splendid Fairy-wren and 27 Red-winged Fairy-wren females banded as nestlings that subsequently became breeders, 30 per cent and 22 per cent, respectively, did so in their natal territory. In these species, as well as the females that disperse and wander in search of a vacancy, others remain as helpers in their natal territory and may disperse directly to a vacancy that occurs in a nearby territory. Males are even more philopatric, with most inheriting their natal territory; of 81 Splendid Fairy-wren and 48 Red-winged Fairy-wren males banded as nestlings that subsequently acquired a territory, 60.4 per cent and 72.9 per cent, respectively, did so in their natal territory; more opportunities to establish new territories were available for Splendid Fairy-wrens (Russell and Rowley 1993*a* and unpublished data).

Dispersal is also influenced by population density. In Splendid Fairy-wrens on Gooseberry Hill, density increased during a period free of fire (1978–84) when both productivity and survival were high; few vacancies became available for male or female breeders in established territories, there was no room for new territories to be established, and reduced dispersal led to an increase in group size and in the number of female helpers (see Plural breeding in Chapter 5).

Survival during the first year

Since most malurids are resident, the fate of young birds can be followed throughout their first year of life. All the available estimates of survival of juveniles and adults are from resighting and recapture of banded birds in intensive studies, where it is reasonable to assume that if a bird was present it was seen and identified. All nestlings were colour-banded, and their fate after fledging followed by regular census, so that we know how many young survived from fledging to independence, and to the next breeding season (Table 7.8).

Survival varies from year to year; juveniles may be taken by predators, become separated from the family group, or, in poor seasons, succumb to the combined effects of food shortage and extreme weather. Over 17 years, the mean survival of fledgling Splendid Fairy-wrens to independence was 68 per cent, with a range from 55 to 87 per cent. In Red-winged Fairy-wrens in the south-west Karri forests, 76 per cent of fledglings survived to independence, with a range over six years of 63–93 per cent.

Estimates of survival to 1 year are minimum values, because in most species, some dispersal occurs in winter and early spring, especially of females dispersing to fill breeding vacancies. In the Superb Fairy-wren in Canberra, where most juvenile females leave their natal territory in August before they are 1 year old, 22 per cent of fledglings survived to 1 July following their hatch (Rowley 1965); at Armidale, about 25 per cent of juvenile males survived from fledging to the next breeding season (Nias 1987). Many Splendid Fairy-wren females did not disperse from their natal territory in their first year, and many that did disperse did not go very far. We estimated survival of fledgling Splendid Fairy-wrens (male and female) to 1 September as 31 per cent over 17 years, with a range of 11–59 per cent. Survival was generally poorer when vegetation cover was sparse after fires; the highest survival was towards the end of the period 1978–84 which had been free of fire.

The survivorship curves of males and females banded as nestlings (Fig. 7.15) show slightly lower survivorship for females over the first two years of life, probably reflecting female dispersal out of the study area; after that, the curves are parallel, indicating similar mortality rates. The survival of Red-winged Fairy-wren fledglings to 1 year is higher than for other fairy-wrens so far studied, 38 per cent with a range of 22–51 per cent. At least three factors may contribute to this high survival. The climate in the south-west is among the least variable in Australia, with regular, moderate rainfall. Few females attempt to raise two broods, so fledglings receive the undivided attention of their large groups. Females rarely disperse from their natal group until they are well into their second year. This is apparent from the survivorship curves of known age Red-winged Fairy-wrens (Fig. 7.15), where survival to 1 year is identical for males and females, lower for females to 2 years, and thereafter, the two curves are parallel, with similar mortality rates.

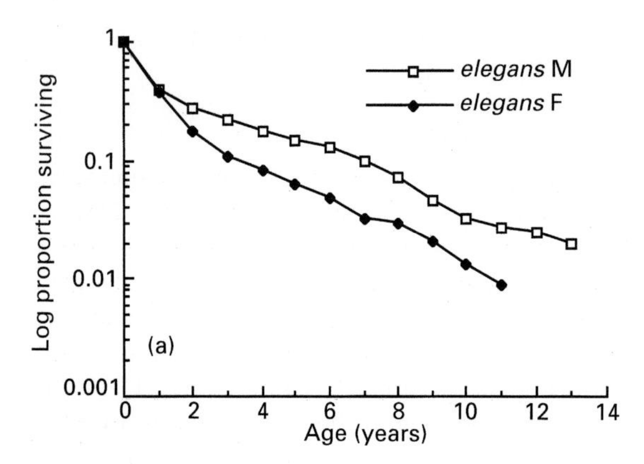

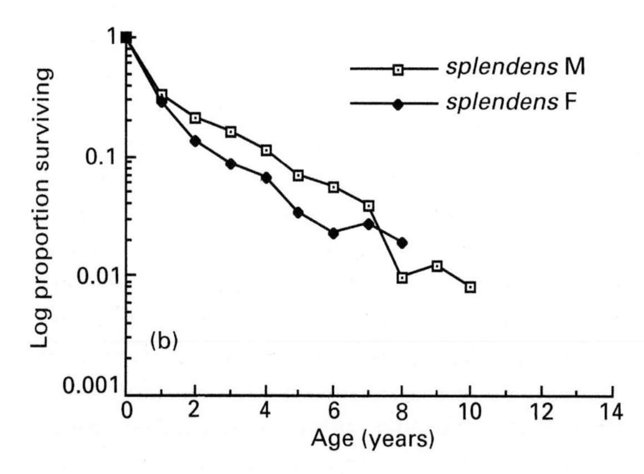

7.15 Survivorship of male (M) and female (F) Red-winged Fairy-wrens (*Malurus elegans*) (a) and Splendid Fairy-wrens (*M. splendens*) (b) banded as nestlings; for both species, the apparent lower survival of female to 2 years reflects dispersal. After 2 years, the curves for male and female are roughly parallel, indicating similar mortality rates. (Data for *M. splendens* from Russell and Rowley 1993*b*, and unpublished data for *M. elegans*.)

In Tidemann's study at Booligal, New South Wales, survival of fledglings to 1 year was poor for all three species studied, although seasonal conditions were good. This suggests that survival in inland areas may be poor, but this needs confirmation from other studies. Our estimate of survival of fledgling White-winged Fairy-wrens in a coastal habitat north of Perth is also low; we suspect this is partly due to undetected dispersal from a small study area, and in part to differences in the social organization, with possibly more movement of birds between territories.

Age at first breeding

Both male and female fairy-wrens are sexually mature and capable of breeding at 1 year old, and this is probably so in all malurid species. Whether or not they do breed is influenced by the social situation at the time. If no opportunities for pairing are available, males and females remain as helpers, generally in their natal territory, and in most species, a significant proportion of males and females do not breed until they are 2 years old or more (Table 7.9). Because the mating system is promiscuous, we do not know at what age males first copulate and actually succeed in fathering young. All we can record is the age

Table 7.9 Age at first breeding in the Splendid Fairy-wren (*Malurus splendens*) and the Red-winged Fairy-wren (*M. elegans*).

Age at first breeding (years)	*Splendid Fairy-wren*				*Red-winged Fairy-wren*			
	Males		*Females*		*Males*		*Females*	
	n	%	*n*	%	*n*	%	*n*	%
1	35	43.2	51	56.7	5	10.4	2	8.0
2	27	33.3	36	40.0	14	29.2	8	32.0
3	9	11.1	3	3.3	9	18.8	8	32.0
4	7	8.6	0	0.0	8	16.7	6	24.0
>4	3	3.7	0	0.0	12	25.0	1	4.0

The table includes all birds of known hatch year that became senior males or laid at least one clutch (female) during the studies.

at which they first achieve the status of senior male in a territorial group or pair. Most female Splendid Fairy-wrens that became breeders did so at 1 year old, but a significant proportion did not breed until they were 2 (Table 7.9). More than 20 per cent of males did not acquire senior status in a territory until they were 3 years old or more. In a period of high survival and stable conditions, few vacancies occurred, and groups accumulated numbers of older helpers, especially males. With increased mortality of breeders, more vacancies were filled by 1-year-old birds. In Red-winged Fairy-wrens, with higher adult survival, the median age at first breeding was 3 years for both males and females, with 28 per cent of females and 42 per cent of males not achieving breeding status until they were 4 years old or more.

Adult survival

The most significant measure of adult survival in a resident species is the proportion of breeding adults surviving from one breeding season to the next (Table 7.8). In fairy-wrens so far studied, this is generally high, particularly when their small size is taken into account: 70 per cent or above for males of four species and females of three. In the south-west forests, Red-winged Fairy-wrens had the highest survival, 78 per cent for breeding males and 77 per cent for breeding females; all 27 senior males present in the 1986 breeding season were still alive at the start of the 1987 season. At Booligal, survival of Superb and Variegated Fairy-wrens was poor, but survival of White-winged Fairy-wrens, the arid zone specialists, was slightly better, and similar to our estimates for the same species in coastal heathland near Perth. Both of these estimates for White-winged Fairy-wrens include all adults, because at present the social system makes it hard to distinguish between breeders and helpers. Even in the very cold winters of Canberra, survival of male Superb Fairy-wrens males is close to 70 per cent; it is interesting that in the very favourable habitat of the National Botanic Gardens annual survival of breeding females was 70 per cent, much higher than the 53 per cent found by Rowley (1965) in an outer urban habitat, or the 44 per cent in somewhat degraded forest bordered by pasture near Armidale, where the favoured blackberry habitat was restricted by spraying (Nias 1987).

With such high survival rates, a few birds attain remarkable ages. Our oldest Splendid Fairy-wren was a male hatched in 1975 who helped and then succeeded as senior male in his natal territory in 1976, dispersed to the next door territory as senior male in 1977, survived the complete burning of his territory in the major fire of 1985, and remained as senior male until late in the 1987 breeding season, a lifespan of over 12 years and a breeding tenure of 12 seasons. Our oldest male Red-winged Fairy-wren was banded in November 1978 as an adult in full nuptial plumage, and has been senior male in his territory since we began to monitor it in November 1981—at least 14 years; when we last saw him in November 1994, he was at least 16 years old; sadly he was not there in November 1995. Such longevity is not exceptional in Red-winged Fairy-wrens; of 28 senior territorial males in 1990, seven were 10 years old or more.

The spread of ages in populations of Splendid and Red-winged Fairy-wrens illustrates the high survival of adults and their persistence in the study area. Figure 7.16 gives a snapshot of each population in the same year, 1987. The large proportion of females of uncertain age reflects the immigration of females into the study area, as replacement breeding females or as helpers. In the Red-winged Fairy-wren, many older males, first banded before the study began in 1980, were still present in 1987. The previous year, 1986, was a year of particularly high survival of juveniles to 1 year, hence the large proportion of 1-year-old birds. The Splendid Fairy-wrens had suffered a major fire 2.5 years previously in January 1985. Breeding success had been low in the 1985 and 1986 breeding seasons, but the persistence of males and females from the good years before the fire is clear.

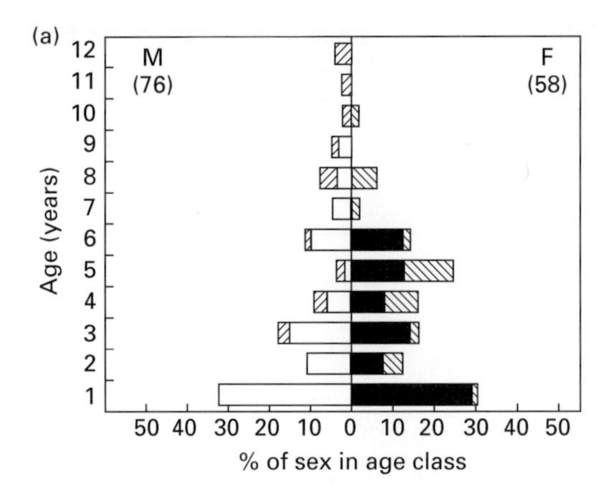

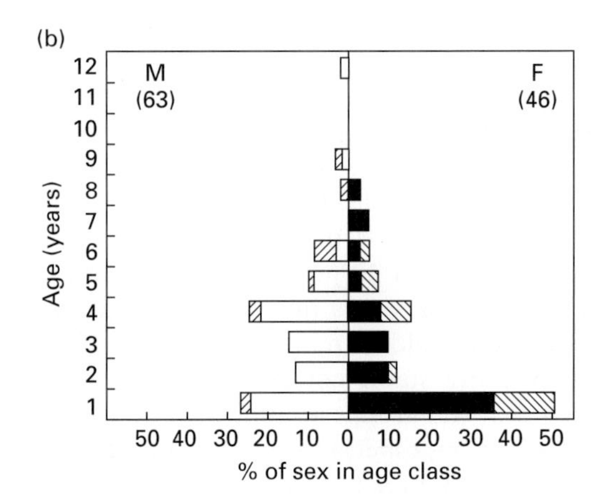

7.16 Age structure of populations of (a) Red-winged Fairy-wren (*Malurus elegans*) and (b) Splendid Fairy-wren (*M. splendens*) in the same year (1987). The bars show the proportion of each sex in different age classes at 1 September each year. In each class, birds of uncertain age are included at their known minimum age (hatched bars). Thus the 6-year-old age class includes birds known to be 6 years old and those at least 6 years old. In *M. elegans* 30% of males and 12% of females, and in *M. splendens* 14% of males and 11% of females, were more than 5 years old. M: male; F: female.

Life history overview

The survival rates of Australian passerines appear to be high. Estimates using older methods which equate survival and recapture rates have been shown to be inappropriate, but there are as yet no estimates of survival using modern estimates from capture–recapture models. At present, the best information comes from population studies of individually marked birds, and the survival estimates are of breeding adults from recapture and resighting in situations where resighting probabilities are high; if a bird is absent, it is reasonable to assume that it is dead. We summarize this information in Table 7.10 for Australian species other than fairy-wrens. All the birds listed are small passerines with body weight less than

Table 7.10 Survival of Australian adult small passerines (<50 g body weight).

Species	*% survival*	*Source*
Brown Thornbill (*Acanthiza pusilla*)	87.0	Bell and Ford 1986
Buff-rumped Thornbill (*A. reguloides*)	58.0	Bell and Ford 1986
Red-browed Tree-creeper (*Climacteris erythrops*)	78.9*	Noske 1991*a*
Brown Tree-creeper (*C. picumnus*)	78.2*	Noske 1991*a*
White-throated Tree-creeper (*Cormobates leucophaea*)	72.6*	Noske 1991*a*
Eastern Yellow Robin (*Eopsaltria australis*)	75.0*	Marchant 1985
White-breasted Robin (*E. georgiana*)	83.0, 73.0	R. and M. Brown, unpublished data
Bell Miner (*Manorina melanophrys*)	67.5, 59.4	Clarke and Heathcote 1990
Golden Whistler (*Pachycephala pectoralis*)	78.0*	Brown *et al.* 1990
Rufous Whistler (*Pachycephala rufiventris*)[1,2]	90, 84	Bridges 1994*b*
Scarlet Robin (*Petroica multicolor*)	87.5, 66.9	Robinson 1990*b*
Flame Robin (*P. phoenicea*)[1]	82.5, 68.8	Robinson 1990*b*
White-browed Scrub-wren (*Sericornis frontalis*)	77.0*	Brown *et al.* 1990

Data from colour-banded populations in which individuals were followed from banding until disappearance (presumed death). Data are for breeding adults unless marked *, where data are for all adults; two figures separated by a comma are survival of breeding males, breeding females.
[1] Migrant
[2] Over-winter survival, April to September

50 grams, whose survival is frequently 70 per cent or above, as is also the case for the fairy-wrens in Table 7.8. This is much higher than the mean survival rates (53 per cent) calculated by Martin (1995) for a series of small shrub- or low-foliage nesting passerines in forest and shrubland/grassland habitats in temperate North America between latitudes 31°N and 47°N. Karr *et al.* (1990) compared survival rates of temperate and tropical species (mostly passerines) using capture–recapture models. For 10 species banded in Maryland, USA, they found mean survival was 54 per cent, and for 25 passerine species banded in Panama, survival was 56 per cent.

The life history of fairy-wrens (and probably other genera within the family) is characterized by low annual fecundity, with the production of two or three small clutches per year, and fairly low breeding success. This low reproductive rate, typical of Australian passerines, is balanced by the high survival of adults (Woinarski 1985; Yom-Tov 1987; Rowley and Russell 1991). Such data as there are for Australian migrants suggest a similar life history (Robinson 1990*a*,*b*; Bridges 1994*b*). For almost all malurids, nests are placed in shrubs or other low vegetation or debris, and, although they are usually enclosed, domed nests, they are subject to the relatively high predation rates typical of nests in such situations (Martin 1995). Once fledged, the survival to sexual maturity is relatively high, and adult survival is high. Most birds do not disperse far from where they were hatched, and breeding is frequently delayed, with many birds not breeding at 1 year old even though they are sexually mature.

We see this as a life history evolved in an environment where large-scale migration is rare and food is not subject to regular large fluctuations from very abundant to scarce as happens in severe northern hemisphere winters. Seasonal peaks in insect abundance occur, but they are not as great as in the northern spring (Woinarski and Cullen 1984; Bell 1985) and not enough to sustain very large broods. Breeding seasons are long, allowing risks to be spread over a number of small clutches. Irregular disturbances occur, drought and fire in particular, when survival and offspring production are threatened, and when the most appropriate strategy for adults is to reduce or forgo breeding attempts and enhance their own survival until conditions improve. Provided an adult survives, there is usually a good year just ahead when two or even three broods can be reared.

8

Conservation

Introduction

Australia, in the short space of 200 years, has changed from a continent peopled by 500 000 nomadic hunter-gatherers to a modern industrialized country of 18 million. For most of this time, the nation has happily 'ridden on the sheep's back', benefiting from the sale of primary products overseas and increasingly supported by exploitation of its forest and mineral resources. Over the past 50 years, it has become obvious that such a state of affairs cannot continue indefinitely. Soil erosion and salination, overgrazing, and weed invasion have all contributed to a decline in production and ultimately to the rise of a rather surprised public conscience. Conservation of biodiversity, sustainable development, and the use of renewable resources are currently popular conservation themes.

For a continent the size of Australia to change from a subsistence primitive society to a major exporter of agricultural products has necessitated enormous changes to the landscape. Seventy per cent of the continent is now used for agriculture, forestry, or grazing stock and although not all this area has been completely cleared, most of the vegetation has been significantly altered as a result of grazing by exotic herbivores and competition from exotic plants. These influences have not been confined to agricultural areas but have spread to unoccupied areas, even the deserts. The boundaries of pastoral settlement were largely determined by 1900. Clearing of land for cropping continued after that, and is still in progress in some states. The spread of feral pests such as rabbits, camels, donkeys, pigs, horses, cats, and foxes, and of widespread weeds has largely taken place in the last 100 years.

Present-day conservation has two broad aims: first, to draw up a national scheme of renewable resource management, such that future generations will be able to benefit from those resources that are at present taken for granted but have been steadily degrading; and second, to try to ensure that viable representative populations of all forms of life persist. To achieve this functional conservation, it is necessary to survey and evaluate those areas of native vegetation remaining in the National Estate and to establish adequate legal security for these places in the form of National Parks and Reserves, with sufficient funds to enable their long-term management and protection. At the moment, it is hoped that most species occur in these remnants. Such optimism is patently unjustified, and so there is a need to re-establish areas of native vegetation that are inadequately represented in the National Estate and this is much more difficult.

Realistic conservation requires the flora and fauna of Australia to be documented. This process is well in hand for certain groups, but in others there is a long way to go; probably at least half the insect species in Australia are still undescribed. Ornithology is further advanced than most other fields. Recent attempts to evaluate the status of birds in

Australia considered 1074 bird taxa (including subspecies), 150 of which were extinct or threatened (23 extinct; 26 endangered; 40 vulnerable; 32 rare; 29 insufficiently known; Garnett 1992*a*,*b*).

Threats to Australian birds

Clearance and fragmentation of habitat

Throughout the nineteenth century large-scale sheep and cattle grazing operations spread from the coasts towards the centre of the continent and occupied all land where there was enough water for livestock. Land clearing for agriculture was widespread until recently; it was still in progress in Western Australia until 1983, and while it has now largely ceased in Western Australia, South Australia, and Victoria, it is still occurring in New South Wales and in Queensland where 450 000 hectares of semi-arid *Acacia* woodland was cleared in 1994. Clearing for urban development is another pressure on available habitat, and is of greatest significance on the east coast.

In the process of clearing, small patches of land were left uncleared, along road verges, as shelter for stock, or because the land was too rocky, swampy, or sandy to be suitable for cropping. Occasionally it was left as a conservation reserve. These fragments of vegetation are now separated by expanses of habitat inhospitable to most birds. Many of these fragments have become very degraded and now contain only very small populations that are probably not viable in the long term. The effects of land clearing and degradation in the wheatbelt of Western Australia have been demonstrated by Saunders and Ingram (1995). Over 90 per cent of the original area of native vegetation has been cleared, more than half of it since 1945. Half of the 195 recorded species have decreased in range and/or abundance since the beginning of the twentieth century; of the seven malurids recorded, one has disappeared (Thick-billed Grasswren (*Amytornis textilis*), always rare in the wheatbelt), five others have reduced range and numbers, and the White-winged Fairy-wren (*Malurus leucopterus leuconotus*) has probably decreased in numbers, but not in range.

Forestry activity

Large-scale logging of eucalyptus forest for timber production has also resulted in the loss of much of the once forested habitat, and the fragmentation or degradation of that which remains. For many years, logging was selective, and non-target species and understorey vegetation were relatively little affected. More recently, clear felling for combined timber and wood-chip extraction has seen the loss of large areas of mature forest. Although these areas will be allowed to regenerate as forest, the consequences of such a large-scale reduction in such a short time are unknown.

In the past, much cleared forest was replanted with soft-wood (*Pinus* spp.), a habitat that is used by very few Australian birds, and very rarely by malurids. Recently large monospecific areas of hardwood (eucalypt) plantations have been established for timber and woodchip production, on land that was already cleared. These will not be of much benefit to any malurids, since little understory is allowed to establish.

Other environmental modifications

Land degradation

As a consequence of clearing and grazing, much agricultural and pastoral land is now seriously degraded by wind and water erosion of top-soil and salination caused by rising water tables. Most of the uncleared remnants of native vegetation in agricultural regions have also become seriously degraded, owing to altered water regimes, fire, loss of understorey after grazing by stock or vermin, or general damage to vegetation by the influx of weeds and the drift of weedkilling sprays used on crops.

Fire

Australia is a very dry contintent, subject to episodes of sustained drought, and fire has always been a significant cause of habitat modification. Fire regimes have changed with the advent of European agriculture. Before then, fires were either occasional, on a large scale, started by lightning in summer, or small scale, low intensity, deliberately lit by Aboriginal people in a process of habitat management for hunting that produced a vegetation mosaic. Since then, fire frequencies have changed in many places. Across northern Australia, summer burning of grasslands including spinifex, to produce green shoots for stock, is a regular practice detrimental to fauna and native vegetation. Further south, human activity has increased the frequency of accidental fires in native forests, and forest management practice is to burn native forests deliberately and frequently, at low intensity, to prevent large-scale uncontrollable, high-intensity accidental fire.

Water

Habitats are threatened by the drainage of wetland areas and swamps, and by changes in water flow by direct use from rivers or their damming to provide water for urban supply and for agriculture. In coastal areas, the drainage of wetland areas for urban development has greatly reduced the amount of habitat available for Southern Emu-wrens (*Stipiturus malachurus*), in particular. In the drier inland, the reduction of flow in rivers threatens the viability of the important floodplain habitat of many species. The vegetation along rivers, their floodplains, and associated billabongs (see Glossary) and swamps are important refuge areas in times of drought. The Grey Grasswren (*Amytornis barbatus*) is one species whose habitat depends on the vegetation supported by the inland flowing rivers, whose floods may be determined by rainfall hundreds of kilometres away, and whose water is coveted by many users before it reaches Central Australia.

Grazing

A very large area of inland Australia, although not cleared for cropping, has been grazed and trampled by introduced herbivores—sheep and cattle in controlled pastoral activities, and also by many feral pests—rabbits, camels, donkeys, goats, horses, pigs, buffalo. All of these have reduced the vegetation cover, especially during drought where forage and water have been short and mouths too many. Much of the reduction of species in the arid and semi-arid areas south of the Tropic of Capricorn was probably due to the devastating effects of stock and rabbits before the advent of myxomatosis.

Weed invasion

In the last 100 years, many introduced plant species have spread uncontrollably at the expense of native vegetation, such as *Acacia nilotica* and *Mimosa pigra* in the north, and blackberry (*Rubus* sp.) in the south. Not all invasive weeds are necessarily hostile to native birds; thick tangles of blackberries are much favoured by the Superb Fairy-wren (*M. cyaneus*) in south-east Australia (Nias 1984).

Predators

Before the arrival of humans, Australia had no large mammalian predators, and only a few small ones. The various so-called Native Cats (*Dasyurus* spp.), weighing 1–2 kilograms, ate insects, lizards, small rodents, and probably some birds, but were probably not major nest predators. The most significant nest predators were other birds, snakes, and large lizards. Humans introduced a range of very efficient mammalian nest predators to Australia. The dingoes that came with Aboriginal humans were less significant than the cats, foxes, and rats that came with the Europeans. Foxes and cats, are able to survive throughout central Australia and are moving into the tropics; they are significant predators of birds, both adults and nests. In urban habitats, it is recognized that hunting by free-ranging domestic cats accounts for the deaths of many birds, and some attempt is now being made to control this. The control of feral foxes, cats, and rabbits over the whole continent poses a far greater problem. Baiting and trapping at local sites are effective, but require continued effort, and they are impossible to apply on a continental scale.

Competition

Several introduced bird species have become well established in Australia, some from aviary escapes, some from deliberate introduction by Acclimitization Societies in the nineteenth century. Many are not insectivorous, and so do not compete directly with any malurid for food, but Common Blackbirds (*Turdus merula*), Song Thrushes (*T. philomelos*), and Common Mynas (*Acridotheres tristis*) are common in south-eastern urban areas where the Superb Fairy-wren occurs.

Demographic problems

The number of birds in a population fluctuates in response to environmental perturbations such as fire, drought, cold winters, or a year of very high brood parasitism. If a very small population is significantly reduced by one such disaster, a second event following too soon after the last may be enough to eliminate the population. It is a well-known tenet of conservation biology that small populations are vulnerable to extinction from such random demographic variation. If small populations are isolated, such losses cannot easily be made good by recruitment from a nearby area. This extinction of small isolated populations has occurred in many of the small remnant patches in agricultural land in Australia (Saunders and Ingram 1995); as the vegetation becomes degraded, shelter is reduced, predation becomes more effective, and food harder to find.

The low reproductive rate of most Australian passerines means that a sudden population decrease is not rapidly made good, especially since the environmental change that brought about the population decline in the first place is likely to have depressed reproductive success and juvenile survival as well. A further complication for the family Maluridae is their co-operative breeding system, as a result of which the actual breeding population is likely to be far less than the total adult population.

A population model developed by Lesley and Michael Brooker (1995) for the Splendid Fairy-wren (*Malurus splendens*) on Gooseberry Hill near Perth tried to capture some of the complexity of the interactions affecting the population processes of a wild species. The model incorporated the effects of fire, nest predation, rainfall, brood parasitism, and the known productivity and survival characteristics from a 17-year study of the population (Rowley *et al.* 1991; Russell and Rowley 1993*b*). The model predicted that fairy-wren populations in small patches (less than 2000 hectares) had a high risk of local extinction if fire frequency was high (more than once in 5 years), as is typical of small urban reserves, but if fire frequency was low (less than once in 20 years), then even a small population (100 breeding females in 400 hectares) may be viable in the long term.

Conservation status of the family Maluridae

Garnett (1992*a*) lists 10 malurid taxa as 'threatened'; this considers only Australian taxa. The 15 taxa in five species endemic to New Guinea are insufficiently known for any realistic attempt to be made to define their conservation status; the Broad-billed Fairy-wren (*M. g. grayi*) is known from only 30 specimens and the subspecies *M. g. campbelli* from only seven (LeCroy and Diamond 1995). Wallace's Wren (*Sipodotus wallacii*) and the Emperor Fairy-wren (*M. cyanocephalus*) both largely depend on forest, and the extent of recent clearing for wood-chips and timber is largely undocumented. The Orange-crowned Wren (*Clytomias insignis*), the only mountain-forest malurid, is unlikely to be endangered, since its habitat is inaccessible. White-shouldered Fairy-wrens (*M. alboscapulatus*) are widespread and are probably the only malurids in

New Guinea to have benefited from clearing, since they thrive in the tall grass of regrowth in overgrown gardens and roadside verges.

Of the 42 Australian taxa that we consider in this book, most have had their distribution curtailed by development that caused the loss of their preferred habitat—generally low shrubby vegetation easy to clear or readily eaten by herbivores (chenopod shrublands). One exception may be the Superb Fairy-wren, which appears to thrive in suburbia, to cope with predatory cats, and even to out-compete the exotic universal colonizer, the House Sparrow (*Passer domesticus*) on the campus of the Australian National University in Canberra. It is hard to explain how its congener in Western Australia the Splendid Fairy-wren, so similar in many respects, appears unable to cope with urbanization, and disappears with settlement.

Current threatened species categories

'Birds to Watch 2' which claims to be the 'official source for birds on the IUCN 'Red List' lists the Maluridae as a subfamily Malurinae and calls them Australian Warblers (Collar *et al.* 1994, p. 170–72). It includes as threatened three bristlebirds (*Dasyornis* spp.), the Slender-billed Thornbill (*Acanthiza iredalei*), two Gerygones (*Gerygone* spp.), the Chestnut-breasted Whiteface (*Aphelocephala pectoralis*), the Yellowhead (*Mohoua ochrocephala*), and the Silktail (*Lamprolia victoriae*). None of these species is considered to be a member of the family Maluridae.

We have assessed the status of all the Australian malurids in terms of the revised criteria recently established (IUCN 1994, Fig. 8.1). Most taxa in the genus *Malurus* are secure except for the Kimberleyan subspecies of the Purple-crowned Fairy-wren (*M. coronatus*). Most forms of the emu-wren *Stipiturus* except those in Tasmania have undergone significant habitat loss, particularly those in the coastal habitats of the south-east and south-west corners that are the subject of so much draining and urban development. One form is endangered and two forms are assessed as vulnerable. Many grasswren *Amytornis* species have suffered the loss or degradation of habitat,

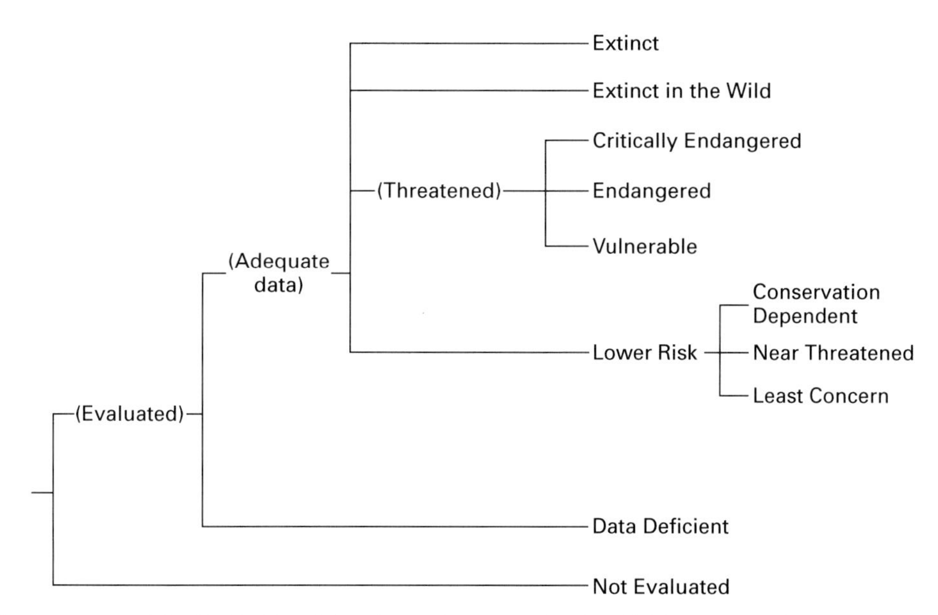

8.1 The structure of the IUCN Red List Categories recognized by IUCN Species Survival Commission, 30 November 1994.

and have undergone a severe reduction in range, but many are rarely seen and little known. We agree with Garnett (1992*a*,*b*) that 10 taxa require watching.

Under the new criteria for categories (IUCN 1994), no member of the Maluridae is known to have become extinct nor is one regarded as Critically Endangered. One taxon is regarded as Endangered, five as Vulnerable and four are regarded as probably Threatened but Data Deficient. Briefly, the most recent criteria for the Endangered and Vulnerable and Data Deficient categories are:

Endangered: a taxon is Endangered when it is not Critically Endangered but is facing a very high risk of extinction in the wild in the near future.

Vulnerable: a taxon is vulnerable when it is not Critically Endangered or Endangered but is facing a high risk of extinction in the wild in the medium-term future (detailed in Table 8.1).

Data Deficient: there is inadequate information to make a direct or indirect asessment of its risk of extinction based on its distribution and/or population status.

Threatened species

The following 10 taxa are potentially threatened.

Endangered

1. Southern Emu-wren (*Stipiturus malachurus intermedius*), in the Mt Lofty Ranges, South Australia. Occupies an area less than 500 km^2 in fewer than 10 localities, and the area of its habitat is decreasing as swampland is drained and due to wildfires. The population is fewer than 2500, severely fragmented, with no one population of more than 250 mature individuals.

Vulnerable

2. Purple-crowned Fairy-wren (*Malurus coronatus coronatus*). The Kimberleyan subspecies has suffered severe reduction of numbers over the past 100 years due to grazing pressure along the river frontage of several rivers—Victoria, Ord, Chamberlain, and Fitzroy. In particular, the loss of *Pandanus aquaticus* and 'Cane-grass' or 'River-Grass' *Chionachne cyanthopoda* very much reduced the habitat available (Rowley 1993); recent attempts to fence off lengths of river frontage (Fitzroy and Victoria Rivers) indicate a change of pastoral attitudes and an increased appreciation of the value of those river frontages. Unfortunately the increase of ecotourism and the repeated firing of the woodland in the northern Kimberleys has led to recent, serious habitat degradation (G. S. Chapman, personal communication). The subspecies occupies an area less than 2000 km^2 with populations severely fragmented and the area and quality of habitat decreasing with grazing. The population is fewer than 10 000, severely fragmented with no one population containing more than 1000 mature individuals.

3. Southern Emu-wren (*Stipiturus malachurus parimeda*) in the extreme south of Eyre Peninsula. The subspecies occupies an area less than 2000 km^2 in fewer than 10 locations and its projected area of habitat is decreasing. The population is fewer than 10 000, severely fragmented, with no population of more than 1000 mature individuals.

4. Mallee Emu-wren (*Stipiturus mallee*). Confined to *Triodia* hummock grassland with low woodland understorey in south-east South Australia and north-west Victoria. Area of occupancy less than 2000 km^2; area and quality of habitat decreasing with fire as is the number of locations where it occurs; little regeneration is taking place. Population estimated to be fewer than 10 000 and declining; severely fragmented, and no one population with more than 1000 mature individuals.

5. Thick-billed Grasswren (*Amytornis textilis textilis*). Distribution has much decreased over the past 100 years, and is now confined to northern Peron Peninsula (*c.* 800 km^2) and the adjacent mainland (*c.* 1500 km^2) in

Plates

Plate 1
Blue and Purple-crowned fairy-wrens, genus *Malurus*

1. Splendid Fairy-wren
Malurus splendens p. 149
Heathland, shrubland, and mallee in south-western, central, and inland eastern Australia. Sexes clearly dimorphic in nuptial plumage; males subspecifically distinct. Outside the breeding season males enter an eclipse female-like plumage; immatures initially resemble females but males gain blue primaries (a) and then black bills before assuming their first nuptial plumage.

(a) *M. s. splendens* of south-western Australia; uniformly cobalt-blue back and belly.
(b) *M. s. melanotus* of inland eastern Australia; black lower back feathering and mid-belly whiteish.
(c) *M. s. callainus* of central Australia; turquoise above, cobalt-blue below, black back.

2. Purple-crowned Fairy-wren
Malurus coronatus p. 155
River margins of northern Australia. Sexes clearly dimorphic in nuptial plumage; outside the breeding season and in times of drought the males enter an eclipse female-like plumage without the distinctive ear-patch of females; immatures resemble females except for brown (not grey) crown, sexes indistinguishable. Two very similar subspecies recognized, separated by 200 km gap, *M. c. coronatus* in north-west Kimberleys and *M. c. macgillivrayi* in the east.

3. Superb Fairy-wren
Malurus cyaneus p. 143
South-eastern Australia. Sexes clearly dimorphic in nuptial plumage; two subspecies recognized. Outside the breeding season males enter an eclipse female-like plumage retaining their black bills. Immatures are indistinguishable and female-like until the bills of the males darken.

(a) *M. c. cyaneus* of Bass Strait Islands and Tasmania; larger and deeper blue.
(b) *M. c. cyanochlamys* of mainland Australia; smaller and paler to the north.

1a ♂
1a ♀
1a imm. ♂
1b ♂
1c ♂
1a ♂ (eclipse)
2 ♀
2 ♂
2 ♂ (eclipse)
3a ♂
3b ♂ (eclipse)
3b ♀
3b ♂

Plate 2
Chestnut-shouldered fairy-wrens, genus *Malurus*

A chestnut-shouldered fairy-wren occurs in nearly every part of Australia except for the extreme south-east. The males of all these species have conspicuous rufous (chestnut) scapulars.

1. Variegated Fairy-wren
Malurus lamberti p. 160
Sexes dimorphic in nuptial plumage; four subspecies recognized, based on two different male plumages (a and b,c,d) and three different female plumages (a,b;c;d). Males of all four subspecies enter an eclipse female-like plumage in the non-breeding season.

(a) *Malurus l. lamberti* of eastern coastal Australia; nuptial male with royal blue back and azure crown; female with bright rufous lores, red-brown bill, and grey-brown back.
(b) *Malurus l. assimilis* throughout inland and western Australia; nuptial male with purple back and variable violet-blue crown; female with bright rufous lores, red-brown bill, and grey-brown back as *M. l. lamberti*.
(c) *Malurus l. rogersi* Kimberley region, north-west Australia; nuptial male indistinguishable from *M. l. assimilis*; female with bright rufous lores, red-brown bill, and grey-blue back.
(d) *Malurus l. dulcis* of Arnhem Land, northern Australia; nuptial males indistinguishable from *M. l. assimilis*; female with white lores, red-brown bill, and grey-blue back.

2. Blue-breasted Fairy-wren
Malurus pulcherrimus p. 168
Sexes dimorphic in nuptial plumage; male with violet-blue crown, back, and ear tufts, and breast navy blue; females very similar to those of the Variegated Fairy-wren (*M. lamberti assimilis*) but lores rufous and bill red-brown. The eclipse female-like plumage of males outside the breeding season is characterized by a white or pale-blue ring of small feathers round the eye.
No subspecies recognized throughout the range across southern Australia.

3. Lovely Fairy-wren
Malurus amabilis p. 165
Sexes dimorphic in nuptial plumage, with tails shorter than those of other chestnut-shouldered fairy-wrens; male with azure crown, back, and ear tufts, and black breast and bill; female with black bill as male, white lores, turquoise ear tufts, and smoky-blue back.
In shrubby forest margins on Cape York Peninsula; no subspecies recognized.

4. Red-winged Fairy-wren
Malurus elegans p. 172
Sexes dimorphic in nuptial plumage; male with silvery-blue crown, back and ear tufts, with navy-blue breast; female grey-brown above with darker scapulars, deep rufous lores not extending behind the eye, and black bill as male. Many males in the season following their hatch do not achieve full nuptial plumage. Eclipse male plumage in the non-breeding season as for female, but retains black lores. In forested south-west corner of Australia; no subspecies recognized.

1c ♀
1a, b ♀
1a ♂
(eclipse)
1d ♀
1b, c, d ♂
1a ♂
2 ♀
2 ♂
2 ♂ (eclipse)
4 ♀
4 ♂
3 ♂
3 ♀
4 ♂

Plate 3
Bi-coloured fairy-wrens, genus *Malurus*

Three species achieve striking male nuptial plumages with only two colours—blue and white, black and white, and red and black. They all lack the erectile ear tufts of the blue and chestnut-shouldered fairy-wrens.

1. White-winged Fairy-wren
Malurus leucopterus p. 176

Sexes dimorphic; three subspecies recognized on colour of nuptial males and size; females of all three are similar drab grey-brown. Unlike other *Malurus*, male White-winged Fairy-wrens do not attain full nuptial plumage until they are in their fourth year. Males in their third year are as (c) 'spotty' males. Most males moult into an eclipse plumage after the breeding season has ended; a very few maintain their nuptial plumage throughout the year.

(a) *Malurus l. leucopterus* nuptial male black with white patch on wing involving scapulars, secondary wing coverts, and innermost (3–6) secondaries; tail deep blue; only on Dirk Hartog Island, Western Australia.

(b) *Malurus l. edouardi* similar to *M. l. leucopterus* in plumage but larger, although tail shorter; only on Barrow Island, Western Australia.

(c) *Malurus l. leuconotus* similar plumage to *M. l. leucopterus* except that black feathering replaced by mid to deep cobalt blue; found in low shrublands across arid and semi-arid Australia with little variation.

2. Red-backed Fairy-wren
Malurus melanocephalus p. 181

Sexes dimorphic with two subspecies recognized, the nuptial males of which differ in the shade of red on the back. This species is the smallest *Malurus*. The females of the two subspecies are similar and a more rufous brown than female White-winged Fairy-wrens (*M. leucopterus*). Outside the breeding season, nuptial males enter an eclipse female-like plumage. As in *M. leucopterus*, males take more than 2 years to attain full nuptial plumage; in their first year, they are indistinguishable in plumage from females but have a cloacal protruberance during the breeding season. In their second year, their bills are darker and in their third year, they may attain some black and red patches as 'spotty' males.

(a) *Malurus m. melanocephalus* nuptial males with fiery orange scapulars and back; eastern Australia from northern New South Wales to base of Cape York.

(b) *Malurus m. cruentatus* nuptial males with scarlet-crimson back and shorter tail; northern tropical Australia.

3. White-shouldered Fairy-wren
Malurus alboscapulatus p. 185

A complicated New Guinea species with three different sorts of female plumage, one of which is black and white like the male. Six subspecies are recognized, three of which are not clearly sexually dimorphic in plumage, while three are. The males of all six subspecies are similarly black and white and do not enter an eclipse plumage. All subspecies typically found in tall grasslands such as frequented by the Red-backed Fairy-wren (*M. melanocephalus*) in northern Australia; avoidance of tall mountains and forests has led to geographical separation of distinct subspecies.

(a) *Malurus a. alboscapulatus* male glossy black except for white scapular tufts; female pied, with head and back black and underparts white; from Vogelkop, West Irian Jaya.

(b) *Malurus a. naimii* as (a), but smaller; from northern to southern central Papua New Guinea.

(c) *Malurus a. aida* male glossy black with white scapulars; female similar to male but duller; northern West Irian Jaya.

(d) *Malurus a. kutubu* as (c) and (e), but larger and south of the cordillera is separated geographically from (c) to the north and (e) to the east.

(e) *Malurus a. moretoni* as (c) and (d), but separated from both by *M. a. naimii* with pied females; lowlands and foothills around eastern end of Papua New Guinea.

(f) *Malurus a. lorentzi* male glossy black except for white scapular tufts; female brown above, creamy white below with cinnamon flanks; from Trans-Fly in south-western Papua New Guinea to the Baliem Valley in Irian Jaya.

1c ♂
1a, b, c♀
1c ♂
1a, b ♂
2b ♂
2a ♀
2a ♂
2a ♂
3a-f ♂
3c-e ♀
3f ♀
3a, b ♀

Plate 4
Four New Guinea malurids

Three genera represented; two species now placed in *Malurus*, and one each in two monospecific genera *Sipodotus* and *Clytomyias*. A fifth New Guinea malurid, *Malurus alboscapulatus*, is shown in Plate 3.

1. Emperor Fairy-wren
Malurus cyanocephalus p. 192

Sexes dimorphic: males blue and black above and below; females crown and face as male but back chestnut, underparts white; white tips to tail feathers. Immature males pass through female-like plumage before achieving adult male plumage.

(a) *Malurus c. cyanocephalus* crown of male with violet cast to the blue; occurs in lowland forests from Vogelkop in the west along northern New Guinea; female similar to those of (b) and (c).

(b) *Malurus c. bonapartii* similar to (a) but crown light cobalt blue; throughout southern lowland forests of New Guinea; female as (a) and (c).

(c) *Malurus c. mysorensis* as (b) but smaller; only found on Biak Island off the north-west of New Guinea; separated from (b) by nominate race (a). Female as (a) and (b).

2. Wallace's Wren
Sipodotus wallacii p. 196

Sexes in this monospecific genus are almost inseparable; males have throat and breast white whereas female is yellowish in these areas; these differences vary over the widespread distribution with two subspecies recognized; rarely cocks tail. Forages in trees rather than undergrowth in foothill rainforest throughout New Guinea. Only *S. w. wallacii* illustrated.

3. Orange-crowned Wren
Clytomias insignis p. 199

Male and female of this monospecific genus virtually indistinguishable; carries tail cocked; two subspecies recognized; the only malurid in mountain forest.

(a) *Clytomyias i. insignis* throat, breast, and upper belly creamy-white in both sexes; confined to the mountains of the Vogelkop in extreme western New Guinea.

(b) *Clytomyias i. oorti* throat, breast, and belly buff, female paler; found in mountain forest each side of the central cordillera of the main part of New Guinea.

4. Broad-billed Fairy-wren
Malurus grayi p. 189

Sexual dimorphism slight, mainly in size (male larger) and belly-colour, which is white in the female and blue in the male; both sexes have pale blue ear tufts. Two subspecies currently recognized, separated by the central cordillera.

(a) *Malurus g. grayi* crown and forehead of male charcoal mottled with pale blue; female with solid charcoal crown and forehead, and white belly; from the Vogelkop along the northern slopes of the central cordillera.

(b) *Malurus g. campbelli* smaller than (a), crown and forehead of male and female black; female otherwise as (a); known only from Mt. Bosavi south of the main cordillera.

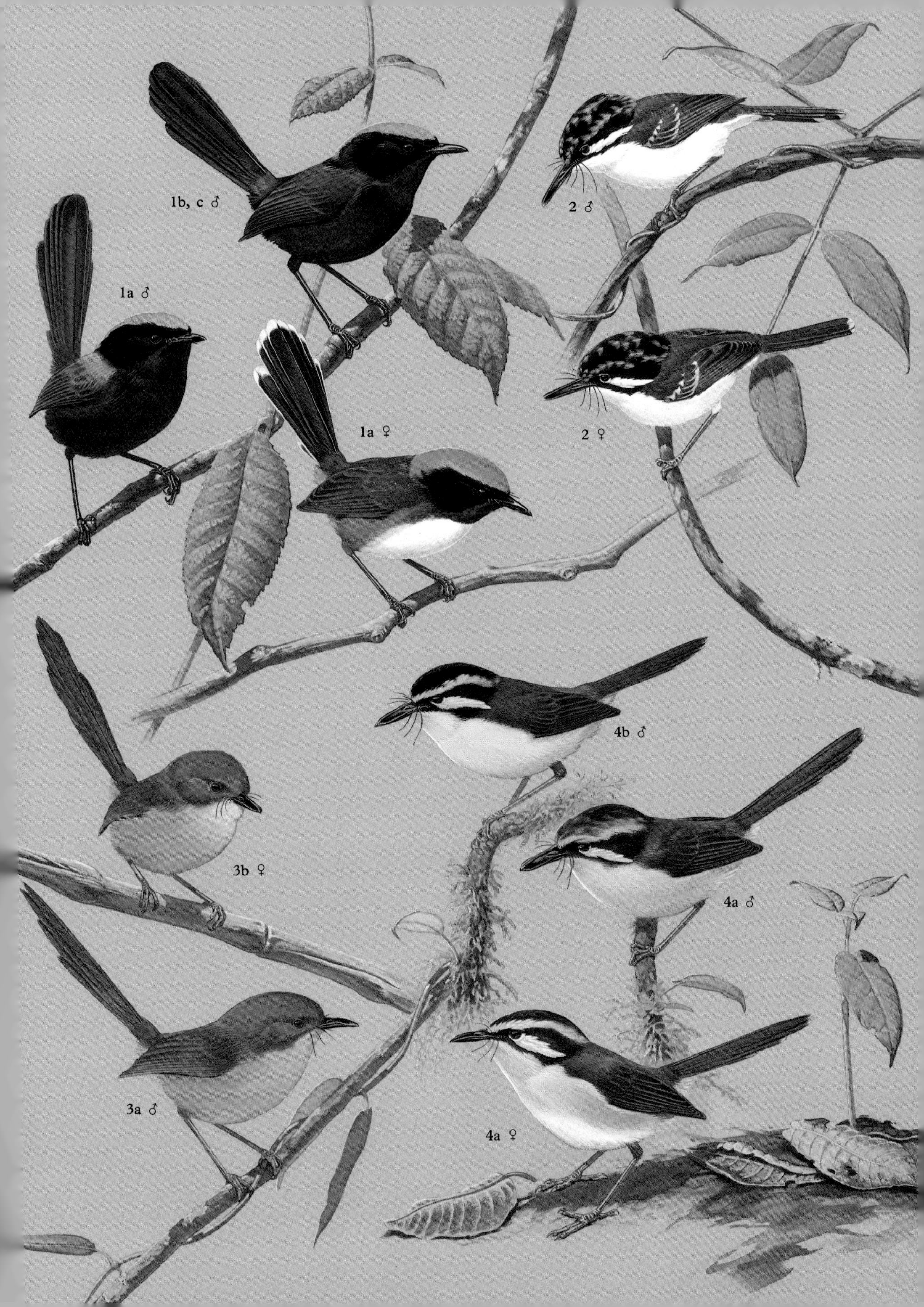
1b, c ♂
2 ♂
1a ♂
1a ♀
2 ♀
4b ♂
3b ♀
4a ♂
3a ♂
4a ♀

Plate 5
Emu-wrens, genus *Stipiturus*

Three species characterized by having only six very long filamentous tail feathers, the central feathers much longer than the outer ones. The barbs of these feathers are sparse and the barbules lack hooks to mesh them together. Sexes dimorphic, males with blue throats, upper breast, and ear coverts contrasting with the plainer females. Juvenile sexes distinct before nestlings fledge, showing incipient male throat colouring.

1. Mallee Emu-wren
Stipiturus malachurus p. 208
Crown plain dull rufous; back olive-brown streaked black; tail one and a half times as long as body; face, ear coverts, and throat sky-blue. Females similar to males without blue colouring and with more streaked crown. Confined to spinifex (*Triodia*) under mallee woodland on each side of the border between South Australia and Victoria, south of the Murray River. No subspecies.

2. Rufous-crowned Emu-wren
Stipiturus ruficeps p. 210
The smallest and brightest of the three emu-wrens. Rufous upper surface, with only slight dorsal streaking; tail shorter, one and one-third times as long as the body, and the vanes more enmeshed than in the other emu-wrens. Females similar to males, but lacking blue face and throat; sometimes show traces of pale blue in ear coverts. Widespread across arid western and central Australia in spinifex (*Triodia*, *Plectrachne* spp.) on sand plains, dunes, and rocky ranges. No subspecies recognized.

3. Southern Emu-wren
Stipiturus malachurus p. 203
The largest of the emu-wrens, rufous brown streaked with black above and plain tawny below, whiter on central belly. Males with sky-blue eyebrow, throat, and upper breast; unlike the other two species the ear coverts are light brown. Tail about twice as long as the body with vanes little enmeshed. Only found in low dense heaths or tussocky swamps in seven discrete geographically isolated subspecies across southern coastal Australia and Tasmania.

(a) *Stipiturus m. malachurus* as described above: in eastern and south-eastern coastal mainland Australia from Noosa (Qld) to the Coorong (South Australia).
(b) *Stipiturus m. westernensis* white-streaked ear coverts, slender tail feathers, olive-grey back with black streaks. South-western Australia, coastal from Jurien Bay to Esperance avoiding forest block.
(c) *Stipiturus m. littleri* similar to (a) but streaking finer and belly all rufous; smaller, with shorter tail. Tasmania.

Not illustrated
Stipiturus m. intermedius Mt Lofty Ranges, South Australia.
Stipiturus m. halmaturinus Kangaroo Island, South Australia.
Stipiturus m. parimeda Eyre Peninsula, South Australia.
Stipiturus m. hartogi Dirk Hartog Island, Western Australia.

2 juv ♂
1 juv ♂
1 ♀
2 ♂
1 ♂
2 ♀
3a ♂
3b ♂
3c ♂
3a juv ♂
3a ♀

Plate 6
Grasswrens, genus *Amytornis*

The eight species of grasswren are the largest malurids but amongst the least known. With their streaked brown plumage, they are well camouflaged for survival in the sparsely vegetated arid habitats in which they live. Only slight sexual dimorphism, with females of most species showing a rufous flank patch.

1. Striated Grasswren

Amytornis striatus p. 227

Medium-sized grasswren with black malar stripe, white throat, and streaked breast, usually found in spinifex (*Triodia* spp. or *Plectrachne* spp.) habitat that separates it from Thick-billed and Dusky Grasswrens (*A. textilis*, *A. purnelli*) where their distributions overlap. Three subspecies currently recognized.

(a) *Amytornis s. whitei* from the Pilbara region of Western Australia; generally redder, with rufous brow and tawny lower breast and belly. Single feather shows detail of shaft streaks: white with fine black margins.

(b) *Amytornis s. merrotsyi* from the northern Flinders Ranges, South Australia; black malar stripe obscured and rufous brow much reduced.

(c) (d) *Amytornis s. striatus* widespread across most of central arid Australia; (c) the north-western birds tend to be more rufous, shaft streaks white with no black margin (see single feather detail), breast often plain white; (d) south-eastern birds browner, shaft streaks white with coarse black margin (see single feather detail), breast buff. Broad black malar stripe and rufous brow of adults is lacking in immatures. Female with rufous patch across each side of lower breast (as in other *Amytornis*).

2. Eyrean Grasswren

Amytornis goyderi p. 233

Deep, blunt heavy bill of this grasswren separates it from the other three grasswren species that occur in the same area. Malar streak vestigial and obscured by streaking; no brow streak. Underside unstreaked, dull white; female with chestnut flanks. Immature browner with streaking less distinct. Found on sand dunes covered with cane-grass (*Zygochloa paradoxa*) in the Simpson and Strzlecki deserts of central Australia. No subspecies recognized.

1a ♂
1b ♂
1c ♂
1d ♀
1d ♂
1d juv
2 ♂
2 ♀
2 juv

Plate 7
Grasswrens, genus *Amytornis*

1. Dusky Grasswren
Amytornis purnelli p. 242

Brownest of all the grasswrens with no brow or malar stripes and with rufous underparts; female with rufous flanks. Found in rocky spinifex-clad hills in central Australia and north-west Queensland. Two subspecies recognized.

(a) *Amytornis p. purnelli* in central Australia, where it is distinguished from the similar Thick-billed Grasswren (*A. textilis*) by more rufous coloration and on the basis of habitat; *A. textilis* prefers saltbush-clad shrubland to the rocky hills of *A. p. purnelli*.

(b) *Amytornis p. ballarae* from the Selwyn Ranges near Mt. Isa, north-west Queensland. Underparts paler grey and back more rufous than (a). Distinguished from the Carpentarian Grasswren (*A. dorotheae*) by latter's white throat and breast contrasting with black head.

2. Thick-billed Grasswren
Amytornis textilis p. 236

Medium-sized and brown, the only grasswren to inhabit the saltbush shrublands of the arid inland plains. Similar to, but not so dark as, the Dusky Grasswren (*A. purnelli*) of the rocky hillsides. Three subspecies recognized.

(a) *Amytornis t. textilis* once widespread through the shrublands of south-western Australia, now found only near Shark Bay on the west coast. Medium brown above, streaked with black and white; tail long, particularly the males; underparts lighter brown.

(b) *Amytornis t. myall* similar to the western subspecies (a) but isolated by the Nullabor Plain to an area around the Gawler Ranges in South Australia. *A. t. myall* is slightly smaller.

(c) *Amytornis t. modestus* at one time known from central Australia through to New South Wales, recently found only within the basin of Lakes Eyre, Torrens, and Frome in South Australia. Smaller than either of the other subspecies, the tail is shorter and the underparts whiter.

3. Black Grasswren
Amytornis housei p. 218

Isolated from all other grasswrens by 500 km, Black Grasswrens live in tumbled sandstone escarpments sparsely covered with spinifex. Rarely flies and, unlike most other malurids, seldom cocks its tail. Head and upper back black streaked white; lower back chestnut, unstreaked; ventral surface of males black streaked white on throat; females, rufous belly and breast, throat as males. Juveniles uniformly dusky, sexes alike.

1a ♂
1a ♀
1b ♂
1a juv
2a ♂
2b ♂
2c ♂
2a ♀
3 juv
3 ♂
3 ♀

Plate 8
Grasswrens, genus *Amytornis*

1. Grey Grasswren
Amytornis barbatus p. 214
Distinguished from other central Australian grasswrens of similar size by the general pale appearance and the black and white facial markings. Two subspecies recognized.

(a) *Amytornis b. barbatus* light fawn-grey overall. Face white with broad white eyebrows meeting in front; a thick black stripe from base of bill through the eyes to side of neck; white ear coverts with black V below, crossing the lower throat. Tail long, narrow, and tapered. Females smaller and duller; juvenile markings less distinct. Confined to lignum-covered swamplands of Bulloo River in north-west New South Wales and northwards into Queensland.

(b) *Amytornis b. diamantina* larger, redder dorsally and less streaked than (a). Five apparently discrete populations separated from nominate subspecies by the Grey Range, on floodplains of Diamantina River, Cooper Creek, and other irregularly flowing inland river systems.

2. Carpentarian Grasswren
Amytornis dorotheae p. 224
Medium-sized grasswren of dissected sandstone ridges some 100 km inland from the Gulf of Carpentaria in south-east Northern Territory and north-west Queensland. Blackish head grading to rufous back with white margined streaks. White throat and breast grading to tawny on belly and flanks. Female as male, but belly and flanks dark chestnut. Juveniles duller. No subspecies.

3. White-throated Grasswren
Amytornis woodwardi p. 221
The largest of the grasswrens, basically similar to the Carpentarian Grasswren (*A. dorotheae*), but darker looking and with distinct black-streaked breastband. Found only on bare rocks and spinifex in the north-west half of the Arnhem Land sandstone plateau, Northern Territory.

1a ♂
1a ♀
1a juv
1b ♂
2 ♂
2 juv
2 ♀
3 ♀
3 ♂

(a) (b) (c) (d) (e) (f)

3.3 Australian vegetation. (a). *Pandanus* lining tropical rivers, Drysdale River, WA (Purple-crowned Fairy-wren *Malurus coronatus*). (b). *Eucalyptus* woodland; Dryandra, WA (Blue-breasted Fairy-wren *Malurus pulcherrimus*). (c). Mallee: many-stemmed eucalypts arising from an underground mass of woody stem, N of Murchison River, WA (Splendid Fairy-wren *Malurus splendens*). (d). Tall Open Forest—Karri (*Eucalyptus diversicolor*) forest in high rainfall regions, Manjimup, WA (Red-winged Fairy-wren *Malurus elegans*). (e). Hummock grassland, 'spinifex', near Giles, WA, covers extensive areas of arid Australia, sometimes as understorey to mallee or Acacia woodland (Rufous-crowned Emu-wren *Stipiturus ruficeps* and grasswrens *Amytornis*, except *A. textilis* and *A. barbatus*). (f). Arid shrubland, Gawler Ranges SA, dominated by shrubs of the family Chenopodiaceae, including *Atriplex* and *Maireana* (White-winged Fairy-wren *Malurus leucopterus* and Thick-billed Grasswren *Amytornis textilis*).

Table 8.1 Criteria for the category Vulnerable used to identify threatened status, the lowest of the threat categories defined in the new IUCN threatened species classification (IUCN 1994).

Vulnerable

A taxon is Vulnerable when it is not Critically Endangered or Endangered but is facing a high risk of extinction in the wild in the medium-term future, as defined by any of the following criteria (A-E):

(A) Population reduction in the form of either of the following:
- (1) An observed, estimated, inferred or suspected reduction of at least 20% during the last 10 years or three generations, whichever is the longer, based on (and specifying) any of the following:
  - (a) direct observation;
  - (b) an index of abundance appropriate for the taxon;
  - (c) a decline in area of occupancy[1], extent of occurrence[2] and/or quality of habitat;
  - (d) actual or potential levels of exploitation;
  - (e) the effects of introduced taxa, hybridization, pathogens, pollutants, competitors, or parasites.
- (2) A reduction of at least 20% projected or suspected to be met within the next 10 years or three generations, whichever is the longer, based on (and specifying) any of (b), (c), (d), or (e) above.

(B) Extent of occurrence estimated to be less than 20 000 km^2 or area of occupancy estimated to be less than 2000 km^2, and estimates indicating any two of the following:
- (1) Severely fragmented or known to exist at no more than 10 locations.
- (2) Continuing decline, inferred, observed or projected, in any of the following:
  - (a) extent of occurrence;
  - (b) area of occupancy;
  - (c) area, extent and/or quality of habitat;
  - (d) number of locations or subpopulations;
  - (e) number of mature individuals.
- (3) Extreme fluctuations in any of the following:
  - (a) extent of occurrence;
  - (b) area of occupancy;
  - (c) number of locations or subpopulations;
  - (d) number of mature individuals.

(C) Population estimated to number fewer than 10 000 mature individuals and either:
- (1) An estimated continuing decline of at least 10% within 10 years or three generations, whichever is longer, or
- (2) A continuing decline, observed, projected or inferred, in numbers of mature individuals and population structure in the form of either
  - (a) severely fragmented (i.e. no sub-population estimated to contain more than 1000 mature individuals);
  - (b) all individuals are in a single subpopulation.

(D) Population very small or restricted in the form of either of the following:
- (1) Population estimated to number less than 1000 mature individuals.
- (2) Population is characterized by an acute restriction in its area of occupancy (typically less than 100 km^2) or in the number of locations (typically less than five). Such a taxon would thus be prone to the effects of human activities (or stochastic events whose impact is increased by human activities) within a very short period of time in an unforeseeable future, and is thus capable of becoming Critically Endangered or even Extinct in a very short period.

(E) Quantitative analysis showing the probability of extinction in the wild is at least 10% within 100 years.

---

[1] Area of occupancy is the area occupied by a taxon within its extent of occurrence.
[2] Extent of occurrence is the area which encompasses all sites of present occurrence of a taxon.

Western Australia. Continuing loss of habitat on mainland due to sheep grazing and fires. Extent of occurrence less than 20 000 km^2 but the population appears to be remarkably dense (2–3 birds per ha) and probably exceeds 10 000 individuals on the north of Peron Peninsula. A continuing decline on the mainland is probable, but the population on Peron

Peninsula would appear secure as it is now a National Park and pastoral grazing has ceased. Nevertheless, a single sub-population may be vulnerable.

6. Thick-billed Grasswren (*A. textilis modestus*). Formerly known from New South Wales and Northern Territory but not seen there in the past 50 years (Garnett 1992*a*). Distribution in saltbush around basins of Lakes Eyre, Torrens, and Frome in South Australia; common where found, but scattered patchily. Estimated population is fewer than 10 000 individuals and the habitat is likely to decrease with continuted grazing; population probably fragmented into small sub-populations of fewer than 1000 individuals.

Data deficient

7. Carpentarian Grasswren (*Amytornis dorotheae*) is known from several widely separated populations on isolated dissected sandstone escarpments inland from the Gulf of Carpentaria in the Northern Territory and Queensland. Lives in mature clumps of *Triodia* hummock grassland in low open woodland with or without shrubby understorey. McKean and Martin (1989) consider that fire as part of pastoral management is the main danger. Occurrence in less than 2000 km^2 with extent and quality of habitat diminished by fire as part of pastoral management. Population estimated to be fewer than 10 000, severely fragmented with no population exceeding 1000 mature individuals.

8. Striated Grasswren (*Amytornis striatus striatus*). Although still widespread in *Triodia* hummock grassland throughout most of arid central Australia, the populations in the east and south-east are now fragmented and declining, following clearing for agriculture that has left little of the original mallee/spinifex vegetation. Insufficiently known.

9. Striated Grasswren (*Amytornis striatus merrotsyi*). Occurs in isolated pockets of the Flinders Ranges in South Australia, where hummock grassland occurs in rocky gullies with an overstorey of trees and shrubs. Occurrence in less than 2000 km^2 with area and extent decreasing due to fires and grazing by goats; populations fragmented, and not containing more than 1000 individuals. Insufficiently known.

10. Thick-billed Grasswren (*Amytornis textilis myall*). Extent of its occurrence less than 20 000 km^2 in the Gawler Ranges and northern Eyre Peninsula, South Australia. Habitat area and quality likely to diminish with continued grazing of the preferred chenopod shrubland. Population unlikely to exceed 10 000 individuals, regarded as severely fragmented. Their range appears to have contracted during the past century, as has that of the other two subspecies of *Amytornis textilis*. Insufficiently known.

Additional considerations

Outside the strict crtieria suggested by the IUCN (1994) we consider that taxa confined to one small island (*Stipiturus malachurus hartogi* and *Malurus leucopterus leucopterus* on Dirk Hartog Island and *M. l. edouardi* on Barrow Island) or to a very specific, localized habitat (*Amytornis barbatus*) require that conservation authorities maintain a concerned watch, even though no obvious decline has been reported recently.

PART II

Species accounts

Genus *Malurus* L. P. Vieillot

Malurus L. P. Vieillot, 1816. *Analyse d'une Nouvelle Ornithologie Elementaire*, p. 69.

Synonyms *Malacurus* Gloger, 1842; *Todopsis* Bonaparte, 1854; *Chenorhamphus* Oustalet, 1878; *Musciparus* Reichenow, 1897; *Hallornis* Mathews, 1912; *Leggeornis* Mathews, 1912; *Rosina* Mathews, 1912; *Ryania* Mathews, 1912; *Nesomalurus* Mathews, 1913; *Devisornis* Mathews, 1917; *Psitodos* Mathews, 1928.

Twelve species with 30 subspecies characterized by marked sexual dimorphism; plumage unstreaked. Males in nuptial plumage have striking blue, black, red, or white feathering with most females mainly brown. Exceptions are three (of six) subspecies of the New Guinean White-shouldered Fairy-wren (*Malurus alboscapulatus*), in which the female and male have similar black-and-white plumages. All species are small insectivores resident in the same place throughout the year. At least one member of the genus occurs in every part of Australia and most parts of New Guinea except the highlands.

Fairy-wrens have small rounded wings and do not fly strongly; P10, the outermost of the 10 primaries, is the shortest; P3–6 are longest, with P1–7 much the same length. Most *Malurus* have long tails which they carry cocked at an angle to the body; the outer of five pairs of feathers are one-third the length of the central three pairs; the fourth pair is 6 mm shorter than the central ones (*M. cyaneus*) giving a fan shape most obvious in *M. amabilis*.

Most foraging is by 'hop-search'—hopping on their long legs over the ground or from branch to branch gleaning as they go. Usually forage, allopreen, and roost as family parties. Breed co-operatively and may produce several broods in an extended season.

In the text that follows, we use the order in Christidis and Boles (1994), except that the first species discussed is *M. cyaneus*, the first species that was found.

Superb Fairy-wren *Malurus cyaneus* (Ellis, 1782)

Motacilla cyanea W. W. Ellis, 1782. *An authentic narrative of a voyage performed by Captain Cook and Capt. Clerke in His Majesty's Ships Resolution and Discovery during the years 1776, 1777, 1778, 1779, and 1780.* Vol. **1**, p. 22.

PLATE 1

Polytypic. Two subspecies. *Malurus cyaneus cyaneus* (Ellis, 1782); *Malurus cyaneus cyanochlamys* Sharpe, 1881.

Description

M. c. cyaneus

ADULT NUPTIAL MALE: crown, ear tufts, and lower back, iridescent azure blue; collar, eyestripe, upper back, scapulars, lower back, and upper tail coverts, black; wings, dark grey-brown; tail, dusky-blue, tip white, becomes worn; breast, blue-black edged with black; belly and flanks, greyish white; bill, black; eye, blackish brown; feet, dark brown.

ADULT FEMALE: crown, back, and wings, mid grey-brown; tail, brown with greenish tinge, white tip becomes worn; lores and feathers round eye, light rufous; chin and belly, greyish white; eyes, as male; bill, orange-brown; feet, brown.

ADULT ECLIPSE MALE: similar to female, but bill and lores remain blackish and tail is darker.

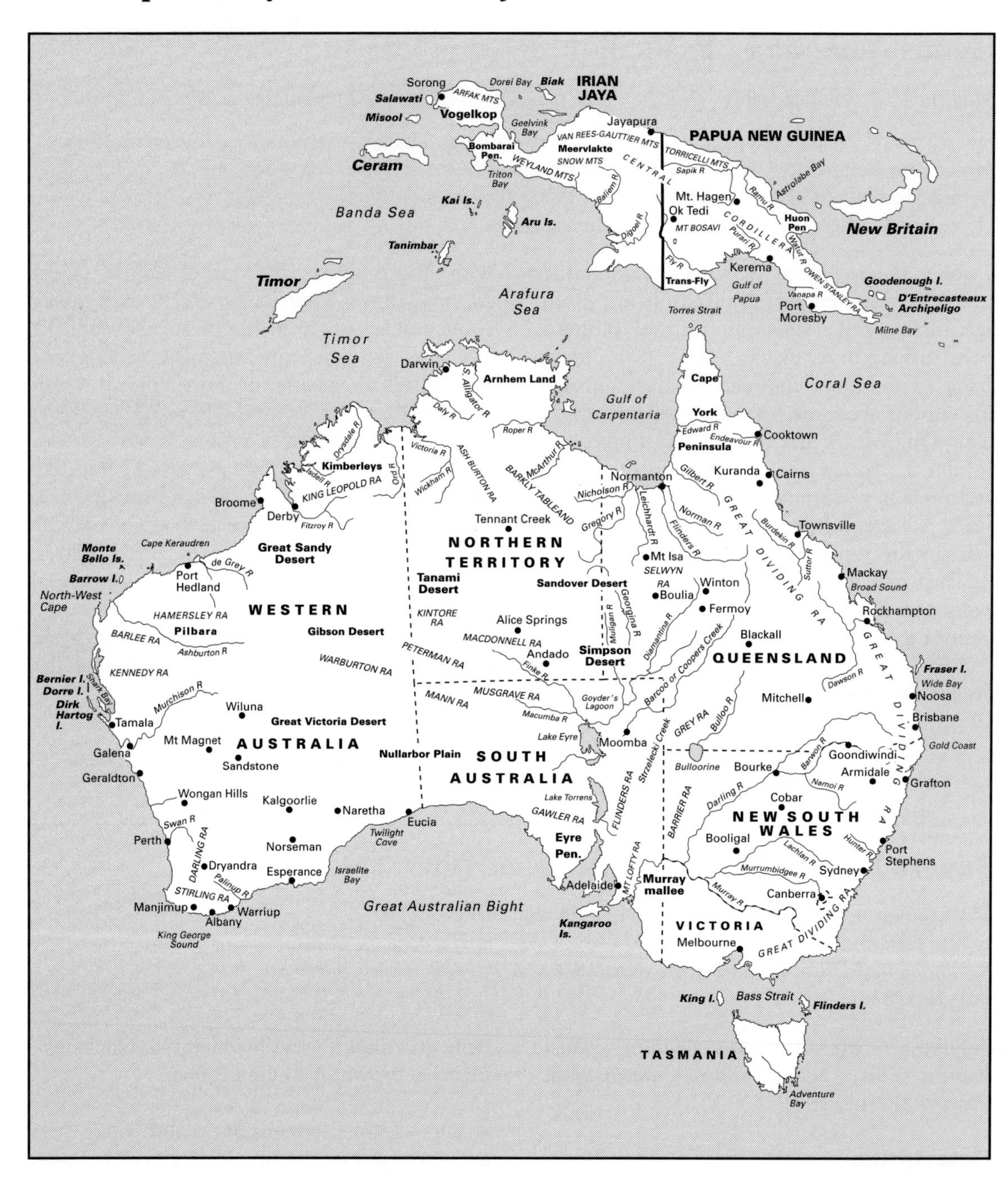

IMMATURE: both sexes indistinguishable from adult female once fledgling has grown its full tail, except for paler lores; males develop darker bills and lores during their first winter.

MOULT: males and females moult all feathers every autumn after breeding when most males enter an eclipse female-like plumage. Males have a second (prenuptial) moult in winter/spring into coloured nuptial plumage. This involves only the body feathers and not remiges or rectrices. Onset varies over 140 days within one population, according to age, status, and condition of individual (see Table; Rowley

Table: Mean date (SD) on which male Superb Fairy-wrens (*Malurus cyaneus*) of different ages began their prenuptial moult

male age (years)	onset of prenuptial moult	
	mean days since 1 Jan	mean date
1	257.6 (14.5)	15 Sept
2	234.3 (15.0)	22 Aug
3	220.0 (18.8)	8 Aug
4	214.0 (20.1)	2 Aug
5 and 5+	192.7 (19.8)	12 July

From data of Mulder and Magrath 1994

1965; Mulder and Magrath 1994). Males may not achieve perfect nuptial plumage at 1 year. Rarely, old males moult directly from one nuptial plumage into the next (5 of 424, Mulder and Magrath 1994).

History and subspecies

In the past, four different species (Campbell 1901*a,b*) and nine subspecies (Mathews 1913) have been named; two subspecies are currently recognized, *Malurus cyaneus cyaneus* (Ellis, 1782) in Tas. and the Bass Strait islands, *M. c. cyanochlamys* Sharpe, 1881 in SE mainland Australia. Tasmanian birds are larger and darker than those on the mainland, with longer tails; males are deeper azure blue on the crown, ear tufts, and mantle, and females are greyer on the breast and flanks (Plate 1). Males from King Is. in Bass Strait are a deeper, more brilliant azure, and birds from Flinders Is. are intermediate between Tasmanian and King Is. birds. Mainland birds are smaller and paler; from SE Qld to Eyre Peninsula in SA, variation is continuous, but birds from the ends of the range differ less from each other than either differs from Tasmanian birds. Birds from Qld are the smallest; pale silvery blue crown, ear tufts, and mantle of male contrast with black breast and white belly. Southwards, there is a trend to richer sky-blue crown, ear tufts, and mantle, blue-black breast, and greyer flanks and belly.

Ellis' original description in 1782 was of the Tasmanian form, as *Motacilla cyanea*. After the First Fleet arrived at Port Jackson (Sydney) in 1788, various accounts illustrated the local form of the Superb Warbler (Fig. 1.1), and in 1790, George Shaw named it *Motacilla superba*. Gould's 1865 Handbook distinguished *Malurus cyaneus* of the mainland east coast from the Tasmanian form which he named *M. longicaudus*, although the form originally described as *cyaneus* was from Tasmania. Sharpe separated the Queensland birds as *M. cyanochlamys* (1881). A. J. North pointed out that *cyaneus* was applied to the Tasmanian form in 1782 (North 1901*b*). Then in 1904, North realized that while Shaw had applied *superba* to Port Jackson wrens in 1790, he had earlier (1789) used it for Tasmanian birds; it was thus not applicable to mainland birds; North proposed *australis*. A. J. Campbell named the darker blue birds of King Is. as a separate species, *M. elizabethae* (1901*b*).

Mathews (1913) treated these species as subspecies, but added more, to a grand total of nine: S. Tas., *cyaneus*; N. Tas., *fletcherae*; King Is., *elizabethae*; Flinders Is., *samueli*; Vic., *henriettae*; SA, *leggei*; Kangaroo Is., *ashbyi*; S. NSW, *australis*; N. NSW and Qld, *cyanochlamys*. In 1930, he reduced this to eight, merging *leggei* with *henriettae*.

Mack's revision (1934) maintained four subspecies, *cyaneus*, Tas.; *elizabethae*, King and Flinders Is.; *australis*, NSW, Vic., SA; *cyanochlamys*, the paler Qld forms. This separation of subspecies by state reflects the separate jurisdictions of the state museums—each interested chiefly in specimens from its own state, so that collections reflected the differences between populations separated by distance, rather than showing that there was a zone of intergradation between two forms. Schodde (1982*b*) demonstrated that variation on the mainland was continuous across putative subspecific boundaries. He grouped all mainland forms in *cyanochlamys* and the Tasmanian and Bass Strait forms in *cyaneus*, although the forms from the three islands differ, reflecting alternate separation and contact as sea levels rose and fell. More objective assessment of the degree of genetic differentiation between apparently different forms may be possible when DNA sequences are known.

Measurements and weights

Malurus cyaneus cyaneus: wing, M (n = 50) 54.0 ± 1.8, F (n = 19) 52.6 ± 1.3; tail, M (n = 50) 61.6 ± 3.1, F (n = 19) 63.8 ± 3.7; bill, M (n = 50) 9.2 ± 0.5, F (n = 19) 8.9 ± 0.6; tarsus, M (n = 50) 24.0 ± 0.9, F (n = 19) 23.7 ± 0.6; weight, M (n = 20) 10.0–13.8 (11.3 ± 0.8), F (n = 31) 9.0–12.2 (10.5 ± 0.8). *Malurus cyaneus cyanochlamys*: wing, M (n = 143) 51.5 ± 1.5, F (n = 21) 49.9 ± 1.6; tail, M (n = 143) 56.6 ± 2.3, F (n = 21) 59.7 ± 3.2; bill, M (n = 143) 8.2 ± 0.6, F (n = 21) 7.8 ± 0.5; tarsus, M (n = 143) 22.8 ± 0.9, F (n = 21) 22.7 ± 0.9; weight, M (n = 46) 8.5–10.5 (9.4 ± 0.5), F (n = 15) 8.0–10.0 (9.1 ± 0.6). (Measurements of birds from S NSW and Vic., not including smaller birds from Qld; weights of *M. c. cyanochlamys* from Hardy 1983.)

Field characters

Male nuptial plumage of blue and black contrasting with white belly and brown wings distinguish *M. cyaneus* from the much bluer Splendid Fairy-wren (*M. splendens*); Variegated Fairy-wren (*M. lamberti*) has distinct chestnut shoulders and longer tail. In inland eastern Australia, several *Malurus* occur, and whereas nuptial males are obviously distinct, males in eclipse plumage and females may be difficult to separate. In eclipse male plumage, the brown wings of *M. cyaneus* distinguish it from *M. splendens* which has blue primaries; *M. lamberti* has a much longer tail, and the face of eclipse males often shows traces of the characteristic blue. Female and immature plumages of *M. cyaneus* are hard to separate; the greenish tinge to the tail of adult females is absent in immatures. Longer, bluer tail and more chestnut lores of female *M. lamberti* distinguish it from female *M. cyaneus*. Female *M. splendens* has a bluer tail than female *M. cyaneus*, but there is little overlap in range.

Voice

Tape-recorded (ANWC). Characteristic *Malurus* reel of song delivered throughout the day by both sexes (more frequently in breeding season), appears to advertise territorial ownership (see sonogram); male song lasts longer; partners may duet in response to reel. Two types of reel identified by Langmore and Mulder (1992): Type I is the typical reel given by both sexes (as in sonogram). Type II given only by male, frequently triggered by calls of potential predators (see Fig. 5.9). Sociable birds, rarely seen alone, they maintain contact with brief 'chet' calls. Alarm call a sharp brief 'chit'; incubating females utter a soft brooding 'purr'; churring threat with threat posture to intruder, including cuckoo.

Range and status

Resident E and SE Australia along coastline and adjacent highlands and islands to inland western slopes and plains. N of Brisbane, absent from coast but continues inland, N to Dawson River and W to Blackall. Throughout Vic. except for NW mallee. Along Murray, Murrumbidgee, Barwon, and Lachlan R to Booligal. Coastal SA to Adelaide, Kangaroo Is., and southern tip of Eyre Peninsula. Rarely found in country cleared for wheat and sheep. Common in suitable shrubby habitat, thrives in suburban Sydney, Melbourne, and Canberra.

Habitat and general habits

Mainly found in areas with annual rainfall greater than 400 mm; much of this country which carried open forest before European settlement is now cleared for farming, so the preferred habitat has been lost; in places, shrubby exotic weeds (Lantana *Lantana camara*, brambles *Rubus vulgaris* and rose briars *Rosa rubiginosa*) provide acceptable substitutes; in cities, exotic shrubs in gardens also suitable. Also persists in drier areas (annual rainfall more than 200 mm), along inland rivers (Murray, Barwon, Lachlan, and Murrumbidgee) and in inland swampy areas such as the Macquarie Marshes and around Booligal. Not in closed

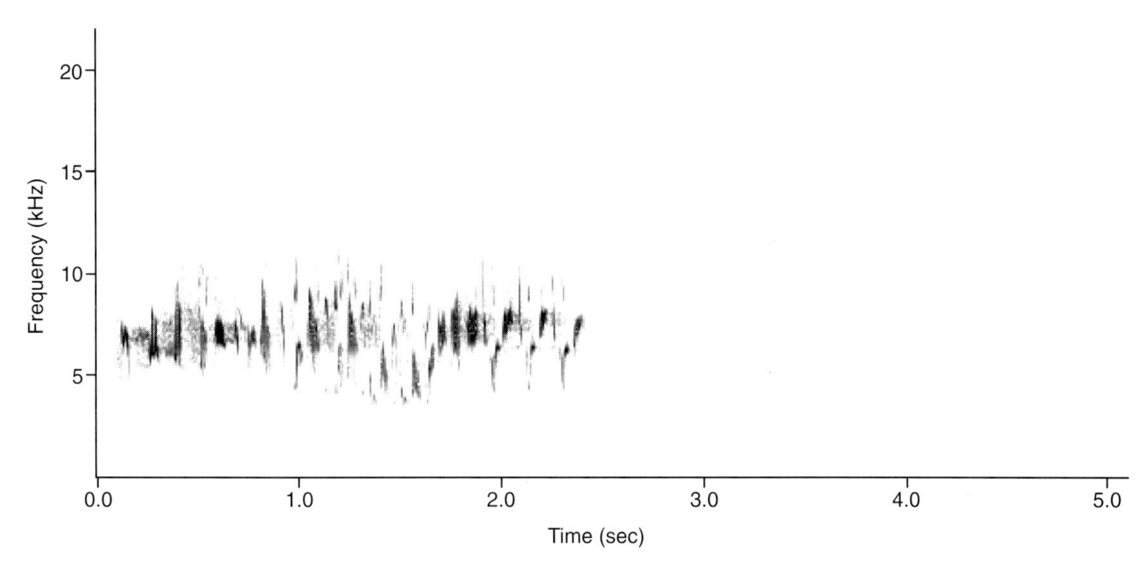

Superb Fairy-wren (*Malurus cyaneus*)
M. c. cyanochlamys, Type I song.
Naomi Langmore, Australian National Botanic Gardens, Canberra.

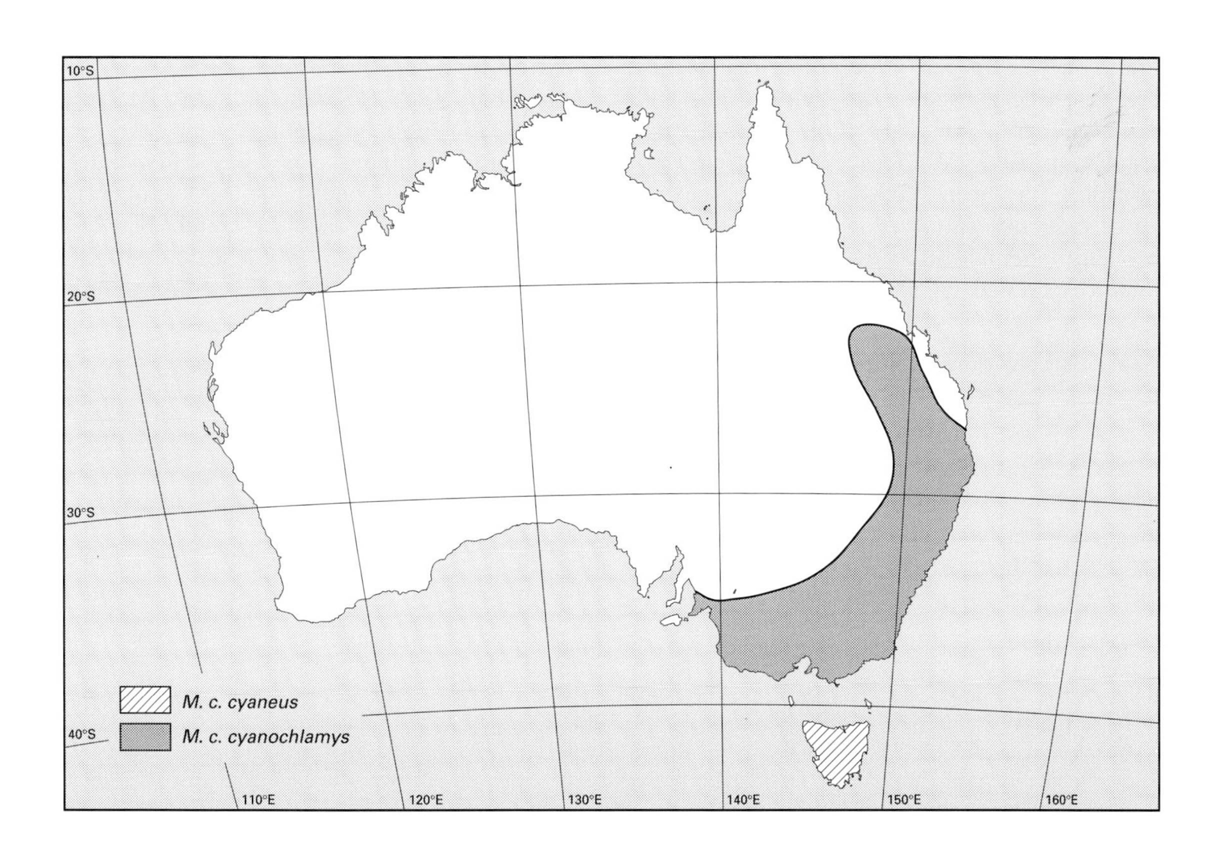

forest nor in sub-alpine and alpine environments. Weak fliers, they seldom venture far from shrubby cover; they forage on the ground and in low shrubs, hopping rapidly on long legs over open areas, dashing to cover when alarmed; tail characteristically held cocked over the back. Diet mainly insects: grasshoppers, shield-bugs, ants, flies, weevils, and various larvae; they also eat a few small seeds (Rowley 1965; Barker and Vestjens 1990). Co-operative breeding species, generally seen in pairs or small groups of 3–5 adults. Studies of colour-banded populations show that groups maintain the same territory throughout year, defended by all members (Rowley 1965; Mulder 1992; Nias and Ford 1992). Group members regularly allopreen; they roost together in dense cover, lined side by side on a horizontal branch (Rowley 1965).

Displays and breeding behaviour

Socially monogamous but sexually promiscuous, they remain paired while both partners survive; divorce is rare (4.2%, Mulder 1992); males seek extra-pair copulations (EPC) but return to social partner and help to feed and raise young in her nest (76% of offspring result from EPC, Mulder *et al.* 1994); no egg-dumping has been recorded. Males seeking EPC frequently carry yellow petals as they approach the female (Rowley 1991). A courting male displays to female with Face Fan; may give Blue-and-Black display with blue feathers at rear of crown and black feathers of nape erect, so that crown appears flat (Pickett 1995). Female may crouch and flutter part-opened wings in invitation display; copulation is rarely seen. Wing-fluttering is also an appeasement display given by either sex. Territory defence is by song battle; persistent trespassers are chased by all group members; a retiring intruder may use 'Impeded Flight' with the black feathers of the back raised (Rowley 1965). Stationary threat postures are directed at intruders (including cuckoos), with body feathers part erected, wings held out from body, head and tail lowered, sometimes with churring call, followed by physical attack (Rowley 1963). Described using Rodent-run distraction display (Rowley 1962; see Fig. 5.4).

Breeding and life cycle

Analysis of 677 nest records found no eggs were laid in April, May, or June; most were laid from Sept. to Jan (see Fig. 7.1(b)). In drier areas (rainfall 200–400 mm) breeding may start earlier (July) and continue into March in favourable years (Tidemann and Marples 1987). Variation in timing occurs in SE Australia; in Canberra, over 5 years, first clutch laid between 14 August and 19 October (a drought year), and the last clutch between 3 January and 4 February (Mulder 1992). Territorial throughout year (0.5–2.0 ha). Only

Table: Aspects of the life history of Superb Fairy-wrens (*Malurus cyaneus*) at four different sites

	Canberra (Rowley 1965)	Canberra (Mulder 1992)	Booligal (Tidemann 1983)	Armidale (Nias 1987)
latitude	36°05′S	36°05′S	33°43′S	30°35′S
longitude	149°15′E	149°15′E	144°47′E	151°44′E
territory size (ha)	—	0.6	1.6	1.4
mean clutch size	3.4	3.2	3.1	3.4
% eggs hatch	66.3	63.3	67.0	63.0
% eggs fledge	56.0	32.6	46.0	39.0
% nests—cuckoo*	3.4	3.0	17.6	13.0
flg/gp/year	*c.* 4	3.9	2.6	2.6
% flg to 1 year	22.0	28.1	11.0	44.0
% ad survival	64.0	69.0	51.0	45.0

* Horsfield's Bronze-Cuckoo (*Chrysococcyx basalis*)

the female builds the domed, oval nest (120 mm high × 80 mm wide) 0.2–1.3 m above ground and well hidden at the edge of dense cover. The initial framework is of spider's web; the outer structure is of coarse grass and fine roots, with an inner layer of finer grasses, lined with soft materials (feathers and fur). Clutch usually 3 (2–4), eggs (16–19 × 12–14 mm, n = 42) tapered oval, pinkish white speckled with red-brown at the broader end; laid daily, incubation by female starts with the last egg. The eggs hatch within 24 hr, 14 days later; altricial nestlings are fed and faecal sacs removed by all group members for 10–14 days. Hatchlings are blind, naked, and red. They darken rapidly as the feathers develop; primaries emerge day 2/3; eyes open day 5/6; young are fully feathered by day 10, but the wings and tail do not complete growth until day 20. Fledglings become independent of adult feeding by *c.* day 40, but young birds remain in the family group long after independence; groups may raise two broods (rarely three) in one season; early hatched young may feed later siblings (Rowley 1965). Males sexually mature at 1 year, usually remain in natal group for 1 or more years as helper before inheriting the senior position, or moving next door to a vacancy. Helpers feed nestlings, care for dependent fledglings, defend territory, and repel predators or cuckoos; also obtain some EPCs. Female helpers are rare; females may breed when 1 year old; some disperse in autumn, most in early spring (Mulder 1995). See Table for comparison of life cycle features from four studies of marked individuals at three different localities.

Splendid Fairy-wren *Malurus splendens* (Quoy and Gaimard, 1830)

Saxicola splendens Quoy and Gaimard, 1830. In J. S. C. Dumont d'Urville, *Voyage de la corvette l'Astrolabe, execute par ordre du Roi pendant les annees 1826, 1827, 1828, 1829, etc.* Vol. **1**, Zoologie, p. 197.

PLATE 1

Polytypic. Three subspecies. *Malurus splendens splendens* (Quoy and Gaimard, 1830); *Malurus splendens melanotus* Gould, 1841; *Malurus splendens callainus* Gould, 1867.

Description

M. s. splendens

ADULT NUPTIAL MALE: head, back, scapulars, and ventral surface, iridescent, deep cobalt blue; wings, grey-brown, edges of primaries turquoise; tail, cobalt blue; black lores and stripe through eye to black collar; ear tufts, oval, sky-blue; black pectoral band (3–5 mm broad); bill, black; eyes, dark brown; feet, black.

ADULT FEMALE: crown, back, and wings, grey-brown; tail, greyish blue-green; lores and feathers round eye, light rufous; chin and belly, greyish white; bill and feet, orange-brown; eyes, dark brown.

ADULT ECLIPSE MALE: grey-brown dorsally and white below as female; tail, cobalt blue; primaries, edged turquoise; bill and lores, black.

IMMATURE: resembles female until first winter when male develops turquoise primaries; later bill and lores darken.

MOULT: both sexes moult all feathers every autumn after breeding (Feb–Apr), when most males enter eclipse female-like plumage; males have second moult in winter/spring (body feathers, not remiges or rectrices) into coloured nuptial plumage. Onset varies according to age, status, and condition of individual; the older the bird, the earlier the nuptial moult. Immatures acquire blue primaries before their bills darken; males may not achieve perfect nuptial plumage at 1 year; their bellies may remain greyish-white. Rarely, a few old males remain 'blue' all year, moulting directly from one nuptial plumage to next (Rowley 1981*a*).

History and subspecies

Three subspecies currently recognized (Schodde 1982*b*). Birds from SW Australia *M. s. splendens* have uniformly cobalt blue back and scapulars, and long, slender tail; birds from central Australia *M. s. callainus* have black band across lower back and black scapulars, blue coloration turquoise above and cobalt below; black pectoral band 3–6 mm broad. Birds from further east *M. s. melanotus* uniformly blue above and below as in nominate race but black backs and scapulars as *M. s. callainus*; black pectoral band is only 1–3 mm broad, mid-belly whitish. Females of all three subspecies similar, but tail of *M. s. splendens* longer.

Originally described as three separate species. The first specimens were taken at King George Sound in SW Australia by Quoy and Gaimard in October 1826, and described as *Saxicola splendens* in 1830. In 1833, John Gould described specimens from the Swan River colony (now Perth, WA) as *M. pectoralis*, emphasizing the black breast-band, but by 1841, he recognized that his *pectoralis* was the same as Quoy and Gaimard's *splendens*; his vernacular name, Banded Superb Warbler, remained in common use. At the same time, he described *M. melanotus*, the Black-backed Superb Warbler, collected from mallee scrub near Mannum on the Murray R. in SA, and differing from *splendens* in being a lighter blue overall, with the lower back and scapulars black; *splendens* is uniformly cobalt blue over the back.

In 1865, Gould received from Samuel White specimens of a third brilliant blue bird collected on the western shore of Spencer Gulf, near the site of present-day Whyalla. He described it in 1867 as *M. callainus*, with the vernacular name of Turquoise Wren, for its turquoise head and upper parts, and pale turquoise ear tufts; the breast was cobalt and the lower back black as in *melanotus*.

The common names of these three forms, Banded, Black-backed, and Turquoise Superb Warblers or Wrens, were in general use for more than 100 years, although debate about whether they were separate species or races of one species continued. Confusion over the origin of specimens led Campbell (1902) to describe a fourth species *whitei* intermediate between *melanotus* and *callainus*. Mathews (1912*a*) treated *callainus* as a subspecies of *melanotus*, and described another subspecies *victoriae* from Victoria. He recognized variation in *splendens* in WA with *M. s. riordani* from northern specimens. Subsequently (1913) he recognized only two species, *M. melanotus* with two subspecies *M. m. melanotus* and *M. m. callainus*, and *M. splendens*, with *M. s. splendens* and *M. s. riordani*. Later, (1922–3, 1930), he maintained these two species, while the 1926 RAOU Checklist maintained three, as did Mack's 1934 revision. Mack recognized no subspecies of *melanotus*, but resurrected Campbell's *whitei* as a subspecies of *callainus* for specimens from near Port Augusta, SA. In *splendens*, he created a new subspecies *aridus* for inland specimens from the Stirling Ranges to Wiluna, WA, and included in it Mathews' *riordani*.

Differences between the three forms were readily apparent in the areas where most people encountered them. It was not until the areas of potential overlap of the three distributions were explored that intermediate (hybrid) forms were recognized. It became clear that some of the variation that gave rise to the subspecies of Mathews and Mack occurs in specimens from areas where *melanotus* and *callainus* or *callainus* and *splendens* intergrade. Specimens from desert areas of WA (Great Victoria Desert, Gibson Desert, and near Wiluna) have shown that areas geographically intermediate between the known distributions of *splendens* and *callainus* have populations with intermediate characteristics, and that *splendens* near Wiluna are paler than typical *splendens* and affected by gene flow from *callainus* (Serventy and Whittell 1967; Ford 1975; Schodde 1982*b*).

The *melanotus–callainus* intergradation is more complex; *M. s. callainus* occurs on the western side of the Flinders Ra. and *melanotus* occurs to the east. Reid *et al.* (1977) examined birds from the Flinders Ra. where the two might be expected to merge. They compared *callainus* from the Flinders Ra. with 'typical' *callainus* from further west on Eyre Peninsula and from near Lake Torrens, and Flinders Ra. *melanotus* with specimens from the Murray. They found 'no evidence of contact'. Schodde (1982*b*) demonstrated that the zone of intergradation between *callainus* and *melanotus* is broad, and that the 'typical' *callainus* that Reid *et al.* used as their standard were, in fact, hybrids; the three forms are clearly not isolated from each other. In the 1975 RAOU Interim List, Parker treated the three forms as subspecies of *M. splendens*.

Measurements and weights

Malurus splendens splendens: wing, M (n = 35) 52.4 ± 1.3, F (n = 19) 51.4 ± 1.6; tail, M (n = 35) 60.4 ± 2.2, F (n = 19) 61.4 ± 3.1; bill, M (n = 35) 8.7 ± 0.6, F (n = 19) 8.3 ± 0.5; tarsus, M (n = 35) 23.5 ± 0.9, F (n = 19) 22.7 ± 0.9; weight, M (n = 43) 9.2–11.1 (10.6 ± 0.4), F (n = 43) 8.6–10.2 (9.6 ± 0.4). *Malurus splendens callainus*: wing, M (n = 63) 50.8 ± 1.9, F (n = 22) 49.5 ± 1.3; tail, M (n = 63) 56.6 ± 2.5, F (n = 22) 56.1 ± 2.7; bill, M (n = 63) 8.4 ± 0.5, F (n = 22) 8.0 ± 0.5; tarsus, M (n = 63) 22.1 ± 1.0, F (n = 22) 21.4 ± 1.0; weight, M (n = 20) 8.0–10.0 (9.0 ± 0.8), F (n = 4) 7.6–9.0 (8.5 ± 0.6). *Malurus splendens melanotus*: wing, M (n = 62) 51.1 ± 1.4, F (n = 15) 49.1 ± 1.3; tail, M (n = 62) 56.2 ± 2.5, F (n = 15) 56.2 ± 2.5; bill, M (n = 62) 8.3 ± 0.4, F (n = 15) 8.1 ± 0.6; tarsus, M (n = 62) 22.0 ± 1.0, F (n = 15) 21.2 ± 0.6; weight, M (n = 18) 7.9–10.3 (9.0 ± 0.8), F (n = 1) 8.7. (Weights from museum specimens and field measurements by authors).

Field characters

Loud, relatively harsh song reel is unmistakeable. Blue underparts and wings of the male separate *M. splendens* from Superb Fairy-wren (*M. cyaneus*). Nuptial male of *M. splendens* and chestnut-shouldered fairy-wrens (*M. lamberti*, *M. pulcherrimus*, and *M. elegans*) clearly separable. In eclipse, blue primaries and bluer tail distinguish *M. splendens* from other *Malurus*. Female *M. splendens* very similar to *M. cyaneus* but greyer above; blue-green tail distinguishes from other *Malurus*. Female chestnut-shouldered fairy-wrens have richer chestnut lores and longer tails than *M. splendens*. In sympatry with Red-winged Fairy-wren (*M. elegans*), *M. s. splendens* occurs in less dense habitat at forest margins. In arid areas, *M. s. callainus* occurs in thick mulga scrub with shrubby understorey, Variegated Fairy-wren (*M. lamberti*) in more variable scrub, and White-winged Fairy-wren (*M. leucopterus*) in specialized, low chenopod and *Zygochloa* habitat. In E, *M. s. melanotus* occurs in mulga scrub and mallee.

Voice

Tape-recorded (ANWC, Baker 1995). Song is a typical *Malurus* reel, with a series of loud trills preceded by a variable number of softer brief introductory trills or whistles (see sonogram). Given by both sexes (breeders and helpers), the song identifies individuals, maintains contact within the group, and advertises territory, especially in boundary disputes. The reel is louder and harsher than *M. cyaneus*, *M. elegans*, *M. lamberti*, and *M. leucopterus*, varies in length, loudness, and structure between and within individuals (Payne *et al.* 1988). Foraging groups maintain contact with soft single 'trrt', sharpening to 'tsit' in alarm; churring threat with threat posture to intruder (Fig. 5.3), including cuckoo (Payne *et al.* 1985). Softer and shorter version of reel also used as contact call; incubating female utters a soft brooding 'purr'.

Range and status

Resident throughout arid and semi-arid inland shrublands in all states except Tas.; in WA and SA reaches coast. Replaced by *M. cyaneus* at S tip of Eyre Peninsula, and by *M. elegans* in SW forests of WA; extends N to Shark Bay in WA, to N edge of Macdonnell Ra. in centre, to Winton in Queensland. Not found in Simpson Desert (Blakers *et al.* 1984). Schodde (1982*b*) suggests that ancestral stock split into three

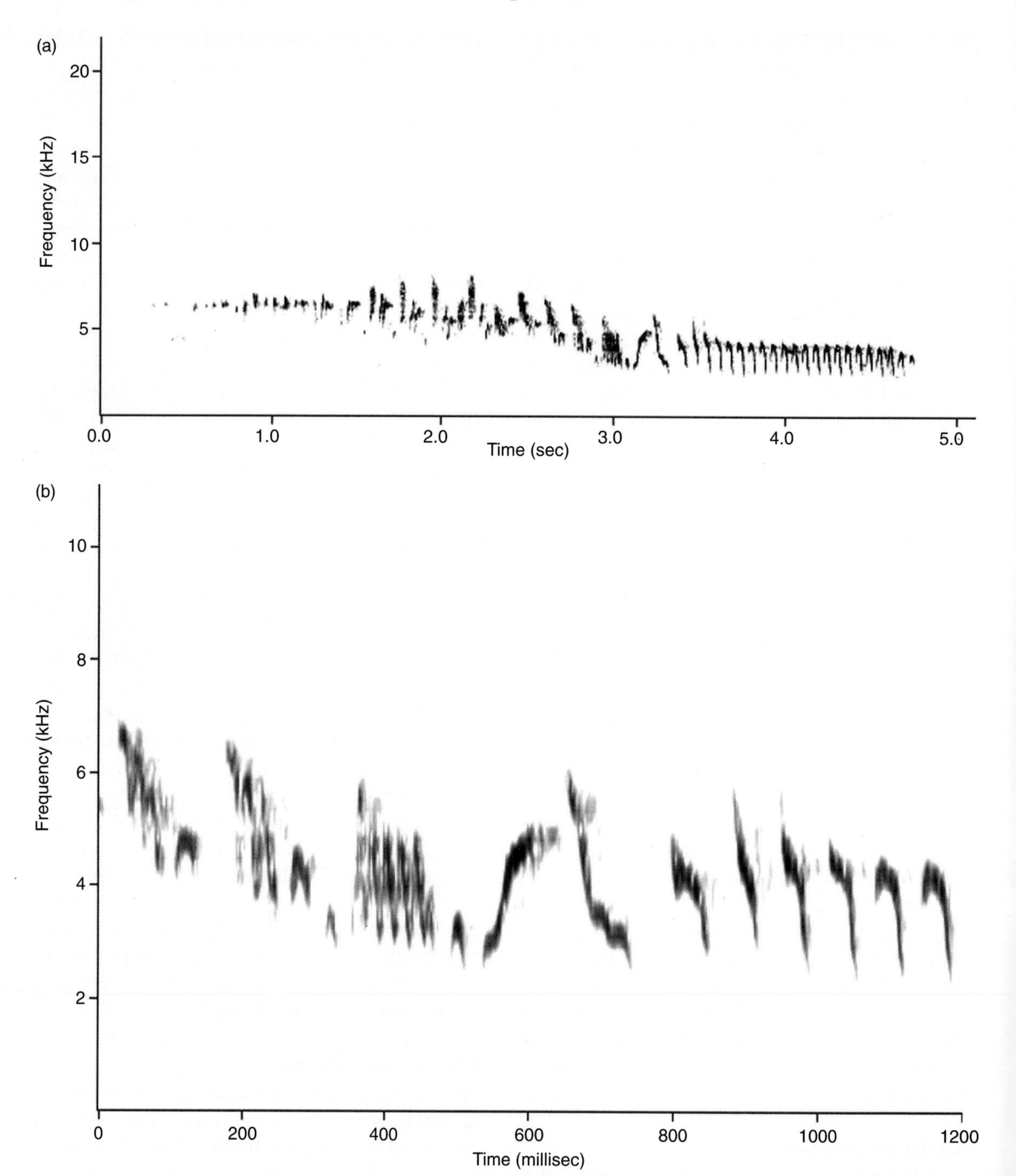

Splendid Fairy-wren (*Malurus splendens*)
(a) *M. s. splendens*, whole song, lasting more than 4 sec (b) detail of song (1.2 sec), at ca. 3 sec from the start of (a), with expanded time and frequency scales.
John Hutchinson, Balingup, Western Australia.

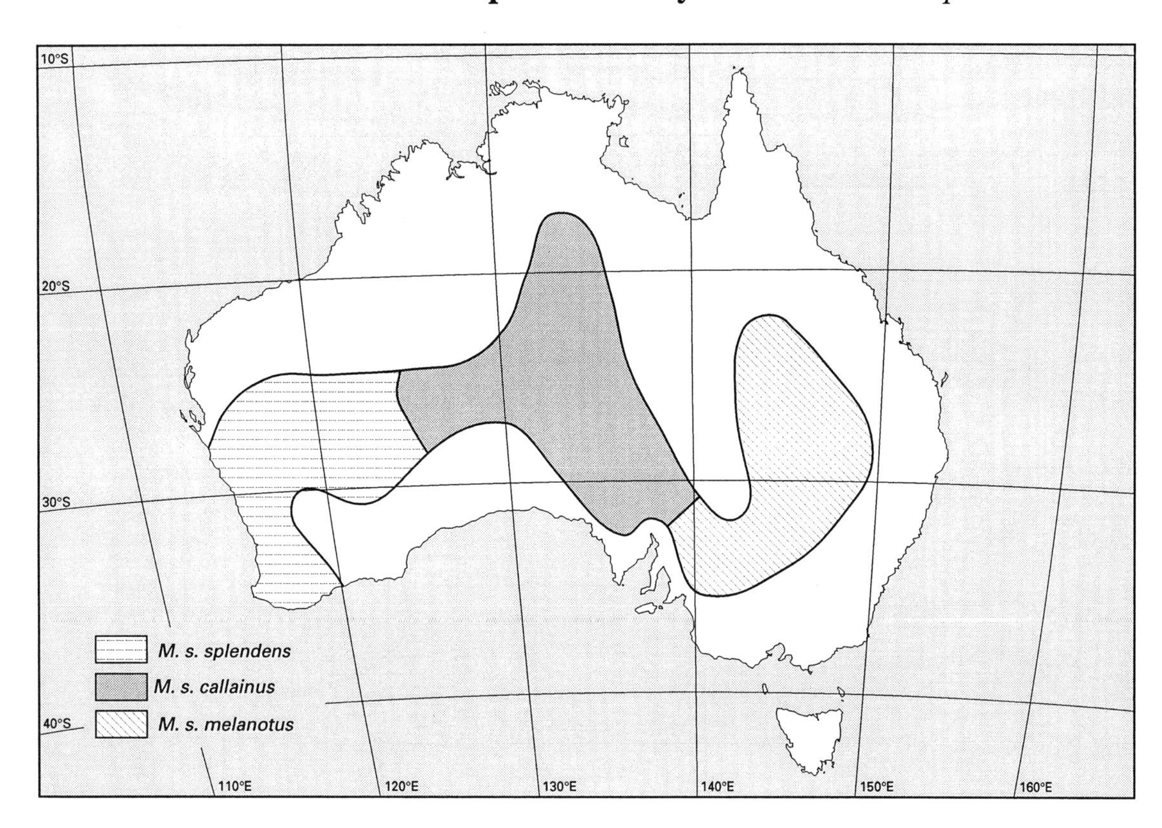

during last arid glacial period, and that these evolved into the three forms known today which have spread and now hybridize where they meet. Common in suitable habitat.

Habitat and general habits

Extremely wide and varied distribution, with populations throughout the arid interior of the continent and the mallee woodland. Live in heathy vegetation overtopped by trees or shrubs throughout drier, largely temperate parts of W and S Australia. Stronger fliers than most other fairy-wrens; versatile foragers over open areas and shrubs by hop-search, pounce, glean, and occasional tower flight in pursuit of swarming termites. Sometimes forage for insects in canopy of flowering eucalypts. Main food: ants, grasshoppers, crickets, spiders, and bugs (Barker and Vestjens 1990; Rowley *et al.* 1991). Breeds co-operatively, and is generally seen in pairs or small groups; studies of colour-banded populations show that groups maintain the same territory throughout year, defended by all members. Group members regularly allopreen, and roost together in dense cover, lined side by side on a horizontal branch. Groups (mean size ± SD 3.3 ± 1.2, range 2–8) remain in their territories from year to year (Rowley 1981*a*).

In woodland–heath habitat, the mean size of contiguous territories was 4.4 ha (range 1.0–8.7; Brooker and Rowley 1995). As vegetation became denser after fire, population density increased, more territories were established, and mean territory size became smaller. Territory size also varied from year to year according to the number of males in the group. When a group grew in size as males matured and did not disperse, the territory increased in area at the expense of adjacent territories. If group size decreased due to death or dispersal of one or more males, territory area contracted.

Displays and breeding behaviour

Socially monogamous but sexually promiscuous, individuals remain paired while both partners survive; divorce is rare (3.4%, Russell and Rowley 1993*a*); males seek EPCs but return to social partner and help feed and raise young in her nest, although more than 66% of offspring result from EPC (Brooker *et al.* 1990). No egg-dumping recorded. Males seeking EPC frequently carry pink or violet petals as they approach the female (Rowley 1991). Courting male displays to female by Face Fan (Fig. 5.14) and Lizard display. Philandering males leaving a territory that they visited may use Sea Horse Flight. Displays by philandering males most frequently seen near nest-building females. Females may crouch and flutter part-opened wings in an invitation display; copulation rarely seen. Wing-fluttering also used as an appeasement display given by either sex. Territory defence usually by song battle; persistent trespassers are chased by all group members; threat postures directed at intruders, including cuckoos: body feathers part erected, wings held out from body, head and tail lowered (see Fig. 5.3), sometimes with churring call, followed by physical attack (Payne *et al.* 1985). Rodent-run distraction display (see Fig. 5.4) by both sexes when nest or young are threatened (Warham 1954).

Breeding and life cycle

Nesting late Aug–Jan; heavy rain in Aug may delay onset; few eggs are laid after late Dec, depends on season and whether female is preoccupied with an earlier brood (Rowley *et al.* 1991, from 774 nesting attempts) (Fig. 7.1(a)). Mean length of breeding season 113 days (range 57–142) between initiation of first and last clutches. Nest built by female, escorted by male. Initial framework of cobwebs, oval domed structure of grass lined with bast-fibre in WA (*Zamia* sp.) and feathers, side entrance near top; base 782 mm (300–2500) above ground, mean size 125 × 64 mm (n = 115), set in *Acacia pulchella* (28%), *Hakea* spp. (31%), and *Dryandra* spp. (11%). Nests built in Sept–Oct placed with their base a mean of 667 mm above ground level, while nests in Nov–Dec were 973 mm a.g.l. Clutch 2.9 ± 0.3 (range 2–4, n = 308), does not vary during season. Eggs (15–18 × 11–13 mm, n = 45) tapered oval, pinkish-white with red-brown spotting in zone around blunter end; laid on successive days usually before 0800 hours. Incubation by female only, 14–15 days (range 12–13 to 16–17; n = 35) tends to be longer in the first half of breeding season. Generally all eggs hatched, brood and clutch sizes similar. Fertility 97% (309 eggs), but this fell to 87.5% (322 eggs) after a major wildfire. Only female broods, but nestlings are fed and faecal sacs removed by all group members for 10–13 days. For 24 nests, one fledged after 9–10 days, 10 after 10–11, eight after 11–12 and five after 12–13 days (Rowley *et al.* 1991). Fledglings remain cryptic for a week; they are fed for a month, and young birds remain in the family group long after independence, sometimes for several years. Males are sexually mature at 1 year, and usually remain in their natal group for 1 or more years as a helper before inheriting the senior position, or moving next door to a vacancy. Helpers feed nestlings, care for dependent fledglings, defend the territory, and repel predators or cuckoos; they may also obtain some EPCs. Some yearling females remain as helpers in their natal territories, others disperse; they can breed at 1 year old. Most dispersal by either sex is only as far as a neighbouring group (Russell and Rowley 1993*a*). Over 16 years, 49% of eggs laid and 49% of nests laid in produced fledglings; 69% of fledglings reached independence, 31% survived to 1 year. Females lay 1–5 (mean 2.0) clutches/year, and re-nest after failure; 36.3% of females re-nested after successfully raising one brood; when first brood had survived to independence, the mean interval between the laying of two successive clutches was 60 days (range 33–87); 20% of nests were parasitized by Horsfield's Bronze-Cuckoo (*Chrysococcyx basalis*), other failures were due to predation (reptiles (see Fig. 7.11), birds, cats, rodents; Rowley *et al.* 1991); because cuckoos were most active in late Oct–Nov , early and late nests were less likely to be parasitized; for nests initiated from Sep to mid-Oct, the mean chance of success was 58%, for nests from mid- Oct to end Nov, the chance of success was 46%, and for nests in December, the chance of success rose again to above 50%. Each year, a mean of 84.5% of groups produced fledglings, 3.2 per group, 2.2 of which reached independence and 0.9 survived to 1 year. Over 18 years, mean sex ratio 1.32 male : 1 female (see Fig. 6.2); mean annual survival of adult males 70%, breeding females, 59% (Russell and Rowley 1993*b*).

Purple-crowned Fairy-wren *Malurus coronatus* Gould, 1858

Malurus coronatus Gould, 1858. *Proceedings of the Zoological Society of London*, **1857**, p. 221.

PLATE 1

Polytypic. Two subspecies. *Malurus coronatus coronatus* Gould, 1858; *Malurus coronatus macgillivrayi* Mathews, 1913,

Description

M. c. coronatus

ADULT NUPTIAL MALE: crown, purple-mauve with central black oblong; nape, black; scapulars, upper and lower back, wing-coverts, cinnamon-brown; flight feathers, grey-brown; tail, deep blue, tipped white, except centre two; black face extends around eye and side of ear, joining nape; no distinct ear tufts; chin, belly, and breast, creamy white with buff shadings; bill, black; eye, blackish brown; feet, dark brown.

ADULT FEMALE: as male in colour of body, bill, eyes, and feet; crown, mid-grey with variable bluish tinge; rich chestnut ear coverts; white eye brow and eye ring; tail, greenish-blue.

ADULT ECLIPSE MALE: crown, grey; face, variable, a mixture of faded black and grey; white eye-brow and eye-ring variable in extent; bill remains black; collar, brown.

IMMATURE: sexes indistinguishable, resembling female but crown brown; tails longer than adults, with narrower white tips; lower half of eye ring white, lack white eyebrow; ear coverts, less rich chestnut; bill, dull black. Males show black in face at 6–9 months.

MOULT: general moult usually early dry season Apr–June, but variable according to season; most males enter eclipse female-like plumage; males have second pre-nuptial moult in late dry season Aug–Sept (body feathers, not remiges or rectrices) into coloured nuptial plumage. Onset varies according to age, status, and condition of individual; tail moulted throughout year. After a poor wet season, males may not moult into nuptial plumage in following Sept (see colour plate Rowley 1988, p. 3). Males do not achieve perfect purple and black crown at 1 year.

History and subspecies

Two subspecies recognized: *M. c. coronatus* Gould, 1858, from Kimberley region of NW Australia; *M. c. macgillivrayi* Mathews, 1913, along rivers bordering Gulf of Carpentaria in E; smaller, greyer on back, whiter on belly, male's tail greenish blue. Western birds are significantly larger, with bluer tails, browner back in males, and distinctly buff breast and belly. In both eastern and western forms, the crown of immatures (both sexes) is brown; in females, the crown becomes distinctly blue-grey with age. (Schodde 1982*b*; Rowley and Russell 1993 and unpublished data).

The first specimens were taken on the Victoria R. by J. R. Elsey, surgeon-naturalist to A. C. Gregory's 1855–6 Expedition across northern Australia; the material was sent to the British Museum and Gould described *M. coronatus* in 1858. Later, Elsey collected a third specimen from the Robinson R., 500 km further east, as he accompanied Gregory on his overland journey to Brisbane. That specimen was overlooked for more than 100 years until Macdonald and Colston (1966) quoted data from the register of the British Museum. Later, specimens were collected on the Leichhardt River in Queensland and at Borroloola, NT, and on the basis of these, Mathews (1913) described an eastern subspecies *macgillivrayi*. *M. coronatus* is so distinct (Plate 1) that, apart from Mathews' creation of the monotypic genus *Rosina* for it (Mathews 1912*c*), the main taxonomic debate has concerned the delineation of subspecies. Because it occurs in what were remote, inaccessible regions of Australia, it was little known from the field until recently; specimens were few, and immature plumages not well known. The eastern and western populations are isolated by an area of unsuitable habitat *c.* 200 km wide, and have probably been isolated for at least 10 000 years. Schodde (1982*b*) recognized two subspecies, *M. c. coronatus* (western) and *M. c. macgillivrayi* (eastern), and we agree.

Measurements and weights

Malurus coronatus coronatus: wing, M (n = 14) 56.2 ± 1.2, F (n = 9) 52.8 ± 1.4; tail, M (n = 14) 71.7 ± 4.9, F (n = 9) 69.5 ± 2.7; bill, M (n = 14) 11.3 ±0.3, F (n = 9) 10.9 ±0.6); tarsus, M (n = 14) 24.9 ± 0.7, F (n = 9) 23.9 ± 0.7; weight, M (n = 41) 10.1–13.0 (11.4 ±0.6), F (n = 38) 9.6–12.6 (10.8 ±0.7). *Malurus coronatus macgillivrayi*: wing, M (n = 11) 53.7 ± 1.0, F (n = 12) 51.6 ± 1.4; tail, M (n = 11) 69.5 ±4.0, F (n = 12) 70.4 ± 4.5; bill, M (n = 11) 10.7 ±0.4, F (n = 12) 10.1 ± 0.5; tarsus, M (n =11) 25.2 ± 0.8, F (n =12) 23.4 ± 0.7; weight, M (n = 27) 9.2–11.5 (10.4 ± 0.5), F (n = 30) 8.7–10.5 (9.5 ± 0.6). (Weights from field measurements by authors.)

Field characters

In nuptial plumage, purple crown of male and chestnut ear coverts of female unmistakeable; eclipse males and immatures harder to recognize. Their size, habitat, and loud calling distinguish them from sympatric Variegated Fairy-wren (*M. lamberti*) and Red-winged Fairy-wren *M. melanocephalus* which are smaller and quieter.

Voice

Tape-recorded (ANWC; Rowley and Russell 1993). Despite the long-held belief that they had no song (which originated from T. H. Bowyer-Bower in Mathews 1918*a*), a loud reel 'cheepa-cheepa-cheepa-cheepa' is given by both sexes, singly and as duet (see Fig. 5.8), and by helpers; individuals differ. Loudness, low frequency, and clear separation between elements make reel easily heard and distinguishable from other species. Group members maintain contact when foraging, with 'chet' calls, singly or in series; alarm call a sharp 'zit'.

Range and status

Resident restricted to riverside vegetation across the Wet–Dry Tropics of N Australia between 400 and 1000 mm rainfall isohyets. Two distinct populations separated by 200 km of unsuitable habitat. Kimberleyan subspecies *M. c. coronatus* has declined due to degradation of river frontages since pastoral settlement; hopefully reversible. Relict populations on Fitzroy and Victoria R.; also on Dunham, Isdell, Beuckelman R., and probably others as yet unsurveyed; healthy populations on Durack and Drysdale R. Carpentarian subspecies *M. c. macgillivrayi* less affected by settlement; healthy populations on most rivers draining into SW Gulf of Carpentaria, from Roper R. in N to Leichhardt R. in E; subspecies regarded as secure but requires careful watching. *M. c. coronatus* rated vulnerable (Garnett 1992*a*; Rowley 1993).

Habitat and general habits

Rarely more than 10 m from river (or spring) running for all of most years. Permanent pools at such places support dense fringe vegetation, typically a variety of tall grasses (e.g. *Coelorhachis rottboellioides*) or *Pandanus aquaticus*, which is a large multi-stemmed shrub or tree with leaves 1m or more long and 10 cm broad, fringed with sharp barbs (see Fig. 3.3(a); last page of colour plates). Forage through and under dense fringe vegetation, eating mainly insects (grasshoppers, bugs, beetles, weevils,

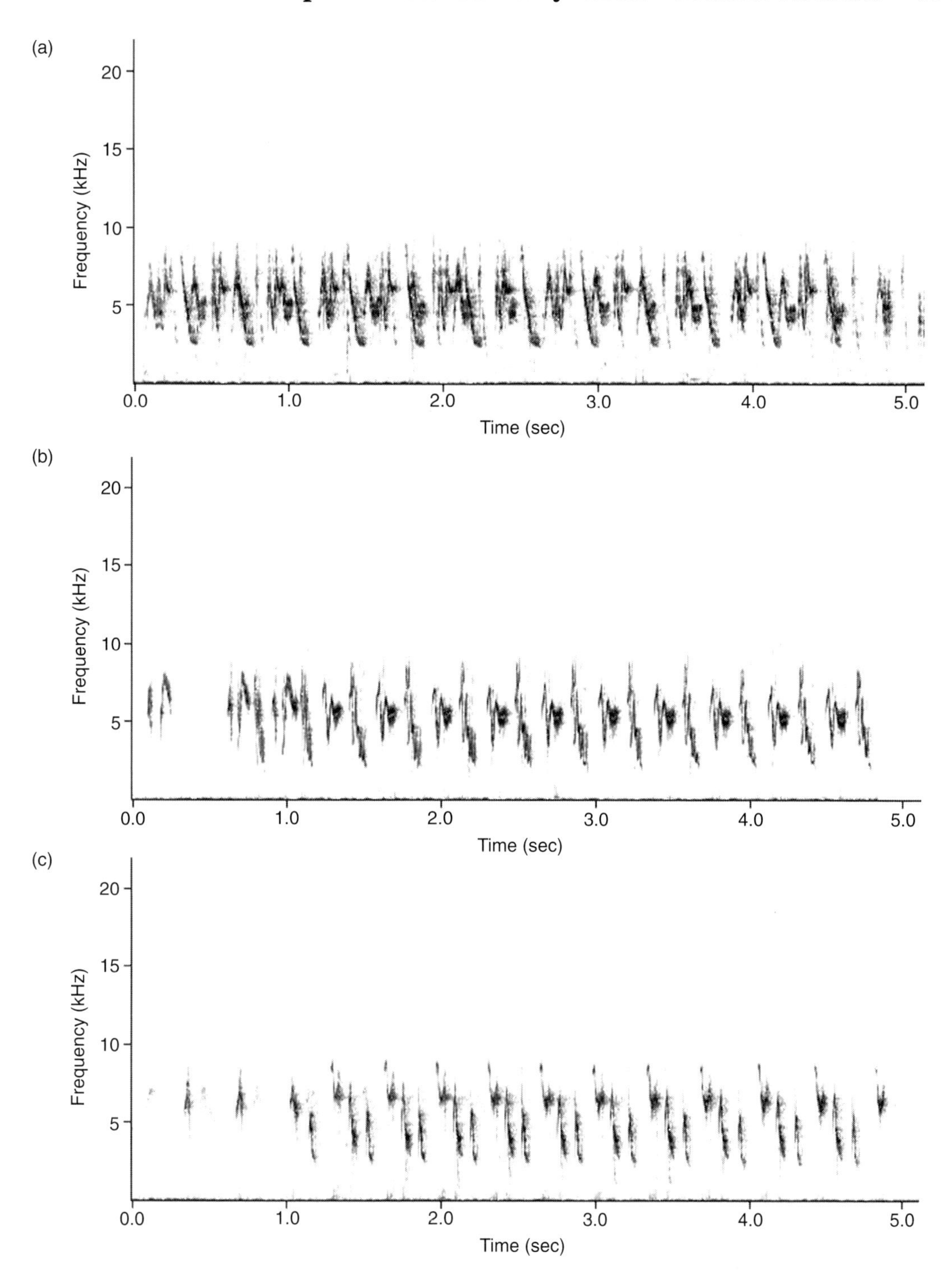

Purple-crowned Fairy-wren (*Malurus coronatus*)
Song. (a) Male *M. c. coronatus*, Drysdale R. (b) Female *M. c. coronatus*, Drysdale R.
(c) Male *M. c. macgillivrayi*, Riversleigh. Scales differ from other *Malurus* sonograms.
E. M. Russell, Drysdale R., Western Australia and Gregory R., Riversleigh Station, Queensland.

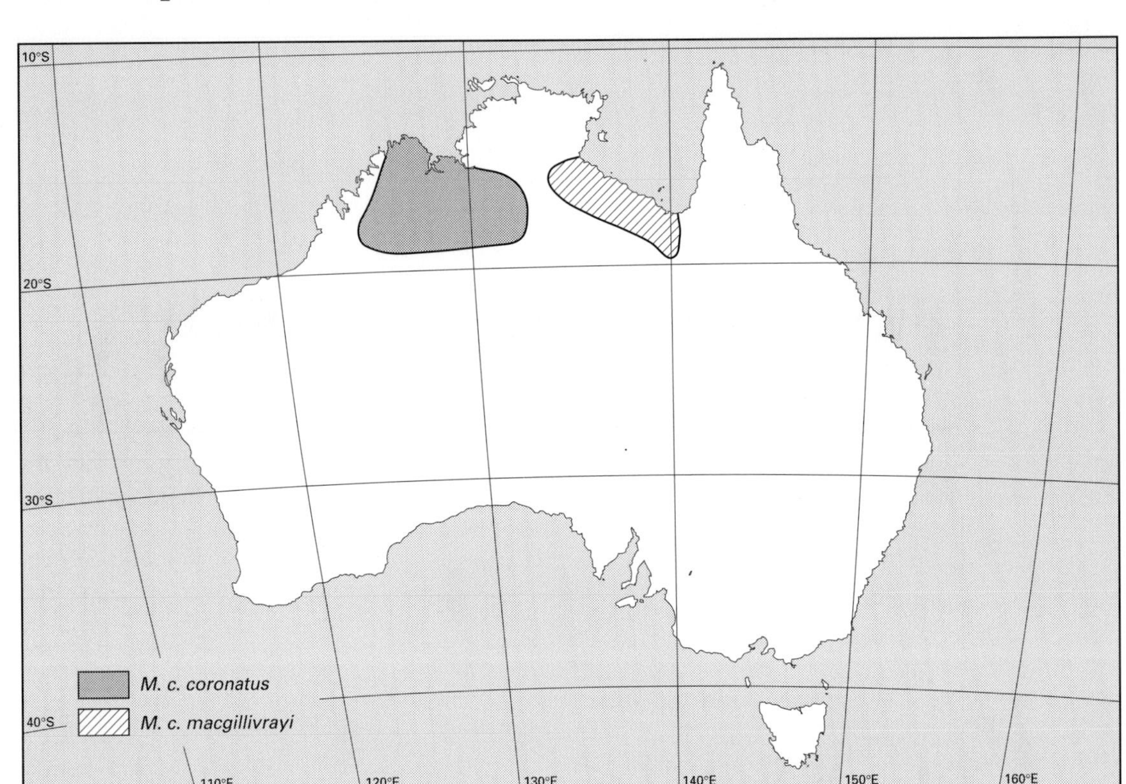

ants), spiders, and worms (Rowley and Russell 1993). Large feet enable them to grasp *Pandanus* leaves, to forage amongst the litter accumulated in leaf axils and to roost perched on the leaves. Co-operative breeding species, generally seen in pairs (83% of 174 groups) or small groups (17%); studies of colour-banded populations show groups maintain the same territory throughout year, defended by all members, who regularly allopreen; roost together in dense cover, lined side by side on a horizontal branch. Groups (mean size 2.3 ± 0.7, range 2–7) remained in the same territories from year to year. Territories are dispersed linearly along rivers, occupying both banks along 200–300 m of river (Rowley and Russell 1993). Duetting by male and female of a breeding pair is an obvious reply to a neighbour's song; both fly to a high point in the open and sing loudly and vigorously together (see Fig. 5.8). Other group members may join with excited chattering, but only the breeding pair sing together (see sonogram of duet in Rowley and Russell 1993).

Displays and breeding behaviour

Socially monogamous but probably sexually promiscuous (as other *Malurus*), remain paired while both partners survive; divorce is rare (1.7%, 174 pair-years) and 51.7% of pairs were intact from one year to next. Frequent duetting reinforces the pair-bond and emphasizes territorial boundaries. Does not have the erectile ear patches or scapulars that other *Malurus* use in display; no courtship displays or Petal-carrying recorded; no information on EPC; Rodent-run distraction display by all group members easily provoked when nest or young threatened.

Breeding and life cycle

Laying recorded in most months with two peaks Apr and Aug (see Fig. 7.1(h)).The tropical Wet–Dry climate has two main seasons, the Wet from Dec to Mar and the Dry from June to Sept, with two transition periods of varying length. In our 5-year study, above average wet season rainfall was followed by laying Apr–May and again in the following dry season. After poor wet seasons, females did not nest and males remained in eclipse plumage. Nest built by female; oval, bulky, domed (mean size 152 × 100, *n* = 25); outer layer of rootlets, grass leaves, and bark, lined with fine grass and rootlets, 24 of 25 described nests were placed in the leaf axils of *P. aquaticus*, where they resembled the clusters of litter that accumulate there after floods; mean height above ground 880 mm (250–2900 mm, *n* = 25); side entrance near top. Mean clutch 2.6 (2–3, *n* = 9), eggs (15.7–16.9 × 12.4–13.7 mm, *n* = 7, Schodde 1982*b*) tapered oval, dull white, with sparse spots and fine blotches of red-brown, concentrated at blunter end, laid on successive days. Nests parasitized by Brush Cuckoo (*Cacomantis variolosus*) (see Fig. 7.10). Incubation by female, for at least 14 days, nestlings hatch within 24 hr of each other, are brooded initially by female; female may be fed on the nest by mate and helpers. Nestlings remain in nest about 10 days, and are fed and their faecal sacs removed by all adults. Fledglings remain cryptic for a week, then travel with the foraging group, fed for *c.* 1 month; young birds remain in the family long after independence; may help to raise later siblings. Both sexes are probably sexually mature at 1 year old, but both tend to stay and help in their natal group for *c.* 12 months after which they take any vacancies available. Although tolerated in their natal group, no female helped more than 2 years. Many males helped at least 3 years before a vacancy occurred nearby or they inherited their natal group. Females dispersed more and further than males. Wet season rainfall has an important influence on annual productivity; females may rear more than one brood per year; from census in June, a mean of 0.78 independent young (range 0.26–1.22) were added to groups in the Kimberley population each year; for the Carpentarian population, the figure was 1.04 (0.22–1.95). The mean annual survival of adult males was 70%, and of breeding females, 70% (Rowley and Russell 1993).

Chestnut-shouldered Fairy-wrens

Four species of chestnut-shouldered fairy-wren with seven distinct forms are currently recognized: Variegated Fairy-wren (*Malurus lamberti*), Lovely Fairy-wren (*M. amabilis*), Red-winged Fairy-wren (*M. elegans*), and Blue-breasted Fairy-wren (*M. pulcherrimus*) (Plate 2). In the past, these have been assigned various taxonomic ranks, and as many as 15 different forms have been described. Mathews (1913) placed them in a separate genus *Leggeornis*. Mack (1934) retained this as a subgenus, but Schodde (1982*b*) does not use it and nor do we.

All the nuptial plumaged males in this group share the same basic colour pattern: blue-violet head and ear tufts, black rump, a black collar, black lores, and they all have scapular patches of long chestnut-rufous feathers. The breast is variously black, blue-black, or violet-black. Females, immature males, and eclipse males are variable shades of brown or blue-grey. In some forms, blue-grey females have white lores and eyering. Some member of the complex is found in every part of Australia except in the extreme SE and Tas. With such wide distribution, considerable variation could be expected, and it is this that has led to the confused taxonomic history of this group. The use of modern molecular techniques will increase our understanding of the similarities and differences in this complex group.

Table: Differences and similarities of chestnut-shouldered fairy-wrens

	Species						
	1a	1b	1c	1d	2	3	4
Males							
crown	azure	----------- violet blue -----------			azure	violet blue	silver blue
ear tufts	------------ lighter than crown ------------				azure	same as crown	same as crown
back	blue	purple	purple	purple	azure	same as crown	silver blue
breast	black	black	black	black	black	indigo	indigo
Females							
lores	rufous	rufous	rufous	white	white	rufous	rufous
around eyes	rufous	rufous	rufous	white	white	rufous	grey
shoulder	grey brown	grey brown	blue grey	blue grey	blue grey	grey brown	rufous
upper surface	grey brown	grey brown	blue grey	blue grey	blue grey	grey brown	grey brown
tail	very long	long	long	long	short	very long	very long
bill	---------------- red-brown ----------------				black	dark brown	black

1a *M. l. lamberti*
1b *M. l. assimilis*
1c *M. l. rogersi*
1d *M. l. dulcis*
2 *M. amabilis*
3 *M. pulcherrimus*
4 *M. elegans*

Variegated Fairy-wren *Malurus lamberti* Vigors and Horsfield, 1827

Malurus lamberti Vigors and Horsfield, 1827. *Transactions of the Linnean Society of London*, **15**, 221. PLATE 2

Polytypic. Four subspecies. *M. l. lamberti* Vigors and Horsfield, 1827; *M. l. assimilis* North, 1901; *M. l. dulcis* Mathews, 1908; *M. l. rogersi* Mathews, 1912.

Description

M. l. lamberti

ADULT NUPTIAL MALE: crown, mid-azure, slightly darker than ear coverts, separated by black collar from mid royal blue upper back; ear tufts, long, lanceolate, extend back from narrow blue ring round eye; lower back, black; scapulars, rufous; wings, deep greyish-brown; tail, turquoise-grey with narrow white tip; lores, throat, and breast, black; sides of breast may be edged violet; belly, creamy-white; flanks, white to tawny grey; bill, black; eye, dark brown; feet, dark grey-brown.

ADULT FEMALE: upper surface and wings, greyish-brown; tail, turquoise-grey, narrowly white-tipped; lores and feathers round eye, rufous; chin, throat, and belly, creamy-white; flanks, white to tawny grey; bill, red-brown; eyes and feet, as male.

ADULT ECLIPSE MALE: as female but bill and lores black.

IMMATURE: Fledglings like female; male shows white eyering, black lores and bill by 6 months (Rowley and Russell, unpublished data).

MOULT: general autumn moult by both sexes, when males enter eclipse plumage; male pre-nuptial moult during winter/spring.

History and subspecies

Four subspecies currently recognized (Schodde 1982*b*): male *M. l. lamberti* on E coast has a royal blue back whereas males of other races have violet-purple backs differing little from each other. Females differ strongly: *M. l. rogersi* (NW Australia) and *M. l. dulcis* (Arnhem Land) have grey-blue upper surfaces and tails, lores and feathers round eye rufous (*rogersi*) or white (*dulcis*); female *M. l. lamberti* and *M. l. assimilis* (inland and W coastal) both have light grey-brown upper surfaces and rufous lores. *M. l. assimilis* generally smaller and lighter in colour in N (both sexes), but variation at any one location precludes further subdivision.

The first chestnut-shouldered fairy-wren encountered was the Variegated, on the coast of NSW, E of the Dividing Ra. Initially thought to be a colour variant of *M. cyaneus*, Vigors and Horsfield (1827) recognized it as distinct, and described it as *M. lamberti*. As settlement spread from Port Jackson, chestnut-shouldered fairy-wrens were encountered over much of inland Australia, but their variability was not formally recognized until North (1901*a*) distinguished the inland forms as *M. assimilis*. His vernacular name Purple-backed Superb Warbler emphasized the main character distinguishing it from *M. lamberti*: the purplish-blue crown, sides of head, and back, as against cobalt in *lamberti*.

The picture became more complex as collectors visited remote parts of the continent. From Bernier Is. off the coast of WA, Ogilvie-Grant (1909, p. 72) described *M. bernieri*, a new form, 'the feathers round the eye and the ear-coverts are bright purplish-blue (not bright cobalt-blue) and contrast but little with the darker purple-blue colour of the crown which is the same in both species'. Schodde (1982*b*) points out that this is the same distinction North had made between *assimilis* and *lamberti*; it illustrates the change in colour of *assimilis* across the continent, from violet-blue on the head and back in the inland SE to lighter blue in the N and W, where ear tufts are similar in colour to those of the E coast *lamberti*.

This variation in *assimilis* led to a proliferation of subspecies. Mathews always maintained *assimilis* as a subspecies of *lamberti* for birds from inland NSW, but in 1930 added six subspecies: *bernieri* (Bernier Is., WA), *mastersi* (NT), *occidentalis* (mid- and NW WA), *morgani* (SA), *dawsonianus* (inland Qld) and *hartogi* (Dirk Hartog Is.). Mack (1934) reduced these to four: *lamberti* (E coast), *assimilis* (Victoria, NSW, SE Qld), *mastersi* (the rest of inland Australia) and *bernieri* (Bernier Is.). The differences between *mastersi*, *assimilis*, and *bernieri* were slight, and the distinction was not maintained by later authors (1975 Interim List, Schodde 1982*b*). Mack was uncertain about the separation between the nominate coastal *lamberti* and inland *assimilis* forms in SE Queensland. Schodde (1982*b*) identified the area of intergradation as a broad tract of country *c.* 700 km long, from Goondiwindi–Wide Bay–Rockhampton–Emerald, where the Dividing Ra. is broad and not very high.

In the 20 years after 1900, two other forms of chestnut-shouldered fairy-wrens were described in which females were blue-grey, *M. dulcis* (Mathews 1908*b*) from Arnhem Land and *M. amabilis rogersi* (Mathews 1912*a*) from the Kimberleys. For a long time, these were regarded either as forms of the Lovely

Wren (*M. amabilis*) from Cape York or as forms of a separate species *M. dulcis* (see the discussion of *M. amabilis* below). Among the specimens collected by the Harold Hall Australian Expedition were some from the Roper R., NT, where Arnhem Land *dulcis* meets the widespread *assimilis*. Harrison (1972) suggested that these greyish females were intermediates between *dulcis* and *assimilis*, an opinion with which Schodde agreed. Parker included *dulcis* and *rogersi* with *assimilis* as subspecies of *M. lamberti* in the 1975 Interim List. Assessment of the overlap between *assimilis* and *rogersi* in the Kimberleys found clear evidence of intergradation, characteristic of a hybrid zone (Ford and Johnstone 1991). Although females of *rogersi* and *dulcis* differ markedly from female *assimilis*, males are very similar. From his analysis of variation in *M. lamberti*, Schodde (1982*b*) recognized four subspecies, *lamberti*, *assimilis*, *dulcis*, and *rogersi*.

Measurements and weights

Malurus lamberti lamberti: wing, M (n = 36) 48.1 ± 1.7, F (n = 8) 46.8 ± 1.9; tail, M (n = 36) 71.9 ± 4.6, F (n = 8) 69.0 ± 3.7; bill, M (n = 36) 9.0 ± 0.7, F (n = 8) 7.9 ± 0.5; tarsus, M (n = 36) 21.9 ± 0.9, F (n = 8) 22.3 ± 0.5; weight, M (n = 49) 7.0–11.0 (8.2 ± 0.7), F (n = 47) 7.0–10.0 (8.1 ± 0.7).
Malurus lamberti assimilis (north of 22^0): wing, M (n = 53) 47.5 ± 1.3, F (n = 33) 46.2 ± 1.5; tail, M (n = 53) 64.5 ± 3.7, F (n = 33) 62.8 ± 3.5; bill, M (n = 53) 8.9 ± 0.5, F (n = 33) 8.6 ± 0.6; tarsus, M (n = 53) 20.6 ± 0.9, F (n = 33) 19.9 ± 0.9; weight, M (n = 38) 6.8–10.0 (7.8 ± 0.6); F (n = 17) 5.9–8.1 (7.3 ± 0.6).
Malurus lamberti assimilis (south of 32^0): wing, M (n = 37) 48.3 ± 1.1, F (n = 7) 47.6 ± 1.3; tail, M (n = 37) 68.7 ± 4.4, F (n = 7) 68.5 ± 3.1; bill, M (n = 37) 8.6 ± 0.4, F (n = 7) 7.7 ± 0.5; tarsus, M (n = 37) 21.2 ± 0.8, F (n = 7) 21.3 ± 0.8; weight, M (n = 17) 6.5–9.0 (7.9 ± 0.8), F (n = 27) 6.0–9.0 (7.6 ± 0.8).
Malurus lamberti dulcis: wing, M (n = 11) 49.1 ± 0.8, F (n = 6) 47.3 ± 1.2; tail, M (n = 11) 62.6 ± 3.0, F (n = 6) 62.5 ± 2.3; bill, M (n = 11) 9.8 ± 0.6, F (n = 6) 9.1 ± 0.5; tarsus, M (n = 11) 20.4 ± 0.4, F (n = 6) 20.4 ± 0.5; weight, M (n = 15) 6.0–8.3 (7.8 ± 0.7), F (n = 7) 7.4–8.9 (8.2 ± 0.5).
Malurus lamberti rogersi: wing, M (n = 23) 49.4 ± 1.9, F (n = 17) 47.6 ± 1.6; tail, M (n =23) 61.8 ± 3.6, F (n =17) 61.1 ± 2.3; bill, M (n = 23) 9. 8 ± 0.6, F (n = 17) 9.5 ± 0.5; tarsus, M (n =23) 21.1 ± 1.0, F (n = 17) 20.1 ± 0.7); weight, M (n = 13) 8.0–11.5 (8.9 ± 0.9), F (n = 6) 7.7–8.7 (8.1 ± 0.4).
(Weights from Hardy 1983, ABBBS, ANWC, and AM.)

Field characters

Widespread, probably overlapping every other *Malurus* sp. somewhere; at any site, only two or three relevant candidates. Most brown female-like birds hard to identify positively in field; females in N (*M. l. rogersi, M. l. dulcis*, and Lovely Fairy-wren *M. amabilis*) clearly distinguished by blue-grey upper surfaces and by location. Most chestnut-shouldered females have rich rufous lores and rufous feathering around the eye, except *M. l. dulcis*, *M. amabilis*, and Red-winged Fairy-wren (*M. elegans*) (see Table); lores of other female *Malurus* paler. Rufous scapulars of nuptial male distinguish *M. lamberti* from other *Malurus* except *M. amabilis*, Blue-breasted Fairy-wren (*M. pulcherrimus*), and *M. elegans* which also have rufous scapulars. Chestnut shouldered nuptial males separable on colour of crown, ear tufts, and breast (see Table).

Voice

Tape-recorded (ANWC, Baker 1995). Song a typical *Malurus* reel given by all group members, lacks prolonged introductory notes, high frequency, similar to that of *M. pulcherrimus*, but slightly stronger and harsher. Maintain social contact by 'tsst' calls, sometimes sibilant 'seeee' or shorter, softer version of reel; alarm short, sharp 'tsit'; call infrequently.

Range and status

Resident; occurs over more than 90% of Australia, from Arnhem Land and Kimberleys in N to Perth in SW, to Victorian Mallee,

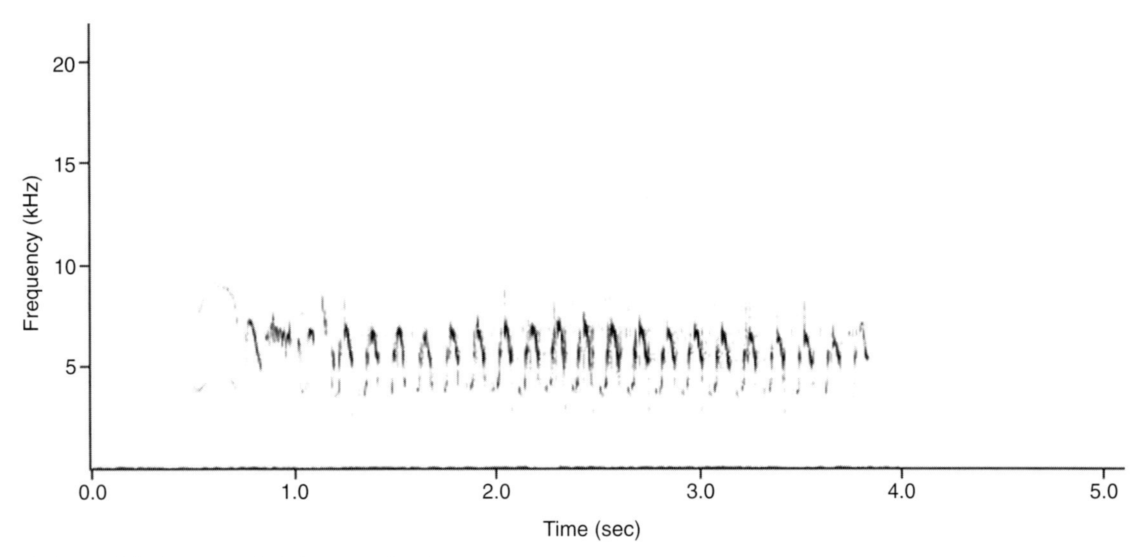

Variegated Fairy-wren (*Malurus lamberti*)
M. l. assimilis, song. E. M. Russell, Peron National Park, Western Australia.

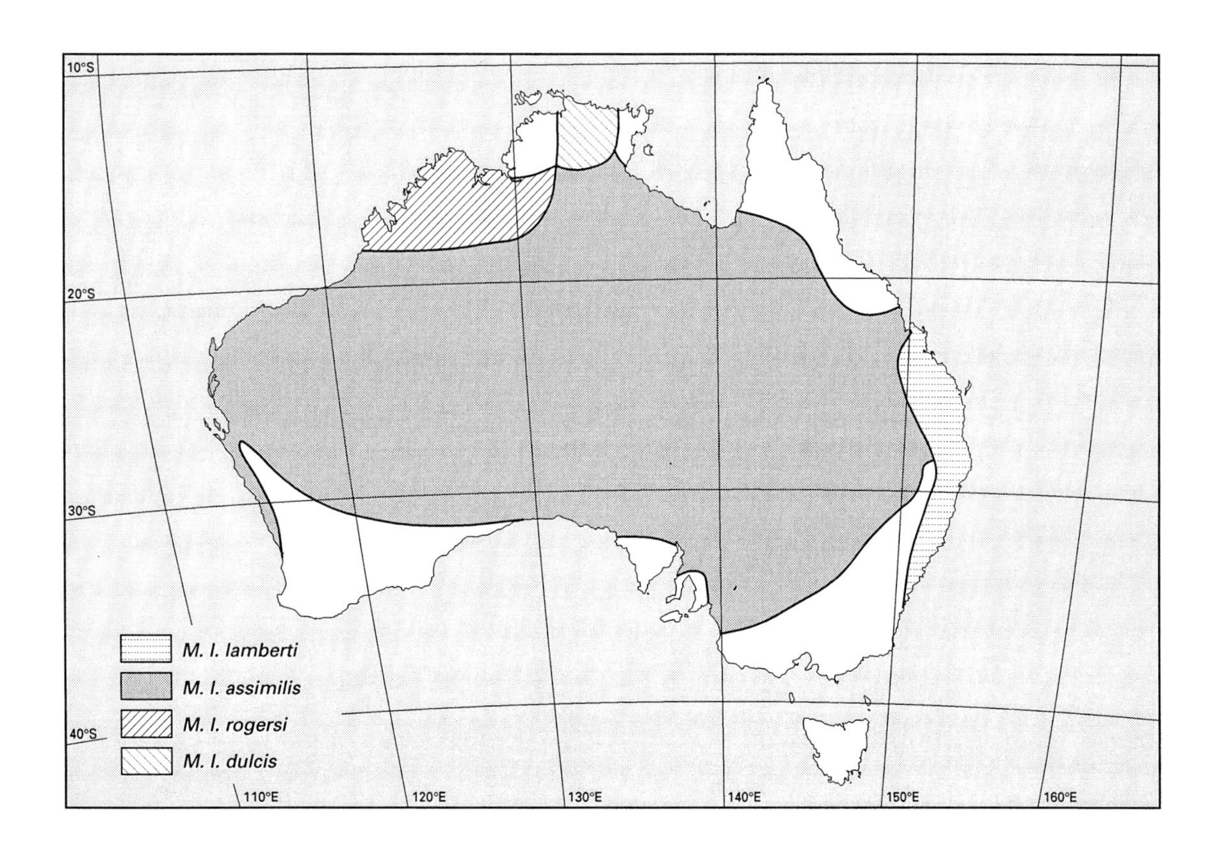

throughout inland areas of all states (except Tas.), coastal from Mackay, Qld S to Wallaga Lake, NSW, absent from Great Dividing Ra. in NSW and Vic. Four subspecies recognized, zones of intergradation identified; E coastal nominate race (*M. l. lamberti*) and inland race (*M. l. assimilis*) meet in SE Qld from Goondiwindi to Rockhampton (Schodde 1982*b*); Kimberley (*M. l. rogersi*) and Arnhem Land (*M. l. dulcis)* races, earlier regarded as separate species, intergrade southward with *M. l. assimilis* (Schodde 1982*b*; Ford and Johnstone 1991). *M. l. assimilis* does not intergrade with *M. amabilis* on Cape York, *M. elegans* near Perth, or *M. pulcherrimus* across SW. Widespread, patchily common.

Habitat and general habits

Lives in dense shrubby vegetation from coastal scrubs to thickets in broken rocky country, creekside shrubbery, and patches of Lignum (*Muehlenbeckia*), *Eremophila*, or *Acacia* across semi-arid and arid interior, through Simpson and Great Sandy Deserts to rocky outcrops in Kimberleys and Arnhem Land. May shelter in mammal burrows to avoid extreme heat (Marchant 1992). Forages through cover, on open ground, and even in tree canopies reached by hopping from branch to branch, since flight is weak. Main food: grasshoppers, bugs, beetles, weevils, flies, caterpillars, wasps, ants, and spiders (Barker and Vestjens 1990). Co-operative breeding species generally seen in pairs or small groups; studies of colour-banded population show groups maintain territories (3–4 ha) throughout year and from year to year (Tidemann 1983; Marchant 1992). Some observations of groups with *c.* 10 birds, including two or three males (White 1910; Wilson 1912; Boehm 1957), led Schodde (1982*b*) to suggest that they leave their territories and form local flocks after the breeding season. However, the groups described by those authors were unbanded and are entirely consistent with a single large group after a successful breeding season; without marked birds, such observations are not sufficient evidence of groups coalescing.

Displays and breeding behaviour

Socially monogamous but probably sexually promiscuous (as other *Malurus*), remain paired while both partners survive; no courtship displays recorded; allopreening by all group members; Petal-carrying by males (yellow petals, Strong and Cuffe 1985) and Rodent-run distraction display by all group members easily provoked when nest or young threatened.

Breeding and life cycle

Breeding mainly July–Dec throughout range, inland birds breed in other months when conditions suitable e.g. after heavy summer rains from tropical cyclones in WA (Carnaby 1954; Robinson 1955) (see Fig. 7.1(f)). Nest built by female; oval, domed, side entrance near top; 11 nests of *M. l. assimilis* at Shark Bay, WA measured by Rowley (unpublished data) were, on average, 97 (90–110) mm tall × 65 (50–90) mm wide; coarser structure than in other *Malurus*, of grass, twigs, and bark, lined with fine grass, plant fibre, fur, and feathers; usually less than 1 m above ground (193mm, 0–450 mm, n = 11), often in dead brush. Clutch 2.8 ± 0.8 (2–4, n = 59), eggs (15–18 × 12–14 mm, n = 54, Schodde 1982*b*) tapered oval, pinkish-white, speckled with red-brown spotting in zone around blunter end, laid on successive days. Incubation by female for 14–16 days depending on temperature, nestlings in nest 10–12 days, fed, and their faecal sacs removed, by all group members. Fledglings remain cryptic 7–10 days, fed by the group for *c.* 1 month, and young birds remain with the group long after independence; some remain as helper in their natal territory for 1 or more years, others disperse and can breed at 1 year old. Females re-nest after failure. *M. l. assimilis* had a very short breeding season in the population studied by Tidemann (1983), with no sign of attempted second broods. Studies of *M. l. lamberti* in coastal SE Australia indicate that females had time to rear two broods, but no definite evidence that they did (Marchant 1992). Females may rear 2–3 broods if season favourable.

Lovely Fairy-wren *Malurus amabilis* Gould, 1852

Malurus amabilis Gould, 1852. *Proceedings of the Zoological Society of London*, **1850**, p. 277

PLATE 2

Monotypic

Description

ADULT NUPTIAL MALE: crown, mantle and oval ear tufts, azure blue; scapulars, bright rufous; wings, dusky-black; tail, blue with broad white tip; lores, nape, lower back, throat, and breast, black; belly, white; eye, brown; bill, black; feet, flesh-grey.

ADULT FEMALE: crown to lower back, smoky blue; ear tufts, turquoise-blue; wings, dark grey; tail, smoky blue with broad white tip; lores and feathers round eye, white; undersurface creamy-white; eye, bill, and legs, as male.

ADULT ECLIPSE MALE: variable, has rufous edges to tertials; may resemble female, or assume grey-brown *assimilis*-like plumage (Forshaw and Muller 1978).

IMMATURE: fledglings as female but duller, with dusky brown bills; full grown tails longer than those of adults.

MOULT: general moult usually Dec–July but variable; tail feathers moulted continually; at first moult, young females acquire adult female plumage. Young males may moult into female-like eclipse plumage or directly into nuptial male plumage (K. A. Muller, in Schodde 1982*b*).

History and subspecies

Previously, two recognized (Mack 1934; Storr 1973), one on Cape York Peninsula and the other south of Cairns, but these differences are not upheld in larger collections (Harrison 1974; Schodde 1982*b*).

Malurus amabilis, one of the least known of the Australian fairy-wrens, was described by Gould in 1852 from 'a single and somewhat imperfect specimen of this bird, bearing the words 'Cape York 1849', ...transmitted by the late Captain Owen Stanley to the Zoological Society of London' (Gould 1865, p. 328). That specimen was a male in full nuptial plumage, and no female was known. Gould was expecting that 'the female of *M. amabilis* will be found to closely resemble that sex of *M. elegans*' (Gould 1865, p. 328), so that when he did receive specimens of female *amabilis* from John Jardine of Cape York, their smoky-blue crown and back, black bill, and white lores led him to describe them as a new species, *M. hypoleucos* (1867), an error that was soon discovered.

Collectors in northern Australia encountered other chestnut-shouldered fairy-wrens with blueish females. North (1904) mentions that De Vis had recorded *amabilis* from Cambridge Gulf in NW Australia. Hartert examined the material collected by J. T. Tunney, and identified a chestnut-shouldered male from the ranges E of the South Alligator R., NT as *M. pulcherrimus*, although he described the blue-grey back, white lores, and red-brown bill of the female and compared them with *amabilis* (1905*a*). When Mathews saw Tunney's specimens in the Tring Museum, he recognized that they were distinct, and described them as *M. dulcis* (1908*b*, 1909). He arranged for J. P. Rogers to collect for him in NW Australia near Wyndham in 1908, and among the birds collected was another form with rufous shoulders and bluish backed females with rufous lores that he named *rogersi* (Mathews 1912*a*). Mathews then grouped the two forms, *dulcis* and *rogersi*, as subspecies of *M. amabilis*, which in his 1913 List and all his later publications appears as *Leggeornis amabilis*. This arrangement was current for many years (RAOU 1926; Cayley 1949).

Mack in 1934 grouped *dulcis* and *rogersi* as *dulcis*, with *amabilis* as a separate species including only *M. amabilis amabilis* and a new southern subspecies *clarus*, with paler, less violet crown and back, giving no reasons for this significant departure from current practice. In a review of the chestnut-shouldered fairy-wrens collected by the Harold Hall Australian Expedition, Harrison (1972) regarded Mack's *clarus* as a synonym of *amabilis*, since variation in *amabilis* was slight and continuous over its range. Harrison described some specimens of female *assimilis* with greyish plumage that he thought were intermediate between *amabilis* and *assimilis* from the Norman R., at the SE corner of the Gulf of Carpentaria, and between *dulcis* and *assimilis* from the Roper R., NT. On this evidence, S.A. Parker included *amabilis*, *dulcis*, and *rogersi* together with *assimilis*, as subspecies of the Variegated Fairy-wren (*M. lamberti*) in the 1975 Interim List. Subsequently Schodde (1982*b*) reinstated *amabilis* as a full species, citing their relatively shorter tails and arboreal habits, the more rounded ear tufts of the male and the bright blue ear coverts and black bill of the female; this is supported by recent biochemical data (Christidis and Schodde, in press).

Measurements and weights

Wing, M (n = 27) 51.6 ± 1.4, F (n = 13) 49.9 ± 2.9; tail, M (n = 27) 54.3 ± 3.0, F (n = 13) 56.0 ± 2.9; bill, M (n = 27) 10.1 ± 0.4, F (n = 13) 9.7 ± 0.5; tarsus, M (n = 27) 21.7 ± 0.6, F (n = 13) 21.7 ± 1.0; weight, M (n = 9) 7.9–9.8 (8.7 ± 0.9), F (n = 4) 8.0–8.9 (8.3 ± 0.6). (Weights from AM, ANWC and QM)

Field characters

Most arboreal of chestnut-shouldered group, seldom foraging on ground, they glean from leaves and branches, flirting their short, broad tails as do fantails *Rhipidura* spp. Tail feathers broadly white tipped; nuptial male has broad oval ear tufts; blue-backed female differs from Variegated Fairy-wren (*M. l. dulcis*) in having blue ear coverts, and from brown-backed female *M. l. assimilis* and Red-backed Fairy-wren (*M. melanocephalus*), their nearest neighbours (see Table for *M. lamberti*).

Voice

Tape-recorded (ANWC). Typical *Malurus* reel, without prolonged introductory notes (see sonogram). Contact calls a single, long drawn-out 'treee', constantly repeated; alarm a sharp, brief 'zit'.

Range and status

Resident, widespread throughout Cape York Peninsula, from Lockerbie Scrub on N tip S

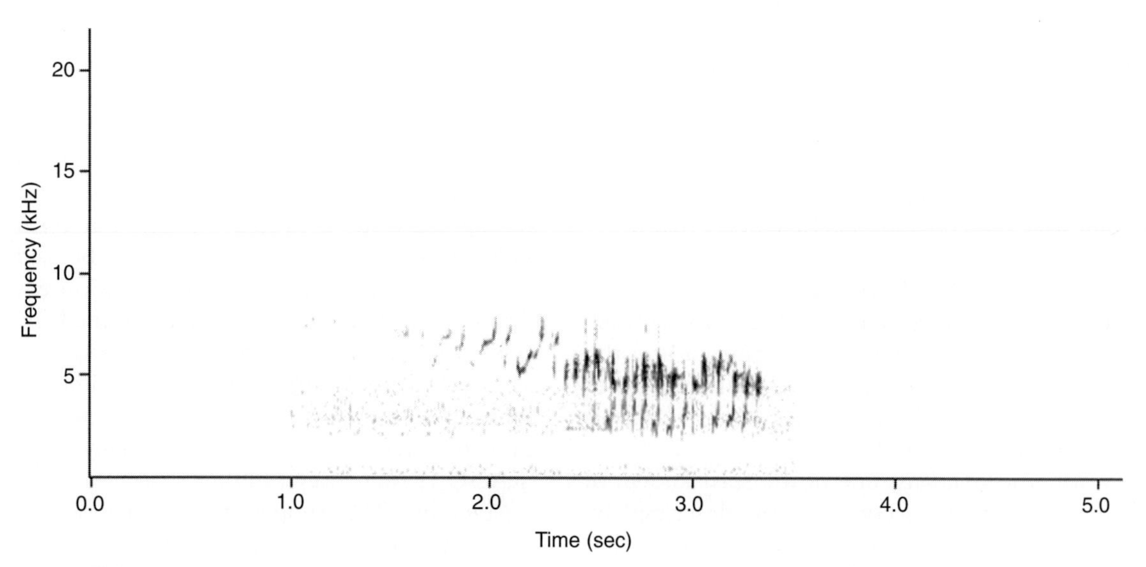

Lovely Fairy-wren (*Malurus amabilis*)
Song. Simon Bennett, Iron Range, Queensland.

to Edward R. on W side and to just N of Townsville on E; mainly coastal, avoiding highlands above 500 m, intergradation with *M. l. assimilis* not established (Schodde 1982*b*). Generally uncommon, common in suitable habitat.

Habitat and general habits

Lives in shrubby margins of closed forest (rainforest) but rarely penetrates more than 50 m into forest; vine thickets, regrowth, heathy margins to mangroves, and *Lantana camara* thickets other favourite areas. Replaces *M. melanocephalus* at edges of grassland. Strong fliers, they glean branches and leaves of trees and shrubs, fanning their stumpy tails, in preference to foraging on ground. Insectivorous: eats a variety of small items hard to identify, including caterpillars (Harrison 1974). Co-operative breeding species, generally seen in pairs or groups of 2–5. No banding studies yet, but appears to be resident throughout year (White 1946). Relationships within group not known; one group recorded with five full-plumaged males (Forshaw and Muller 1978) and another with one nuptial male and three female-plumaged birds (White 1946).

Displays and breeding behaviour

Males seen chasing each other high in shrubbery and also carrying yellow petals (H. B. Gill in Schodde 1982*b*). When alarmed near nest or young, all group members perform Rodent-run distraction display. No other displays recorded. Male seen to feed female but stage of breeding cycle not known (White 1946). Appears socially monogamous, but EPCs probably occur, as in other *Malurus*.

Breeding and life cycle

Nesting recorded in most months (July–Apr); unusual for *Malurus*, male may help female build nest (Macgillivray 1914; White 1946; Tarr 1948, 1949). Two nests found by authors

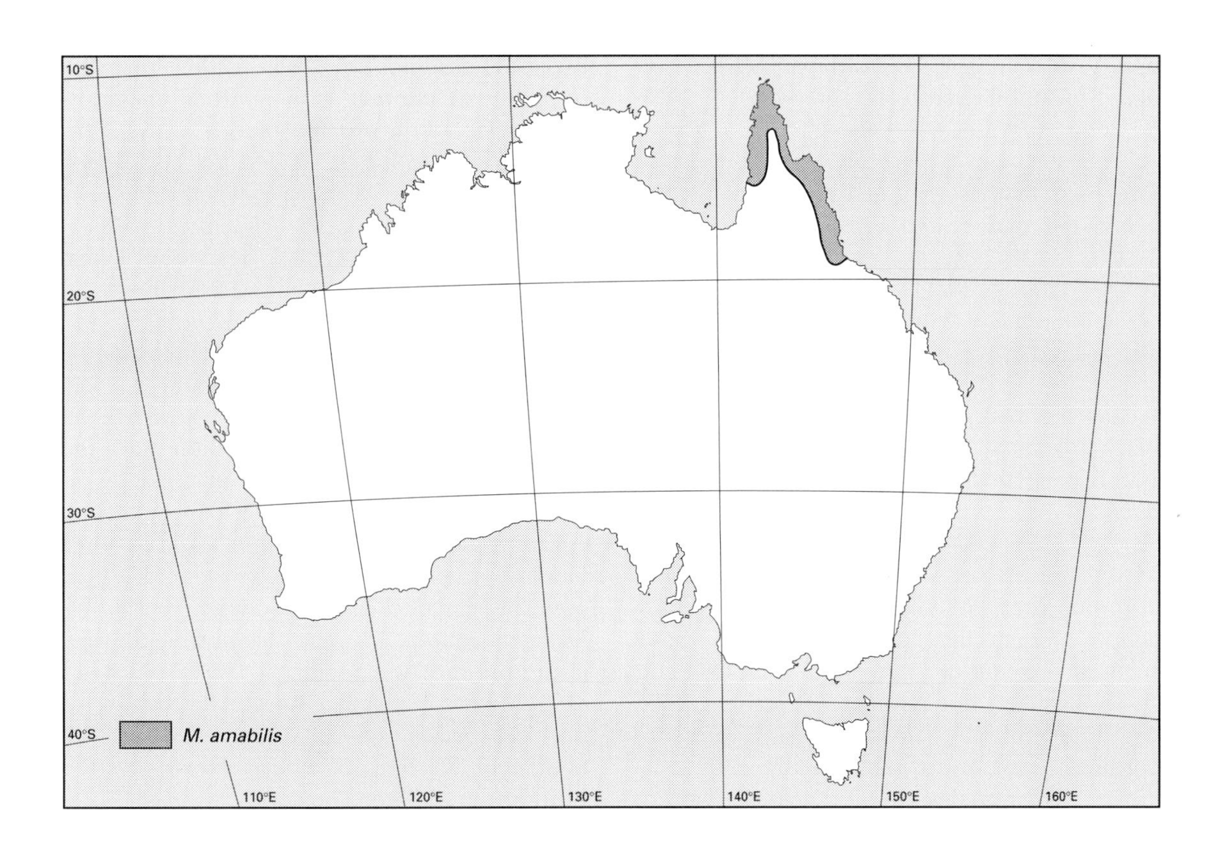

in Lockerbie Scrub (Jan 1992): oval, domed (120–160 × 60–70 mm) with side entrance 45 mm wide and small landing platform; built of twigs, rootlets, bark, and grass with moss in roof; 250 mm and 400 mm from the ground, the nests were suspended from horizontal branches. Clutch 3, eggs (15–17 × 11–13, *n* = 21, Schodde 1982*b*) tapered oval, white with sparse, fine red-brown spotting in zone around blunt end; laid on successive days, incubated by female 13–14 days; nestlings fed and their faecal sacs removed by all group for 14 days (White 1946). Several broods raised in year; independent young probably remain with group and help raise siblings. No colour-banding studies; relationships and survival unknown.

Blue-breasted Fairy-wren *Malurus pulcherrimus* Gould, 1844

Malurus pulcherrimus Gould, 1844. *Proceedings of the Zoological Society of London*, **1844**, p. 106.

PLATE 2

Monotypic

Description

ADULT NUPTIAL MALE: crown, upper back, and ear tufts, violet-cobalt; ear tufts, long, lanceolate, extend back from narrow blue ring round eye; lores, nape, and lower back, black; scapulars, rich rufous; wings, deep grey-brown; tail, dull blue with narrow white tip; throat and breast, dark indigo; belly, greyish white; eye, dark brown; bill, black; feet, grey-brown.

ADULT FEMALE: grey-brown above; rufous lores and feathers round eye; creamy white below; bill, dark brown; eyes and feet, as male.

ADULT ECLIPSE MALE: as female but retains black bill, lores, and trace of pale blue on face and around eye.

IMMATURE: fledglings like female; male shows black in lores at 3 months and whitish feathers round eye (Rowley and Russell, unpublished data).

MOULT: Both sexes moult all feathers every autumn after breeding, most males enter eclipse female-like plumage; males have second pre-nuptial moult in winter/spring (body feathers, not remiges or rectrices) into coloured nuptial plumage. Onset varies according to age, status, and condition of individual. Most males do not achieve perfect nuptial plumage at 1 year. Rarely, old males moult directly from one nuptial plumage to next (Rowley and Russell, unpublished data)

History and subspecies

Gould (1844), when he described specimens from Wongan Hills in SW Australia, distinguished *M. pulcherrimus* from other chestnut-shouldered forms by its indigo-blue throat and breast. He did not know that further inland, it occurred in contact with another chestnut-shouldered form, *assimilis*, and it is the similarity with *assimilis* that led to conflicting opinions for many years as to whether *pulcherrimus* is a separate species or identical with *assimilis*. Thus North (1888) described a nest from the Napier Range (inland from Derby in NW Australia) as that of *M. pulcherrimus* and Hartert identified some specimens collected by J. T. Tunney near the South Alligator R., NT, as *M. pulcherrimus*: 'I suppose these are all *M. pulcherrimus*, though I cannot call the throat and chest deep blue, but consider it, like Dr. Sharpe, to be black' (1905*a*, p. 223).

The reason for the confusion was that since Gould's original description, no-one else had seen the bird until C. P. Conigrave and A. W. Milligan collected it near the Stirling Ranges in the south of WA (Milligan 1903). Mathews (1913) gave vernacular names for several of his subspecies of *Leggeornis lamberti* as Southern, Western, Northern, North-western Blue-breasted Wren; however, he used Blue-breasted Wren for *L. pulcherrimus pulcherrimus*, and *L. p. stirlingi* was the South-western Blue-breasted Wren. Mathews later merged his subspecies *stirlingi* within *L. p. pulcherrimus*.

The ultimate confusion occurred when Mellor described a new subspecies *L. lamberti eyrei*, from Eyre Peninsula, SA, distinguished from *assimilis* by the 'decided dark blue tinge' of the throat and breast (1921, p. 10). Soon afterwards, Ashby (1924) compared Mellor's specimens with one of Conigrave's *pulcherrimus* from WA, and realized that the two were the same. This established the existence of the outlying population of *M. pulcherrimus* on Eyre Peninsula. Mack (1934) did not believe that such a distant population could be derived from *pulcherrimus*, and, although he recognized that the Eyre Peninsula specimens had a blue tinge on the breast, he argued that they must be a variant form of *lamberti* and included them in his subspecies *mastersi*. An extreme solution proposed by Mayr and Serventy (1944), that all the chestnut-shouldered fairy-wrens should be treated as subspecies of *M. lamberti*, was later rejected.

After more specimens had been collected, it became apparent that blue-breasted and black-breasted forms could overlap, with no indication of intergradation (Serventy 1951; Ford 1966). The 1975 Interim List and Schodde (1982*b*) maintain *M. pulcherrimus* as a full species with no subspecies, and this is supported by recent biochemical data (Christidis and Schodde, in press).

Measurements and weights

Wing, M (n = 35) 50.5 ± 1.5, F (n = 14) 49.3 ± 1.4; tail, M (n = 35) 73.0 ± 4.1, F (n = 14) 69.4 ± 3.1; bill, M (n = 35) 9.1 ± 0.5, F (n = 14) 8.6 ± 0.4; tarsus, M (n = 35) 23.0 ± 0.8, F (n = 14) 22.0 ± 0.6; weight, M (n = 29) 8.4–10.3 (9.4 ± 0.5), F (n = 18) 7.8–10.5 (9.0 ± 0.4). (Weights from field measurements by authors.)

Field characters

Colours of nuptial male easily distinguish *M. pulcherrimus* from Splendid Fairy-wren (*M. splendens*) and White-winged Fairy-wren (*M. leucopterus*). Within chestnut-shouldered group, is separated from Red-winged Fairy-wren (*M. elegans*) by the habitat and silvery blue of the latter; *M. pulcherrimus* larger than Variegated Fairy-wren (*M. l. assimilis*), in field separable by more violet crown and indigo breast compared with the latter's bluer crown and black breast; light conditions may make this distinction difficult. In area of overlap, crown and eartufts are the same violet-blue in *M. pulcherrimus*, while *M. l. assimilis* has ear tufts lighter than azure crown. Eclipse male *M. pulcherrimus* are hard to distinguish from *M. l. assimilis* in the field. The rufous lores of female chestnut-shouldered wrens are darker than those of *M. splendens* and *M. leucopterus*; the lores of *M. l. assimilis* are a richer rufous and the bills a brighter red-brown than those of *M. pulcherrimus*. In female *M. elegans*, black bill and rufous lores not extending round eye separate them from female *M. pulcherrimus*.

Voice

Tape-recorded (ANWC). Song typical *Malurus* reel, weaker and higher frequency than *M. elegans*, with fewer, briefer introductory down-slurred notes (see sonogram); maintains contact by single, high brief 'see' calls; single 'tsit' alarm call, with more intense churring alarm rattle. The songs of *M. elegans* and *M. pulcherrimus* are quite similar, having only one or two types of relatively simple repeated elements in their song, but the introductory elements in *M. pulcherrimus* are shorter.

Range and status

Resident, in broad diagonal belt across SW of WA; replaced by *M. elegans* in wetter SW and by *M. lamberti* to N and E; no evidence of interbreeding although so similar in appearance (Ford 1966). In the W, occurs inland from the coast, S of Shark Bay (Tamala), through Galena, Wongan Hills, and Dryandra to Stirling Ra., to S coast at Warriup (70 km E of Albany), and throughout WA wheatbelt in remnant patches of sandplain heath (Ford

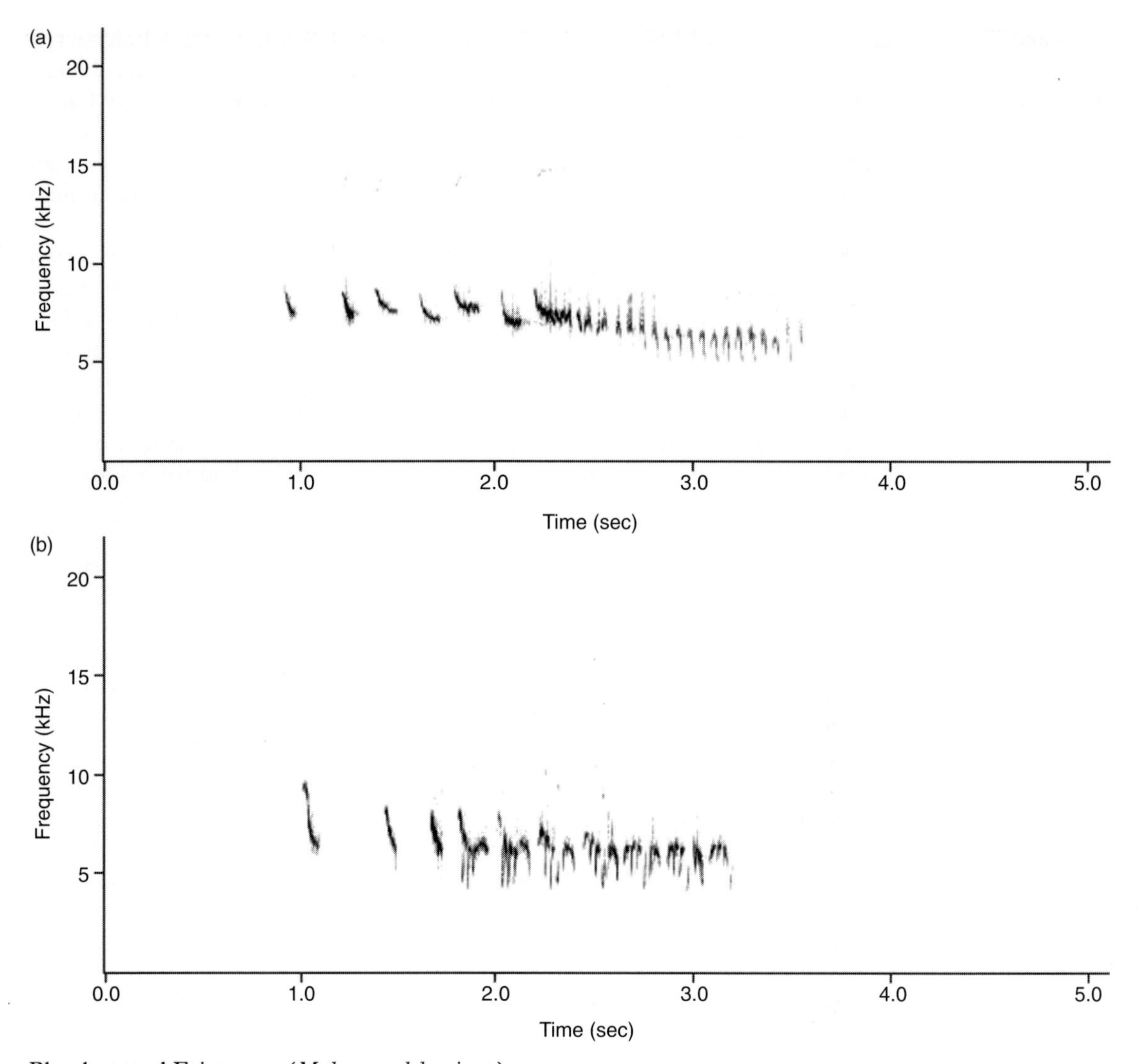

Blue-breasted Fairy-wren (*Malurus pulcherrimus*)
(a) Song. R. B. Payne, Dryandra State Forest, Western Australia. (b) Song, G. S. Chapman, Manmanning, Western Australia.

1969; Rowley 1981*b*), continuing E in coastal heathlands to Eucla. From there, absent for 600 km until the isolated population on Eyre Peninsula, SA up to Gawler Ranges, which is not subspecifically different. Status, uncommon overall, but common where habitat remains suitable. Main threat to survival is clearing of habitat for agriculture; the remaining habitat is very susceptible to fire.

Habitat and general habits

Thrives in tall sandplain heath and eucalypt woodland (*E. wandoo*) with adequate understorey, e.g. poisonbush (*Gastrolobium* spp.). Forage on ground most of winter and in low shrubs throughout spring and summer; reluctant to fly; females may even carry nesting material to nest 'on foot'. Diet: beetles, ants, weevils, flies, wasps, caterpillars (Schodde

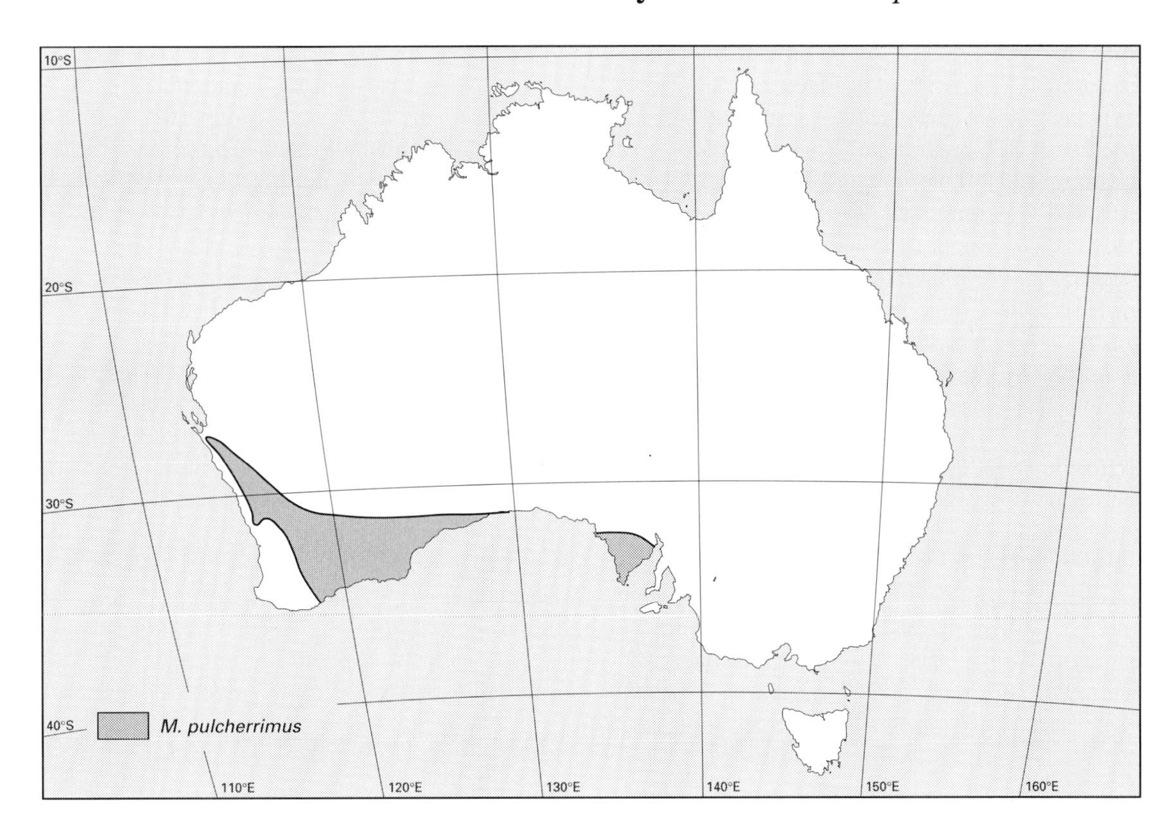

1982*b*; Rowley and Russell, unpublished data). Shy and quiet, probably more common than generally believed. Co-operative breeding species, generally seen in pairs or small groups; studies of colour-banded populations show that groups maintain the same territory throughout the year and from year to year, defended by all members. Groups (mean size 2.6, range 2–7), regularly allopreen and roost together in dense cover. Over 3 years at Dryandra, WA, helpers occurred in 40 % of groups; 26 % of groups contained two or more females; 23 % had two or more males; 48 % of 27 helpers were males, 52 % females (Rowley and Russell, unpublished data.). Helpers feed nestlings, care for dependent fledglings, defend territory, and repel predators or cuckoos; they also probably obtain some EPCs, as do other *Malurus*.

In the dry open woodland of Dryandra State Forest we estimate the territories to be 1–2 ha. Where there are areas of suitable habitat with plenty of dense understorey vegetation, territories adjoin, but the more open areas are left unoccupied so that some territories are not bounded on all sides by neighbours. This species is one of the few that are successfully surviving in the remnants left behind by extensive clearing of heathland for wheat farming (M. G. and L. C. Brooker, unpublished data).

Displays and breeding behaviour

Socially monogamous but possibly sexually promiscuous, as other *Malurus*; remain paired while both partners survive. No courtship displays recorded; orange and yellow petals sometimes carried by males; Rodent-run distraction display given by all members when nest or young threatened (Rowley and Russell, unpublished data).

Breeding and life cycle

At Dryandra WA, of 93 nesting attempts between Sept and Jan, only two clutches laid in Sept, and eight in Jan, the remainder equally distributed Oct–Dec (see Fig. 7.1(c)). Nest built by female; oval, domed (mean size 111 × 66 mm, n = 21), side entrance near top; outer layer of coarser grass, leaves, and small sticks, lined with fine grass, plant down, feathers, and mammal fur; mean height of base 585 mm (100–1500) above ground; change in both substrate and height for later nests. Early nests (Sept–Oct) close to ground, frequently in piles of dead brush; later nests higher, in the shady canopy of a live, green shrub (*Gastrolobium*), better ventilated and more isolated from ground predators. Clutch, mean 2.8 ± 0.4 (2–3, n = 33); eggs (16–18 × 11–14 mm, Schodde 1982*b*) tapered oval, dull cream-white, with sparse spots and fine blotches of red-brown, concentrated in zone around blunter end, laid on successive days. Incubation by female; 14–15 days after last egg laid. Nestlings hatched within 24 hours of each other, brooded initially by female and remain in the nest for 10–12 days; fed, and faecal sacs removed, by all adults of group, but not by young of the year. Fledglings remain cryptic for a week, then travel with the foraging group, fed for *c.* 1 month; young birds remain in the family group long after independence. Both sexes can breed at 1 year but most do not do so until 2 years old; males usually remain in natal group for 1 or more years as helper before inheriting the senior position, or moving next door to a vacancy. Some yearling females help in natal group; some disperse after 1 year and help in another group. Most dispersal by males or females at Dryandra only to nearby group (Rowley and Russell, unpublished data). Extensive studies of isolated remnant populations in wheatbelt by M. G. and L. C. Brooker have shown much more extensive dispersal.

Productivity low despite repeat nesting and multiple broods. Females lay 1–4 (mean 2.3) clutches/year, re-nest after failure, 13% reared second brood (n = 46); 11% of nests were parasitized by Horsfield's Bronze-Cuckoo (*Chrysococcyx basalis*), other failures were due to predation, mainly by reptiles (45% of nests); some nests already parasitized by a cuckoo were taken by predators and since we have attributed that failure to the cuckoo, the proportion of nests taken by predators is higher, reaching 50%. Mean breeding success (fledglings from eggs) in Dryandra Forest was only 33%, with a range over 4 years of 24–63%, but 2 of those years (1993, 1994) were very dry, and levels of predation were always high; 54% of fledglings reached independence, 29% survived to 1 year. Over 4 years, a mean of 74 % of groups (n = 46) produced fledglings, 1.9 per group, of which 1.0 reached independence, and 0.5 survived to 1 year old; mean sex ratio was 1 M : 1 F; mean annual survival of adult males 78%, breeding females, 65% (Rowley and Russell, unpublished data).

Red-winged Fairy-wren *Malurus elegans* Gould, 1837

Malurus elegans Gould, 1837. *A Synopsis of the Birds of Australia and the Adjacent Islands*. Pt 1, pl. 2.

PLATE 2

Monotypic

Description

ADULT NUPTIAL MALE: crown, upper back and ear tufts, iridescent, silvery sky-blue; ear tufts, long, lanceolate, extend back from narrow blue ring round eye; lores, nape, and lower back, black; scapulars, bright rufous; wings, deep grey-brown; tail, dusky blue with narrow white tip; throat and breast, navy blue; belly, greyish white; eye, dark brown; bill, black; feet, dark grey-brown.

ADULT FEMALE: grey-brown above, greyer over crown, scapulars suffused with rufous; lores, deep rufous, *not encircling eye*; eyering, pale grey; chin and breast, white; belly, grey-buff; bill, black; eyes, feet, and tail, as male.

ADULT ECLIPSE MALE: as female, but retains black lores; often with traces of blue.

IMMATURES: fledglings like female; male shows black in lores from 6 weeks (Russell *et al.* 1991).

MOULT: Both sexes moult all feathers every autumn after breeding, most males enter eclipse female-like plumage; males have second pre-nuptial moult in winter/spring (body feathers, not remiges or rectrices) into coloured nuptial plumage; onset varies according to age, status, and condition of individual. 1-year-old males: rarely achieve near-full nuptial plumage (16%), remainder variable: no blue visible (16%), trace blue in face (14%), 'spotty': ear tufts and crown, half sky-blue, half grey; breast, half navy blue, half grey; trace of rufous in scapulars (53%). All have black bill and lores ($n = 90$, known age). Rarely, old males moult directly from one nuptial plumage to next (Russell *et al.* 1991).

History and subspecies

Gould's description of *Malurus elegans* in 1837 gave its locality as the E coast, but with further material from John Gilbert collected in 1839 and 1840 in SW Australia, he soon realized his error, and his 1865 Handbook gave the locality as the Swan R. It has a restricted distribution in SW Australia, with little variation, despite the isolation of populations in the Stirling and Porongorup Ra. for perhaps 5000 years. Mathews in 1916 separated birds from the more southerly Karri forests as a subspecies *warreni* on the grounds that the females were darker, but as Schodde (1982*b*) says, 'they are not', and we agree.

Measurements and weights

Wing, M ($n = 41$) 51.6 ± 1.5, F ($n = 10$) 49.0 ± 0.7; tail, M ($n = 41$) 75.3 ± 3.5, F ($n = 10$) 71.5 ± 3.9; bill, M ($n = 41$) 10.0 ± 0.6, F ($n = 10$) 9.3 ± 0.6; tarsus, M ($n = 41$) 24.5 ± 0.8, F ($n = 10$) 23.4 ± 0.9; weight, M ($n = 468$) 8.5–11.5 (9.9 ± 0.5), F ($n = 323$) 8.0–11.5 (9.3 ± 0.6). (Weights from field measurements by authors and R.J. and M. N. Brown.)

Field characters

Distinguished from adjoining Blue-breasted Fairy-wren (*M. pulcherrimus*) and sympatric Splendid Fairy-wren (*M. splendens*) by silvery blue nuptial plumage of male (especially crown and upper back) and in female, black bill, deep rufous lores not encircling eye, and rufous suffusion of scapulars (see Table, p. 160). Eclipse male of *M. elegans* and *M. pulcherrimus* virtually indistinguishable in the field, unless traces of diagnostic blue in face (silvery or violet blue). However, *M. elegans* usually found in wetter forest environments. Eclipse male *M. splendens* have blue primaries. Long introductory whistles to the song reel unique to *M. elegans*. Reel of *M. splendens* is louder and harsher, and that of *M. pulcherrimus* higher, weaker and shorter.

Voice

Tape-recorded (ANWC; Payne *et al.* 1991; Baker 1995). Song a typical *Malurus* reel, a trill preceded by three or four prolonged down-slurred whistles (see sonogram). It is given by both sexes (breeders and helpers), identifies individuals (Payne *et al.* 1991), maintains contact and advertises territory, especially when boundaries are disputed. When used as a contact call, the trill is softer and shorter. Groups foraging through dense cover maintain contact with soft calls, a series of 3–4 descending 'see-see-see', repeated continually; alarm loud, sharp 'tsit'. The songs of *M. elegans* and *M. pulcherrimus* are similar, having only one or two relatively simple repeated elements, but the introductory elements in *M. pulcherrimus* are shorter.

Range and status

Resident in the higher rainfall forested SW from Moore R. N of Perth to S coast and E to Warriup, 70 km east of Albany; inland to where SW Jarrah (*E. marginata*) forest gives way to

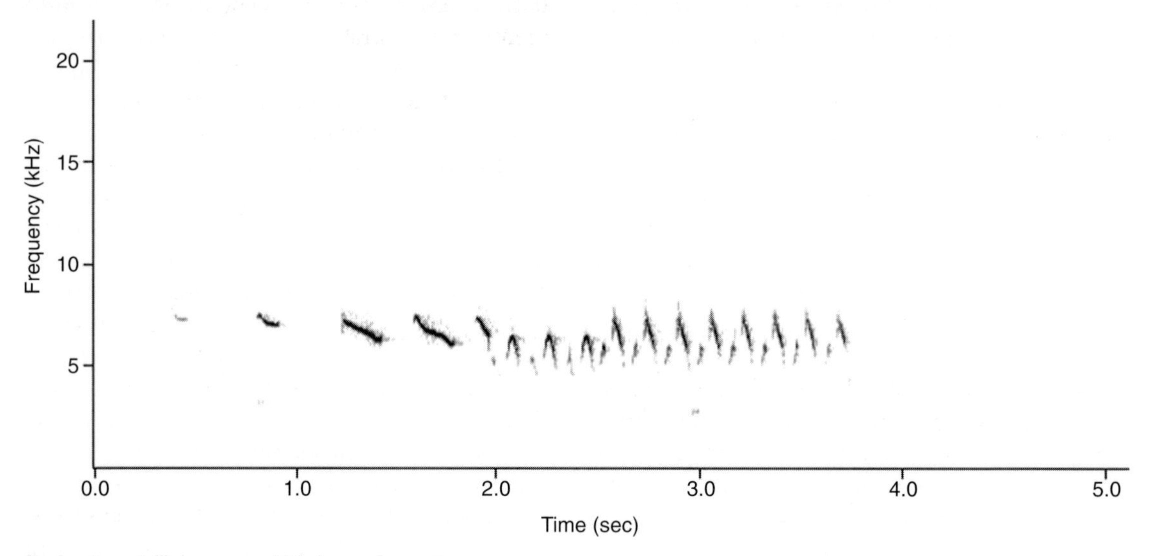

Red-winged Fairy-wren (*Malurus elegans*)
Song. R. B. Payne, Smith's Brook Reserve, Manjimup, Western Australia.

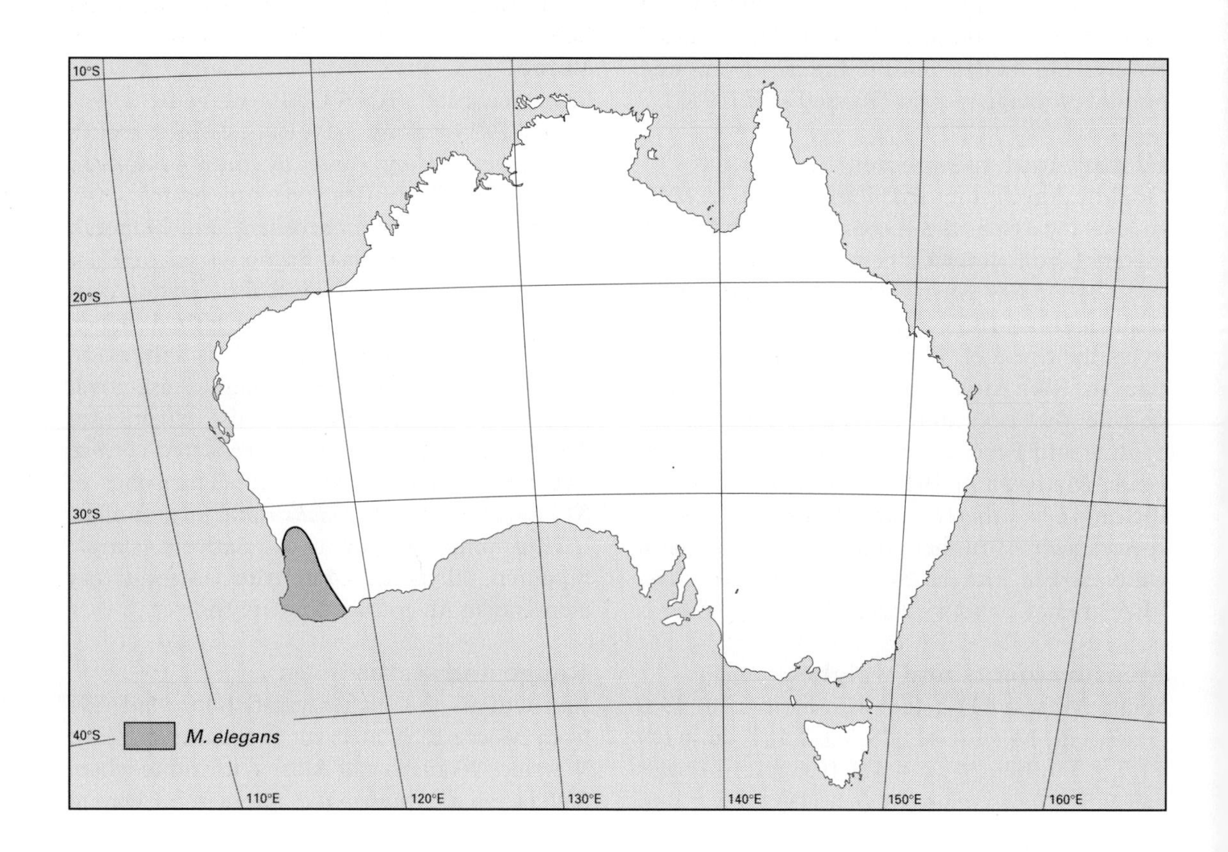

open woodland; replaced by *M. lamberti* to extreme NW and *M. pulcherrimus* to N and E.

Habitat and general habits

Lives in dense cover; common throughout understorey of Karri (*E. diversicolor*) forest, including riverside Cutting Sedge (*Lepidosperma effusum*); in tall, wet coastal heath, and denser, wetter gullies of Jarrah (*E. marginata*). Shy, hard to see most of year; probably more common than hitherto believed (3.2 ± 0.4 birds per ha over 6 years, Rowley *et al.* 1988). Largest of chestnut-shouldered fairy-wrens, with small wings, long tails, and weak flight; forage mostly on or near the ground, by hop-search and pounce in litter and dead brush, by gleaning in low shrubs, bracken (*Pteridium esculentum*), and occasionally in understorey trees to 5 m. Insectivorous: eats ants and beetles throughout year plus spiders, bugs, and caterpillars in breeding season (Wooller and Calver 1981*a*; Rowley *et al.* 1988). Co-operative breeding species, generally seen in pairs or small groups; studies of a colour-banded population show groups maintain same territory throughout year (see Fig. 5.1). In the dense and varied understory of tall Karri forest in SW where the climate is one of the least variable in Australia, occupies territories of 1–2 ha, with boundaries remarkably stable over 10 fire-free years. After an intense accidental autumn wildfire, most territories and their owners remained the same for the following breeding season. In less favourable environments, territories are more restricted, confined to dense stream-side vegetation.

Territories are defended by all members, who regularly allopreen and roost together in dense cover. Groups (mean size 4.0, range 2–9) remain in their territories (mean 1.2, range 0.4–2.4 ha) throughout the year and from year to year. Over 6 years, helpers occurred in 82.5% of groups; 55% of groups contained two or more females but 57% of 337 helpers were males (Rowley *et al.* 1988). Territorial disputes involve all group members, generally males against males and females against females. Vocal identification of individuals is important in recognition of group members and strangers (Payne *et al.* 1991).

Displays and breeding behaviour

Socially monogamous but probably sexually promiscuous (as other *Malurus*); remain paired while both partners survive; divorce is rare; males carry yellow petals (Rowley 1991). Pale blue mantle feathers of males used in display more than in other species. Face Fan and Wing-fluttering displays commonly used. Rodent-run distraction display by all group members easily provoked when nest or young threatened.

Breeding and life cycle

Compared with other *Malurus*, breeding season is short (Oct–Dec; see Fig. 7.1(d)); of 252 nesting attempts between 1980 and 1986 at Manjimup WA, only four clutches laid in Sept and six in Jan; length of breeding season varies, even in one of the least variable Australian environments; over 6 years 1982–6, first clutch laid between 16 Aug and 10 Oct, and last between 10 Dec and 10 Jan. Second broods not common; only 16.6% of females re-nested after successful first brood. Nest built by female; oval, domed (mean size 130 × 77 mm, $n = 154$), side entrance near top; outer layer of coarser grass, leaves, bracken, and small sticks, lined with fine grass and plant down from *Clematis pubescens* or bast from *Banksia grandis*; mean height of base above ground 215 mm (40–1200); 76% of 193 nests were in dead bracken and/or dead brush, half of which were grown through with small shrubs. Clutch 2.4 (1–3, $n = 117$); eggs (16.6–18.7 × 12.6–13.1 mm, $n = 15$) tapered oval, dull cream—white, with sparse spots and fine blotches of brown, concentrated in zone around blunter end, laid on successive days before midday. Incubation by female for 14–15 days after last egg laid. Fertility 94%; nestlings hatch within 24 hr of each other; they are brooded initially by the female and stay in the nest for 11–12 days, where they are fed and their faecal sacs removed by all adults of group, but not by young of the year. The fledglings are still growing (see Fig. 7.9) and remain cryptic for a week, then travel with the foraging group, who feed them for *c.* 1 month; young birds remain in the family group long after independence. Males are sexually mature at 1 year, but usually remain in their natal group for 1 or more years as helper before inheriting the senior position, or moving next door to a vacancy. Helpers feed nestlings, care for dependent fledglings, defend the territory, and repel predators or cuckoos. Because annual survival of breeding females is high (79%), breeding vacancies are scarce; mean age of females at first breeding is 2.8 years (1–6), either in their natal or a neighbouring group. Most young females help in their natal group for a year, then many disperse and help in another group. Most dispersal by males or females only to neighbouring group (Rowley and Russell, unpublished data). Over 6 years, 52.3% of eggs laid and 57.3% of nests laid in (*n* = 117) produced fledglings; 76% of fledglings reached independence, and 42.6% survived to 1 year. Females lay 1–3 (mean 1.4) clutches/year and re-nest after failure; in Karri forest near Manjimup, WA, the level of parasitism was very low, only six of 206 known nests (3%), two Fan-tailed Cuckoos (*Cacomantis flabelliformis*), three Horsfield's Bronze-Cuckoo (*Chrysococcyx basalis*), and one unknown; Horsfield's Bronze-Cuckoo was rarely heard, but Fan-tailed Cuckoos were common; other failures were due to predation (mainly reptiles). Each year, a mean of 76.7% of groups produced fledglings, 2.5 per group, 1.9 of which reached independence and 1.1 survived to 1 year. Over 11 years, the mean sex ratio was 1.2 M : 1 F; mean annual survival of adult males 78%, breeding females 79% (see Fig. 7.15); survival of helpers (M: 73%, F: 58%) reflects greater dispersal by females (Rowley *et al.* 1988).

White-winged Fairy-wren *Malurus leucopterus* Dumont, 1824

Malurus leucopterus Dumont, 1824. *Dictionnaire des Sciences Naturelles*, **30**, p. 118.

PLATE 3

Polytypic. Three subspecies. *Malurus leucopterus leucopterus* Dumont, 1824; *Malurus leucopterus leuconotus* Gould, 1865; *Malurus leucopterus edouardi* A. J. Campbell, 1901.

Description

M. l. leucopterus

ADULT NUPTIAL MALE: black all over except for white shoulder patch of scapulars, secondary wing-coverts, and innermost (3–6) secondaries; primary coverts, black; outer flight feathers, grey-brown edged blue; tail, deep dull blue; bill, black; eye, dark brown; feet, dark grey-brown or black.

ADULT FEMALE: drab grey-brown crown, back, and wings; tail, grey faintly washed with turquoise; lores, feathers round eye, and undersurfaces, off-white; eyes and legs, as male; bill, pinkish brown.

ADULT ECLIPSE MALE: resembles female except for darker bill.

IMMATURE: sexes indistinguishable; once tails full grown, resemble female, but newer plumage, less worn, and tails less blue.

MOULT: both sexes moult all feathers in autumn after breeding; most nuptial and part-nuptial males enter eclipse female-like plumage with second pre-nuptial moult in winter/spring into coloured nuptial plumage (body feathers only, not remiges or rectrices). Rarely, old males moult direct from nuptial to nuptial plumage (Tidemann 1989; Rowley and Russell 1995). Unlike other *Malurus* which may achieve nuptial plumage at 1 year old, males do not do so until their fourth year. After their second year, they progressively approach full nuptial plumage, attained in fourth year. Males entering second year are indistinguishable from females except for slightly bluer tails and a cloacal protruberance. In third year, some males show small patches of white on shoulders with rest of body grey or with patches of blue; bills darken, losing pinkish colour as male becomes blue (Tidemann 1989; Rowley and Russell 1995).

History and subspecies

Three subspecies currently recognized (Schodde 1982*b*). *Malurus leucopterus leucopterus* Dumont, 1824, confined to Dirk Hartog Is., WA: small, with long tail; male black, female drab grey-brown. *Malurus leucopterus leuconotus* Gould, 1865, widespread on mainland: male mid to deep cobalt-blue, occasional individuals may appear black; female drab grey-brown, becoming paler and browner underneath from S to N; both tending smaller S to N. *Malurus leucopterus edouardi* A. J. Campbell, 1901 on Barrow Is., WA: larger than nominate, tail shorter, male as black as nominate, female cinnamon fawn above, creamy-white below.

The first specimens collected were of a distinct island form from Dirk Hartog Is. off the coast of WA near Shark Bay, taken by Quoy and Gaimard in 1818 in the course of the Freycinet expedition, and described by Dumont (1824) from a painting by Arrago. In this form, nuptial males are glossy black, with a white shoulder patch formed by scapulars, secondary wing-coverts, and the 3–6 innermost secondaries. These birds were not seen again for nearly 100 years, until re-discovered on Dirk Hartog Is. (Carter 1917), by which time a similar black-and-white form, *M. l. edouardi*, had been described from Barrow Is., much further north (Campbell 1901*c*).

Quoy and Gaimard's original specimens were lost in a shipwreck, and this caused doubt about the correctness of the original description, since all the white-winged birds resembling *leucopterus* in mainland Australia were a brilliant deep cobalt blue. Gould (1865) doubted that the birds from Dirk Hartog Is. could be the same species as his blue-and-white White-winged Superb Warbler from inland NSW, which was never found near the coast, and for which he proposed the name *Malurus cyanotus* should they prove to be different. He had also described, from a single specimen in a private collection (1865, p. 332), a second blue-and-white bird, the White-backed Superb Warbler (*Malurus leuconotus*), that appeared to have a white back, in addition to white scapulars, wing-coverts, and innermost flight feathers. At the end of the nineteenth century, both names were in use for what were treated as two quite different species of blue-and-white birds. A. J. Campbell, in his treatise on nests and eggs, did not appear to find it at all odd that there should be two such similar species in similar habitats in the same areas (1901*a*). North was 'by no means certain that they are distinct...', especially when specimens of both forms originated from the same place (1904, Vol. 1, p. 217). The compilers of the 1913 Checklist expressed this doubt by listing only *M. cyanotus* and *M. leucopterus*.

There were thus two sources of confusion. First, were the black-and-white forms a separate species from the blue-and-white forms? Second, were the blue-and-white White-winged Wren and the White-backed Wren the same or different? Mathews had no doubt. In his 1913 List, the Blue-and-White Wrens of his genus *Hallornis* were represented by two species, *H. cyanotus* the White-winged Wren and *H. leuconotus* the White-backed Wren. The black-and-white wrens, *leucopterus* and *edouardi*, he placed in the separate genus *Nesomalurus*!

In describing his rediscovery of *M. leucopterus* on Dirk Hartog Is., Carter discussed the question of the White-backed versus the White-winged Wren, having seen many specimens before they were prepared as skins. He explained that all had varying amounts of white on the back, overlaid by long blue feathers, and that 'much depends on the making of a skin as to whether these white feathers show or not.... After examining a long series from various parts of Australia, Mr. G. M. Mathews and myself are agreed that there is but one species...'(1917, p. 592). F. Lawson Whitlock, that experienced observer and collector, agreed. Commenting on Carter's illustration of the skin of a White-winged Wren with white feathers across the back, Whitlock wrote in 1921 'Personally, I have never met with a White-winged Wren showing this characteristic so perfectly; but I could easily make up a skin to do so' (p. 180). By this time, there was general agreement that White-backed Wrens were an artefact.

Completely overturning this agreement, however, a specimen showing a white back turned up from Lyndhurst in South Australia, with the whole back white, as well as the scapulars. It was hailed as a representative of the 'long-lost, probably very local but quite distinct bird, *M. leuconotus*' by Kinghorn and Iredale (1924), and reinstated on the lists of Australian birds. The 1926 Checklist maintained three separate species, *M. leucopterus*, *M. cyanotus*, and *M. leuconotus*, while Mathews (1930) placed *cyanotus* as one of five subspecies of *leuconotus*. Only three confirmed specimens of white-backed males are known, one recorded by Kinghorn and Iredale (1924), another collected in north-western NSW (Boles and Dingley 1977), and a third collected in 1988 from west of Windorah, Qld, and now in the ANWC (R. Schodde, personal communication).

Mack (1934) recognized that occasional variant individuals had a white back while most did not. He maintained the black-and-white Dirk Hartog form as a separate species *leucopterus*, and thought it unfortunate that the name *leuconotus* had priority over the more suitable *cyanotus* for the blue-and-white form. Both forms are variable, with some signs of blue feathers in black plumage, and blue-and-white birds varying from light blue to almost black on the mainland of NW Australia. More recent opinion (Parker 1975; Schodde 1982*b*) places the blue-and-white and black-and-white forms together in the appropriately named species *leucopterus* (white-winged), with subspecies *leucopterus*, *leuconotus*, and *edouardi*.

Measurements and weights

Malurus leucopterus leucopterus: wing, M (n = 14) 43.7 ± 1.5, F (n = 8) 42.4 ± 1.1; tail, M (n = 14) 55.8 ± 2.2, F (n = 8) 57.6 ± 1.1; bill, M (n = 14) 8.5 ± 0.5, F (n = 8) 8.4 ± 0.7; tarsus, M (n = 14) 19.9 ± 1.2), F (n = 8) 19.7 ± 1.2.

Malurus leucopterus leuconotus: wing, M (n = 119) 47.3 ± 1.5, F (n = 40) 45.5 ± 1.2; tail, M (n = 119) 57.4 ± 2.6, F (n = 40) 57.6 ± 1.1; bill, M (n = 119) 8.5 ± 0.5, F (n = 40) 8.4 ± 0.7; tarsus, M (n =119) 19.8 ± 0.8, F (n = 40) 19.7 ± 1.2; weight, M (n = 36) 6.7–8.4 (7.6 ± 0.4), F (n = 26) 6.3–8.2 (7.2 ± 0.6).

Malurus leucopterus edouardi: wing, M (n = 10) 45.4 ± 0.9, F (n = 6) 44.2 ± 1.0; tail, M (n = 10) 54.0 ± 1.9, F (n = 6) 56.3 ± 2.6; bill, M (n = 10) 8.4 ± 0.3, F (n = 6) 8.5 ± 0.6; tarsus, M (n = 10) 19.5 ± 0.8, F (n = 6) 19.3 ± 0.6; weight, M (n = 14) 6.9–8.2 (7.5 ± 0.5), F (n = 4) 7.0–8.1 (7.6 ± 0.4).

(Weights from Rowley and Russell 1995; Serventy and Marshall 1964; and R. D. Wooller, personal communication)

Field characters

Within the genus, only the Red-backed Fairy-wren (*M. melanocephalus*) is smaller. Full nuptial male unmistakeable whether black or blue. Most birds encountered are female-plumaged; small size and plainness, with pale bill and lores, and drab grey upper surface, distinguish them from sympatric Splendid Fairy-wren (*M. splendens*) and Variegated Fairy-wren (*M. lamberti*). Female *M. melanocephalus* have plain brown tails and warmer brown upper surface.

Voice

Tape-recorded (ANWC; Tidemann 1980; Baker 1995). Song a prolonged reel introduced by three to five short notes, lasting up to 4 sec (sonogram). Reel sounds more regular than in other *Malurus*, with frequencies rising and falling almost as sine wave; given by both sexes, reel appears to maintain group cohesion and to advertise territory. Relatively weak, high-pitched song carries a surprising distance above the low vegetation of their habitat. Contact 'tsit' or low intensity reel between group members when closer together. Alarm call an abrupt harsh 'trit'.

Range and status

Resident; occurs throughout arid and semi-arid areas between latitudes 20 and 34°S, including major deserts and Nullarbor Plain; inland from Great Dividing Ra. in E to Barrow and Dirk Hartog Is. in W; the only *Malurus* on Barrow Is. Not present in the wetter SW and SE; replaced by *M. melanocephalus* in N. Abundant in suitable habitat.

Habitat and general habits

Common in low shrubland throughout arid and semi-arid, especially in samphire on salt-pans, chenopod shrublands, desert sandhills with cane-grass (*Zygochloa*), floodplains with Lignum (*Muehlenbeckia*), and extensive spinifex and heathy areas; replaced by *M. lamberti* or *M. splendens* in taller vegetation and in mallee and mulga woodlands. Their small size enables them to forage easily over and through shrubs, gleaning leaves and stems, flying brief sorties to catch flying prey, and hop-searching on ground. Insectivorous; diet includes crickets, bugs, beetles, weevils, ants, caterpillars, spiders, with some seeds and fruits e.g. *Chenopodium*, *Rhagodia*, *Euphorbia*, *Portulaca* (Barker and Vestjens 1990). Co-operative breeding species, generally seen in pairs or small groups (2–11); socially complex and not fully understood. Males in nuptial plumage rare, outnumbered by brown birds (5 of 61, White 1949). Population appears to consist of 'clans', each with a full plumaged nuptial male and usually several breeding females and brown males; in the non-breeding season, they range over a large clan territory maintained from year to year (Tidemann 1980, 1983; Rowley and Russell 1995). On our study area at Pipidinny, WA, clan areas were contiguous and all habitat was occupied by territories. In winter, we regularly found birds where we expected to find them in their known clan areas with their usual associates, and we have no evidence that groups coalesced and wandered.

However, at Booligal, groups foraged together in extra-territorial areas that had remained unoccupied throughout the breeding season (Tidemann 1983). Boehm (1957), in an agricultural area, recorded small flocks with

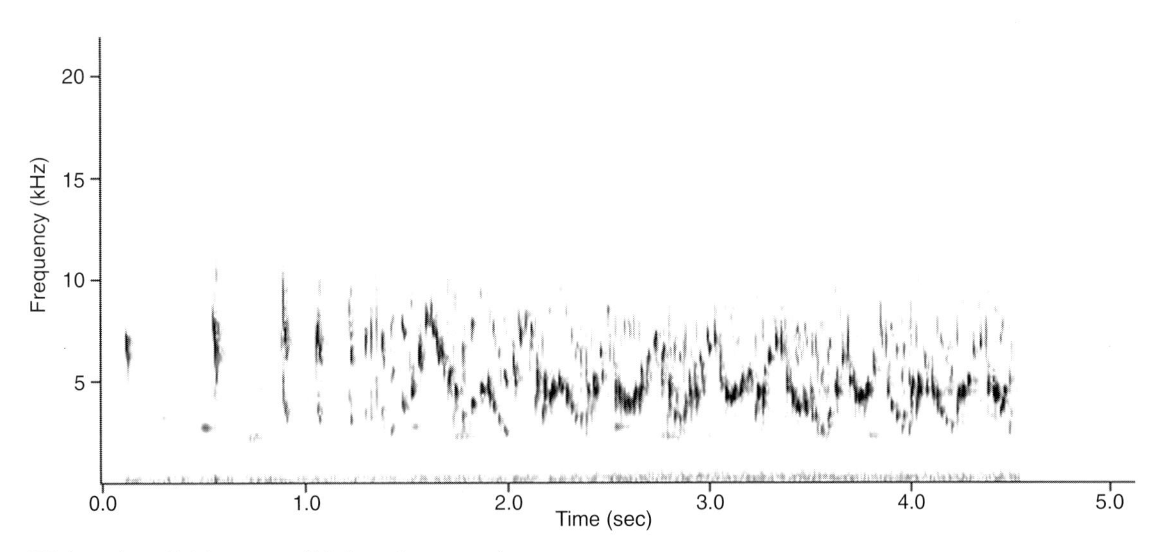

White-winged Fairy-wren (*Malurus leucopterus*)
M. l. leuconotus, song. R. K. Templeton, SW Western Australia.

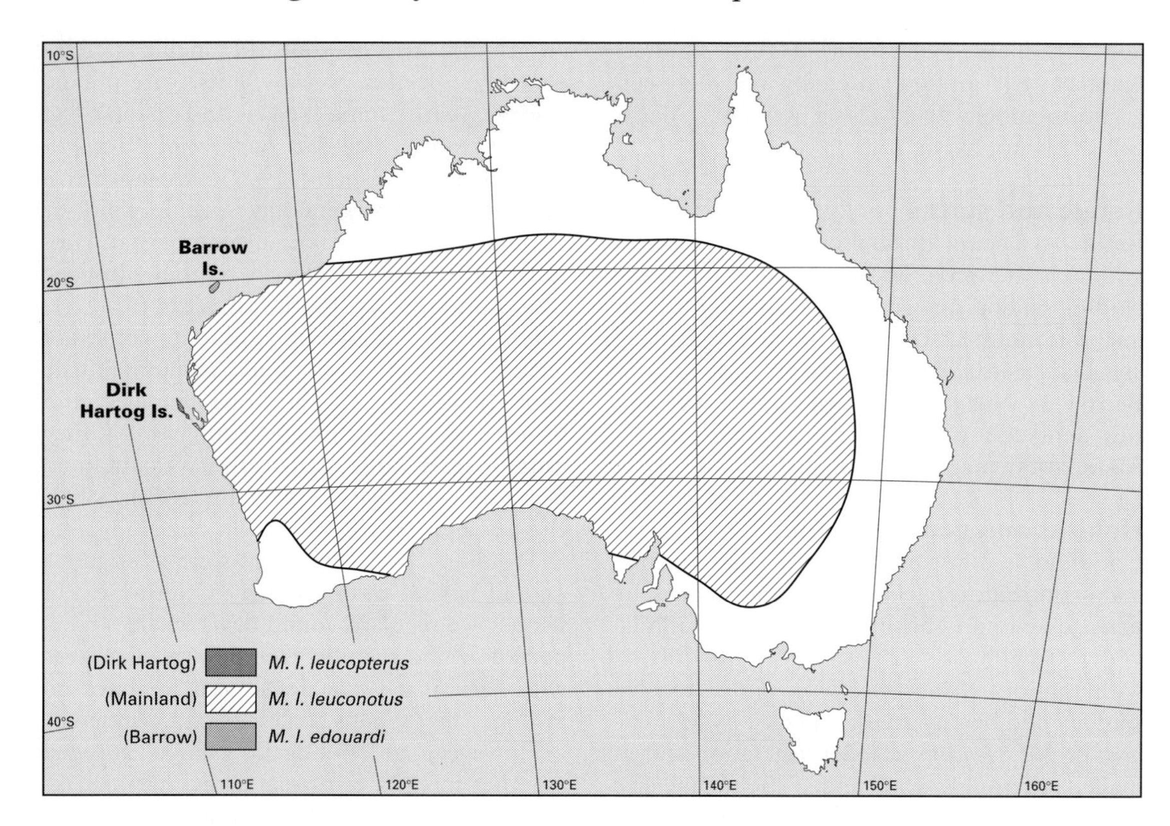

three or four blue males in full plumage in autumn, suggesting that more than one clan had amalgamated and were wandering far from their usual haunts.

Displays and breeding behaviour

Nuptial males carry blue petals in display, usually when intruding into a neighbouring territory; all adults perform the Rodent-run distraction display when nestlings or group members are threatened; allopreen frequently; Wing-fluttering display by both sexes is probably a form of appeasement, as well as sexual invitation. In breeding season, several clan females may nest apart; observations suggest nuptial male attends each female during nest-building and probably mates with her (Tidemann 1983); nuptial male helps feed nestlings of primary female; other females assisted by brown males that tolerate nuptial male visits; other clan members (either sex) may also help. Possible EPC by neighbours, but lack of DNA analysis at present precludes confirmation of relationships of group members.

Breeding and life cycle

Reported laying in every month except June, particularly after good rains but mainly spring and summer (Tidemann and Marples 1987; see Figs 7.1, 7.2). Female builds nest, smallest of any *Malurus*; oval, domed, mean size 105 × 63 mm (90–140 × 60–70, n = 15), side entrance near top; outer layer of fine grasses, lining plant down and feathers; usually less than 1 m above the ground (mean = 237 mm, 120–370 mm, n = 15), deep in a bush. Clutch 3.5 (2–4, n = 75, RAOU NRS; Rowley and Russell 1995); eggs (15–16 × 10–12 mm, n = 57, Schodde 1982*b*), tapered oval, creamy-white, with sparse red-brown spots in zone around blunter end, laid on successive days. Incubation by female, 13–14 days after last egg laid. Fertility 97%, nestlings hatched within 24 hr of each other, and brooded initially by female, remain in nest for 10–11 days, fed and faecal sacs removed by female, her male partner, and sometimes helpers; fledglings remain cryptic for 1 week, fed by parents for *c.* 4 weeks, young birds remain in the group after independence and may help rear young of later broods. Male sexually mature at 1 year, although still brown plumaged, and usually remain in natal group for 1 or more years as helper. Some yearling females remain as helper in natal territory; age at first breeding not known. From study of colour-banded population over 4 years, 81% of eggs laid (n = 93 eggs) and 88% of nests laid in (n = 34 nests) produced fledglings; 50% of fledglings reached independence after *c.* 4 weeks, only 10% of fledglings known to be alive after 1 year, but significant dispersal is likely. Females re-nest after failure and after rearing first brood; 20% of females (n = 39) reared two broods (Rowley and Russell 1995). No brood parasitism in this study, but Horsfield's Bronze-Cuckoo (*Chrysococcyx basalis*) recorded as a nest parasite by Brooker and Brooker (1989*a*); at Booligal, NSW, Tidemann (1983) found 36% of 29 nests parasitized by Horsfield's Bronze-Cuckoo. Annual survival of all adult males, 54% (n = 41), of nuptial males, 64% (n = 14), and 45% of banded breeding females (n = 35) survived to breed 1 year later (Rowley and Russell, unpublished data).

Red-backed Fairy-wren *Malurus melanocephalus* (Latham, 1801)

Muscicapa melanocephalus Latham, 1801. *Supplementum Indicis Ornithologi*, London, p. 52.

PLATE 3

Polytypic. Two subspecies. *Malurus melanocephalus melanocephalus* (Latham, 1801) and *Malurus melanocephalus cruentatus* Gould, 1840.

Description

M. m. melanocephalus

ADULT NUPTIAL MALE: body, satin-black except for dusky-brown flight feathers (black on innermost 3–5), and fiery orange of scapulars, and upper and lower back; tail, black; eye, dark brown; bill, black; feet, light brown.

ADULT FEMALE: upper surfaces, warm mid-brown; wings and tail, dark grey-brown; lores and eye rim, pale buff; underneath, whitish; throat, thighs, and flanks, tinged cinnamon-buff; eye, dark brown; bill, pinkish brown; feet, light brown.

ADULT ECLIPSE MALE: indistinguishable from female except for darker bill.

IMMATURE: sexes indistinguishable; once tails full grown, resemble female, but newer plumage, less worn.

MOULT: in temperate areas, main body moult in autumn; both sexes moult all feathers in autumn after breeding; appear to moult tail twice a year (Schodde 1982*b*). Most nuptial and part-nuptial males enter eclipse female-like plumage with second pre-nuptial moult in winter/spring into coloured nuptial plumage (body feathers only, not remiges or rectrices). Onset varies according to age, status, and condition of individual. Rarely, old males moult direct from nuptial to nuptial plumage. Moult probably more variable in tropics. Unlike most other *Malurus* which may achieve nuptial plumage at 1 year old, males usually delay until their fourth year. Young males in third year may show traces of black and red, primaries remain dark grey-brown as female; moult into female-like eclipse after breeding season; no long-term studies of colour-banded, known-age birds.

History and subspecies

Two subspecies recognized: *M. melanocephalus melanocephalus* (Latham, 1801), from coastal E Australia. *M. m. cruentatus* Gould, 1840, across monsoonal N Australia: scarlet to crimson back, shorter tail.

Malurus melanocephalus was described several times from different specimens. It was one of several species described by Latham in 1801 in a supplement to his *General Synopsis of Birds*, from the series of paintings known as the Watling Drawings (Hindwood 1970). Latham based his description of *Muscicapa melanocephalus* on a partly coloured male with black head, orange back, and white belly, probably from near Port Stephens, *c.* 250 km N of Sydney. Lewin's *Birds of New Holland* in 1808 included a painting and description of a fully coloured male from further north, as *Sylvia dorsalis*. On the page following their account of *M. melanocephalus*, Vigors and Horsfield (1827) described a third species, *Malurus brownii*, from specimens collected on the coast of central Qld during Matthew Flinders' voyage. Finally, Gould received from Surgeon Benjamin Bynoe some birds collected during surveys of NW Australia by HMS Beagle and he described a crimson-backed, fully plumaged male as *Malurus cruentatus* (1840). By 1865, Gould had realized that *dorsalis* and *brownii* were forms of *melanocephalus* and so maintained only two species, *melanocephalus* and *cruentatus*. This usage was current at the beginning of the twentieth century (Campbell 1901*a*; North 1904). Mathews (1913) placed *cruentatus* and *melanocephalus* in a monospecific genus *Ryania*, along with *pyrrhonotus* from N Qld and *melvillensis* from Melville Is. The 1926 Checklist and Mack (1934) reinstated *melanocephalus* in *Malurus* and Mack recognized three subspecies, *melanocephalus* (the orange-backed form), *cruentatus* (the scarlet-backed form), and *pyrrhonotus* (intermediate), from near Cairns in N Qld. Schodde (1982*b*) explains that specimens referred to *pyrrhonotus* are from the hybrid zone between *melanocephalus* and *cruentatus*, bounded by the Norman, Burdekin, and Endeavour R. in NE Qld.

Measurements and weights

Malurus melanocephalus melanocephalus: wing, M (n = 52) 44.0 ± 1.3, F (n = 16) 43.1 ± 1.9; tail, M (nuptial) (n = 40) 48.9 ± 2.6, M (eclipse) (n = 12) 53.9 ± 4.7, F (breeding) (n = 9) 53.3 3.4; F (non-breeding) (n = 7) 58.2 4.7; bill, M (n = 52) 8.6 ± 0.6, F (n = 16) 8.6 ± 0.4; tarsus, M (n = 52) 20.0 ± 1.1, F (n = 16) 19.7 ± 1.4; weight, M (n = 21) 6.0–10.0 (7.9 ± 1.0), F (n = 29) 5.0–10.0 (7.9 ± 1.2). *Malurus melanocephalus cruentatus*: wing, M (n = 80) 42.7 ± 1.5, F (n = 32) 41.6 ±1.5; tail, M (nuptial) (n = 63) 40.8 ± 3.4, M (eclipse) (n = 17) 51.6 ± 4.0, breeding F (n = 20) 46.9 ± 6.3 ; non-breeding F (n = 12) 52.9 ± 3.2; bill, M (n = 80) 8.7 ± 0.5, F (n = 32) 8.5 ± 0.5; tarsus, M (n = 80) 19.1 ± 0.8, F (n = 32) 19.0 ± 0.9; weight, M (n = 14) 5.7–7.5 (7.1 ± 0.7), F (n = 3) 6.3–7.0 (6.6 ± 0.4). (Weights from museum specimens, ABBBS, and S. Blaber personal communication.)

Field characters

The smallest fairy-wren. Red and black of nuptial male unmistakeable; russet tinge of body colour in female and brown male, complete lack of blue and diminutive size separates them from Variegated Fairy-wren (*M. lamberti*); small size, preference for tall grass, and softer, higher, and less strident calls distinguish them from Purple-crowned Fairy-wren (*M. coronatus*). Rarely, if ever, overlap with White-winged Fairy-wren (*M. leucopterus*).

Voice

Tape-recorded (ANWC); soft reel with weak introductory notes (sonogram); predominant frequencies high, descending a little during the reel. Contact calls quiet, one-syllable 'ssst', given constantly between foraging group members, barely audible at more than 10–15 m. Alarm call louder, short 'zit', much higher and softer than alarm of larger congeners.

Range and status

Resident, locally nomadic; *M. m. melanocephalus* originally as far S as Port Stephens on

E coast (Latham 1801), N to Burdekin R. and inland to Great Dividing Ra. *M. m. cruentatus:* in W, throughout Kimberleys S to Cape Keraudren; N of *c.* Lat. 20°S, E to Charters Towers, including offshore islands—Melville, Bathurst, Groot Eylandt, and Sir Edward Pellew. Across the base of Cape York Peninsula, the two subspecies intergrade between the Burdekin, Norman, and Endeavour R., with some intermediate forms and some typical of each subspecies. Schodde (1982*b*) regards this as evidence that the two races were separated during arid Pleistocene glacial periods *c.* 12 000 years ago and came together again only relatively recently.

Habitat and general habits

Replaces *M. leucopterus* where northern open woodland (savannah) and summer rainfall replace the shrublands and spinifex grasslands of the arid inland. Their preferred habitat, ephemeral tall grass with summer growth and winter drying off, is liable to grazing and burning, which may preclude permanent territories; they prefer unburnt to recently burnt vegetation (Woinarski 1990). Forage as group through tall grass, usually by hop-search, sometimes reaching to low canopy. Insectivorous; eat cockroaches, grasshoppers, bugs, beetles, weevils, ants, spiders. Social system probably similar to *M. leucopterus*, with male nuptial plumage relatively rare because it is not achieved until the fourth year. Thorough study remains to be done; from our limited study over three exceptionally dry years in S Qld, most nestings were by pairs, with males in nuptial or brown plumage. Fledged young led away from nest area and may help raise later broods. When breeding ceases, groups tend to leave territories, coalesce, and wander over larger areas in flocks of up to 30 birds (Macgillivray 1914). Evidence to date suggests that pairs may return to same area to breed in subsequent years, sometimes accompanied by surviving progeny. (S. Blaber, Rowley and Russell, unpublished data.)

Displays and breeding behaviour

Spectacular 'Puff-ball' display by nuptial plumaged male, probably to rivals or intruders; erects scarlet scapulars and back feathers over wing; displaying males give Sea Horse Flight. Petal-carrying (red petals/berries) by male

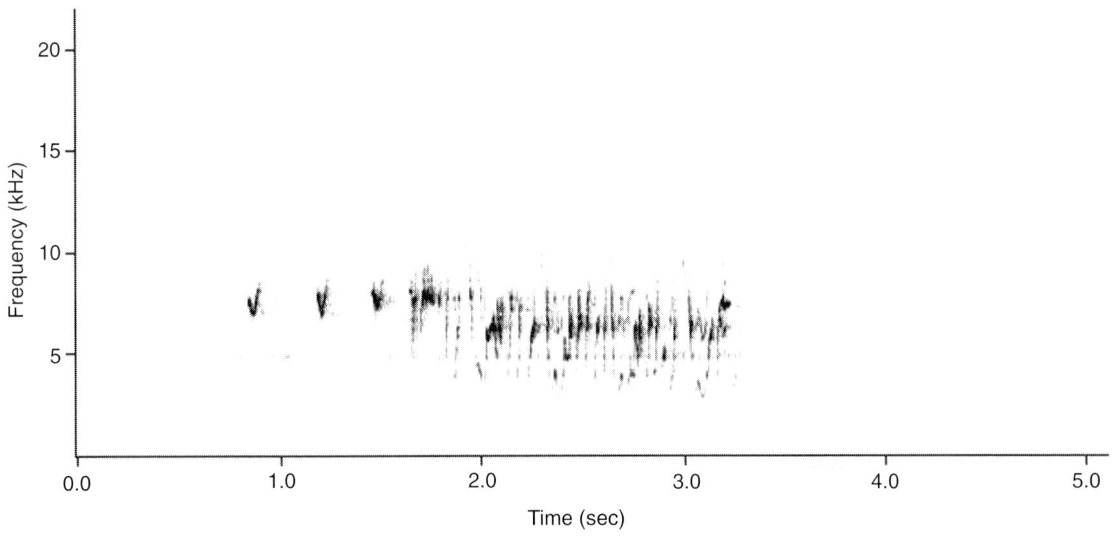

Red-backed Fairy-wren (*Malurus melanocephalus*)
M. m. cruentatus, song. E. M. Russell, Drysdale R., Western Australia.

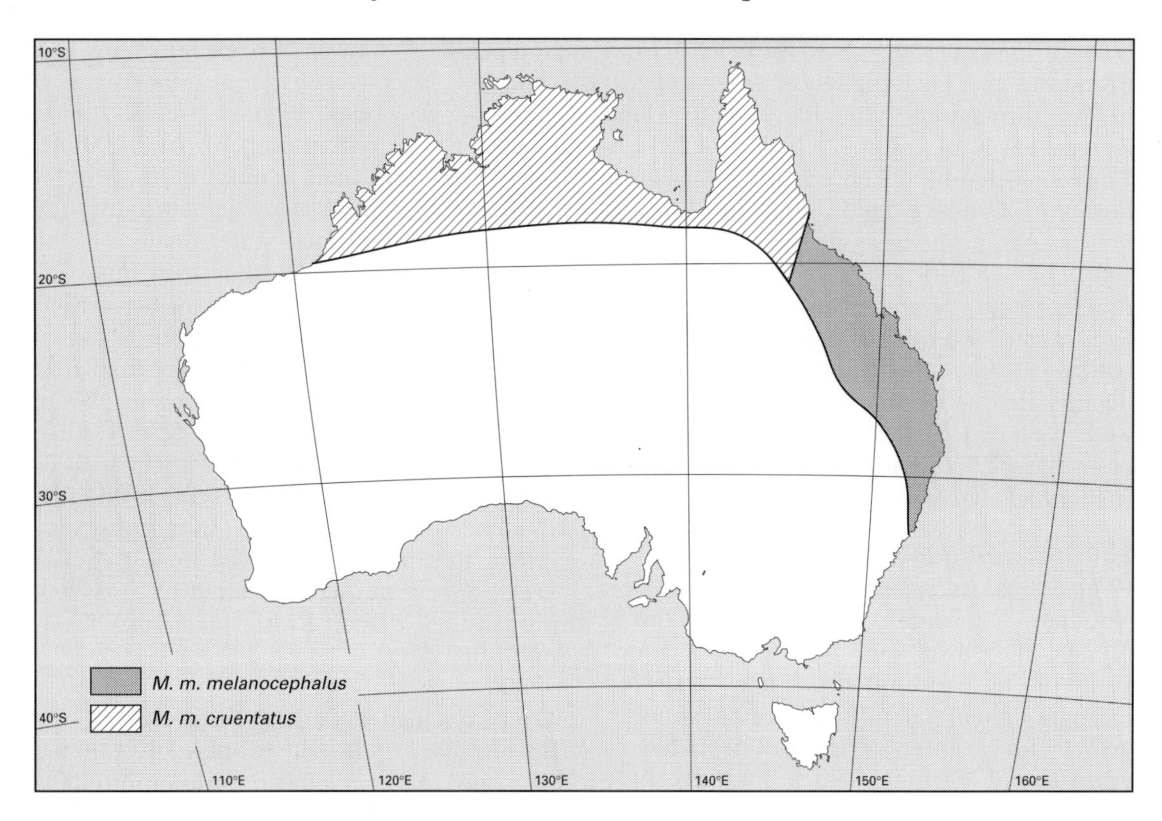

probably seeking EPC. Also 'Wing-shrinking' display to female, maximizes exposure of red feathers (W. T. Cooper *in litt.* Dec 1993; see colour photograph by N. and E. Taylor, p. 444 in Schodde and Tidemann 1988). Both sexes give Rodent-run distraction display when alarmed near nest.

Breeding and life cycle

In the tropics nesting may occur in all months but mainly in wet season (summer and autumn); in sub-tropical areas, breeding largely from Oct to Jan (Macgillivray 1914; Lavery *et al.* 1968) (see Fig. 7.1(g)). Nest built by female, male may help, but not all do so (Harvey and Harvey 1919; Favaloro 1931; Rowley and Russell, unpublished data); nest oval, domed (size 100–130 × 60–80 mm), side entrance (30–50 mm) near top; outer layer of cobwebs and grass, lined with fine grass, rootlets, and plant down; usually less than 1 m above ground, in shrubs or tall grass tussocks. Clutch 3–4, eggs (14.5–17 × 10–13 mm, n = 66, Schodde 1982*b*) tapered oval, dull white, with sparse spots and fine blotches of red-brown, concentrated in zone around blunter end, laid on successive days. Incubation by female, 12–13 days after last egg laid; nestlings hatch within 24 hr of each other, brooded initially by female, then remain in nest 11–12 days and are fed, and faecal sacs removed, by both parents. Fledglings led away from nest area by parents, remain cryptic for a week, then join parents foraging; fed for about a month; they remain in family group after independence and early broods may help raise later siblings. Two records from N coastal NSW of recent fledged young attended by a full plumaged male *M. cyaneus*: brood fed by parents and male *M. cyaneus* (Cooper 1969); two recently fledged young perching each side of male *M. cyaneus*, and being preened by him (Kay Lloyd in record 541/12 of RAOU NRS).

White-shouldered Fairy-wren *Malurus alboscapulatus* A. B. Meyer, 1874

Malurus alboscapulatus A. B. Meyer, 1874. *Sitzungsberichte der Mathematische-Naturwissenschaftlichen Classe der Kaiserlichen Akademie der Wissenshaften, Wien*, **69**, p. 496.

PLATE 3

Polytypic. Six subspecies. *Malurus alboscapulatus alboscapulatus* A. B. Meyer, 1874; *Malurus alboscapulatus naimii* Salvadori and d'Albertis, 1875; *Malurus alboscapulatus moretoni* De Vis, 1892; *Malurus alboscapulatus lorentzi* van Oort, 1909; *Malurus alboscapulatus aida* Hartert, 1930; *Malurus alboscapulatus kutubu* Schodde and Hitchcock, 1968.

Description

ADULT NUPTIAL MALE: all six subspecies, entirely glossy black except for white scapular tuft; tail, short, rounded; bill, black; eye, dark brown or black; feet, black or dusky brown.

ADULT FEMALE: Six subspecies are currently accepted on basis of three different female colorations and separation of populations (see Map and Table). 1. Female **pied**: *M. a. alboscapulatus* and *M. a. naimii*, separated by 1000 km: head and back, black; white scapular tufts; narrow white brow and broken white ring around eye; underparts white with variable amounts of black flecking; feet, brown; eye and bill as male. 2. Female **black**: *M. a. aida*, *M. a. kutubu*, and *M. a. moretoni*: female as male, with white scapular tufts, browner wings and duller overall, faint white brow. *M. a. moretoni* much larger. 3. Female **brown**. *M. a lorentzi:* crown, grey-brown; back, mid-brown; no white scapular tufts; narrow white brow and broken eye-ring; under surfaces, creamy-white; flanks, cinnamon; feet, brown; eyes and bill as male.

JUVENILE: fledglings variable, moult to immature plumage after a few months (see table); *alboscapulatus* and *naimii*: blackish-grey above (including scapulars), whitish below; *aida*: sooty brown all over; *moretoni:* sooty black or

Table: Subspecies of *Malurus alboscapulatus*: diagnostic features of females, immatures, and nestlings.

	alboscapulatus	*naimii*	*aida*	*moretoni*	*kutubu*	*lorentzi*
Adult female	Black above with white scapulars as male; white abdomen, variable black on flanks, thighs, breast, white brow and eye-ring		Sooty black above and below, less glossy than male, white scapulars			Brown, with white underparts, white brow and broken eye-ring
Immature	Unknown	Like female, male = female	Mainly black, with some white on chin and abdomen	Whitish abdomen and chin	Like female	Paler than female, white face less distinct
Nestling, fledgling	Blackish grey above (including scapulars), whitish below		Sooty brown all over	Sooty black, white chin	Unknown	As immature

grey with white chin; *kutubu*: undescribed; *lorentzi*: like adult female; face paler, white markings less distinct.

IMMATURES: resemble adult female, but are duller and lack white scapulars which are gained at first moult in subspecies where they occur.

MOULT: appear to moult annually, over many months. No eclipse plumage.

History and subspecies

Six subspecies recognized (Schodde 1982*b*): 1. *M. a. alboscapulatus* A. B. Meyer, 1874; female pied; separated by 1000 km from 2. *M. a. naimii* Salvadori and d'Albertis, 1875; female pied. 3. *M. a. aida* Hartert, 1930; female black; separated by central cordillera from 4. *M. a. kutubu* Schodde and Hitchcock, 1968; female black. 5. *M. a. moretoni* De Vis, 1892; female black; much larger; separated from *M. a. kutubu* by *M. a. naimii*. 6. *M. a. lorentzi* van Oort, 1909 ; female brown.

Populations vary in three main characters: body size, the length of the tail, and the plumage of females, which may be black, pied, or brown like that of typical Australian *Malurus* females. These differences in female plumage led to the description of many separate species—*M. alboscapulatus* Meyer, 1874 (pied F), *M. naimii* Salvadori and d'Albertis, 1875 (pied F), *M. moretoni* De Vis, 1892 (black F), *M. tappenbecki* (as *Musciparus*) Reichenow, 1897 (pied F), and *M. lorentzi* van Oort, 1909 (brown F).The confusion was increased by the variable plumage of sub-adult males. Collections were made on behalf of Dutch, German, and British museums, and the overall pattern of variation was not appreciated for many years. Mathews (1917) placed Meyer's *Malurus alboscapulatus* from the Arfak Mountains in a separate genus *Devisornis* Mathews, 1917, and subsequently, placed *naimii, tappenbecki, lorentzi,* and *alboscapulatus* in the same genus, for which the name by priority became *Musciparus*. Mathews (1930) listed the form with a black female as *Nesomalurus moretoni* (a genus in which he also placed the black forms of the White-winged Fairy-wren, *M. leucopterus*), with all other forms in *Musciparus*. At last, after examining a large series of specimens collected by Ernst Mayr, and after discussion with Stresemann, Hartert (1930) treated the various known forms as subspecies of *M. alboscapulatus* , and described yet another, *M. a. aida*. A further three subspecies were described from collections by various Archbold expeditions (*M. a. balim, M. a. dogwa, M. a. mafulu*, Mayr and Rand 1935; Rand 1940) and *M. a. kutubu* by Schodde and Hitchcock (1968).

In his treatment of the *M. alboscapulatus* species complex, Schodde (1982*b*) attributes differences in size and relative tail length to habitat differences and suggests that the plumage phases alone are evidence of speciation, explained by chance genetic shifts in isolated populations. He suggests that the present distribution, in which populations with black females occur close to populations with either pied or brown females, has arisen by dispersal along river valleys that converge in mountain areas from lowland sites that were originally far apart. Schodde recognized only six subspecies, on the basis that they are isolated from each other: *alboscapulatus, moretoni, aida* (including *randi*), *kutubu, lorentzi* (including *balim* and *dogwa*), and *naimii* (including *tappenbecki* and *mafulu*). Coates (1990) maintained *M. a. naimii, tappenbecki,* and *mafulu* as distinct.

Measurements and weights

Malurus alboscapulatus alboscapulatus: wing, M (n = 5) 50.6 ± 0.6, F (n = 5) 48.8 ± 0.8; tail, M (n = 5) 41.3 ± 1.3, F (n = 5) 44.3 ± 2.2; bill, M (n = 5) 11.0 ± 0.4, F (n = 5) 10.5 ± 0.3; tarsus, M (n = 5) 21.2 ± 0.5, F (n = 5) 21.1 ± 0.3; weight, M (n = 4) 10.5–12.0. (From Hartert 1930, only range given.)
Malurus alboscapulatus lorentzi (from Trans-Fly; originally *M. a. dogwa*): wing, M (n = 7) 43.4 ± 1.0, F (n = 6) 42.3 ± 1.3; tail, M (n = 7) 38.9 ± 3.7, F (n = 6) 45.0 ± 1.8; bill, M (n = 7) 9.0 ± 0.3, F (n = 6) 9.1 ± 0.5; tarsus,

M (*n* = 7) 20.4 ± 1.1, F (*n* = 6) 20.0 ± 0.5. (No weights recorded for this subspecies.)

Malurus alboscapulatus lorentzi (from Grand Baliem Valley; originally *M. a. balim*): wing, M (*n* = 6) 50.1 ± 1.2, F (*n* = 6) 48.3 ± 1.2; tail, M (*n* = 6) 46.8 ± 3.9, F (*n* = 6) 53.0 ± 2.5; bill, M (*n* = 6) 9.3 ± 0.4, F (*n* = 6) 9.3 ± 0.2; tarsus, M (*n* = 6) 22.3 ± 0.7, F (*n* = 6) 21.5 ± 0.3; weight, M (*n* = 1) 15.0. (Specimens of original *M. l. lorentzi* from lowland SW New Guinea and from Grand Baliem Valley not significantly different in size; weight from Ripley 1964.)

Malurus alboscapulatus naimii (lowland forms): wing, M (*n* = 11) 46.8 ± 1.5, F (*n* = 10) 44.8 ± 1.4; tail, M (*n* = 11) 40.3 ± 2.2, F (*n* = 10) 43.3 ± 1.9; bill, M (*n* = 11) 10.7 ± 0.5, F (*n* = 10) 10.0 ± 0.5; tarsus, M (*n* = 11) 20.9 ± 1.0, F (*n* = 10) 20.3 ± 0.6; weight, M (*n* = 8) 7.0–8.5 (7.6 ± 0.6), F (*n* = 2) 6.5–9.0. (Weights from ANWC.)

Malurus alboscapulatus naimii (highland forms): wing, M (*n* = 16) 50.1 ± 1.9, F (*n* = 18) 48.1 ± 1.4; tail, M (*n* =16) 46.3 ± 3.2, F (*n* = 18) 49.0 ± 3.0; bill, M (*n* = 16) 10.5 ± 0.9, F (*n* = 18) 9.3 ± 0.7; tarsus, M (*n* = 16) 22.6 ± 0.6, F (*n* = 18) 21.6 ± 0.6; weight, M (*n* = 10) 9.3–12.3 (11.1 ± 1.0), F (*n* = 4) 9.7–11.0 (10.6 ± 0.6). (Weights from Diamond 1972.)

Malurus alboscapulatus aida (lowland form, originally *M. a. aida*): wing, M (*n* = 10) 48.9 ± 2.4, F (*n* = 6) 47.5 ± 0.6; tail, M (*n* =10) 39.4 ± 2.6, F (*n* = 6) 41.8 ± 2.6; bill, M (*n* = 10) 10.4 ± 0.3, F (*n* = 6) 10.1 ± 0.4; tarsus, M (*n* = 10) 21.3 ± 0.5, F (*n* = 6) 21.4 ± 0.7. (No weights recorded.)

Malurus alboscapulatus aida (highland form, originally *M. a. randi*): wing, M (*n* = 6) 53.2 ± 2.0, F (*n* = 4) 51.5 ± 1.3; tail, M (*n* = 6) 44.2 ± 2.9, F (*n* = 4) 46.0 ± 1.8; bill, M (*n* = 6) 10.4 ± 1.0, F (*n* = 4) 10.6 ± 0.4; tarsus, M (*n* = 6) 22.5 ± 1.2, F (*n* = 4) 22.5 ± 1.3; weight, M (*n* = 2) 10.0–10.5. (Weights from Gilliard and Le Croy 1961; Ripley 1964.)

Malurus alboscapulatus kutubu: wing, M (*n* = 3) 53.0 ± 0.0, F (*n* = 4) 52.5 ± 1.3; tail, M (*n* = 3) 46.3 ± 0.4, F (*n* = 4) 48.5 ± 3.5.; bill, M (*n* = 3) 11.0 ± 0.5, F (*n* = 4) 10.1± 0.5; tarsus, M (*n* = 3) 23.9 ± 0.9, F (*n* = 4) 22.8 ± 0.8; weight, M (*n* = 2) 11.5–12.0, F (*n* = 3) 9.0–11.0 (9.8). (Weights from Schodde and Hitchcock 1968.)

Malurus alboscapulatus moretoni: wing, M (*n* = 33) 47.9 ± 2.0, F (*n* = 20) 46.0 ± 1.1; tail, M (*n* = 33) 41.8 ± 3.2, F (*n* = 20) 44.3 ± 2.7; bill, M (*n* = 33) 10.0 ± 0.5, F (*n* = 20) 9.8 ± 0.5; tarsus, M (*n* = 33) 21.6 ± 0.9, F (*n* = 20) 21.1 ± 1.1; weight, M (*n* = 20) 8.0–11.0 (9.0 ± 1.1), F (*n* = 13) 7.0–9.0 (8.3 ± 1.2). (Weights from ANWC.)

Field characters

Cocked tail, white scapulars, and small size distinguish them from all other malurids and thornbills in New Guinea; typically in tall grasslands (as Red-backed Fairy-wren *M. melanocephalus* in N Australia); seldom seen above 2 m, mainly in dense cover. (See series of colour photographs by B. Coates in Coates 1990, p. 93–6.)

Voice

Tape-recorded (see sonogram). Typical malurid reel 'the song is a burst of high-pitched reeling twittering lasting 3-4 seconds. The (contact) call is a short low dry monotone twitter or chipping of 2-3 notes, repeated' (Coates 1990. p. 94). 'Very like... Australian Red-backed Fairy-wrens, alarm call loud scolding version of contact chirp' (Schodde 1982*b*). Regular dawn chorus, they are called 'alarm clock' birds in some parts of New Guinea (Stringer 1979).

Range and status

Resident in New Guinea mainland, Yule and Samarai Is. (close inshore), common throughout lowlands, rapidly colonize new clearings up to 2300 m. Less frequent in western extremities e.g. Vogelkop. Subspecies (six currently recognized) separated largely by avoidance of mountainous areas and unbroken tracts of rainforest.

Habitat and general habits

Prefers grassland, cane-grass, overgrown village gardens, regrowth, and roadside verges.

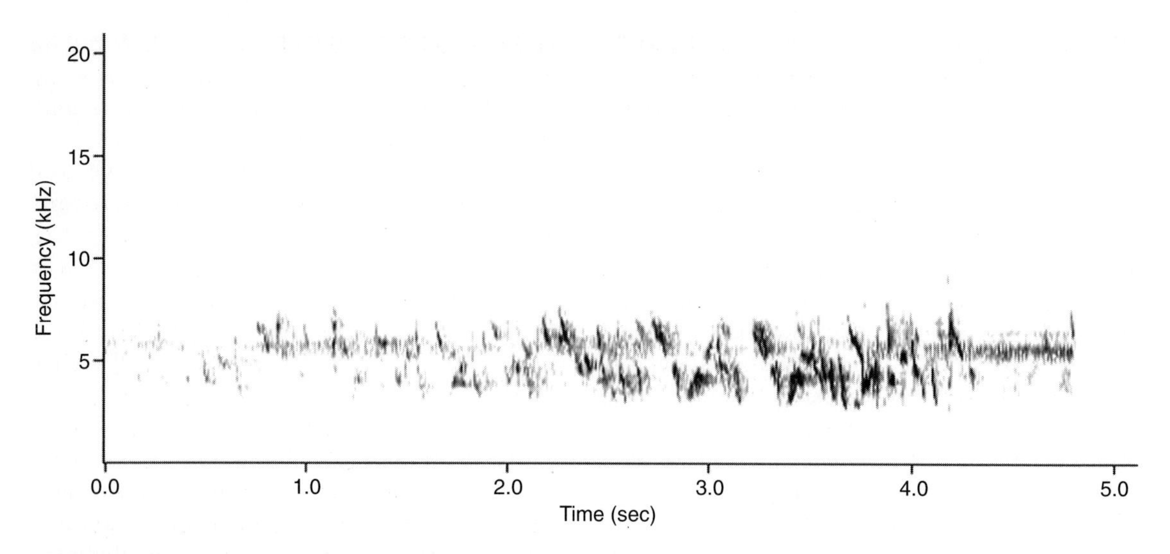

White-shouldered Fairy-wren (*Malurus alboscapulatus*)
M. a. moretoni, song. H. Crouch, nr Port Moresby, Papua New Guinea.

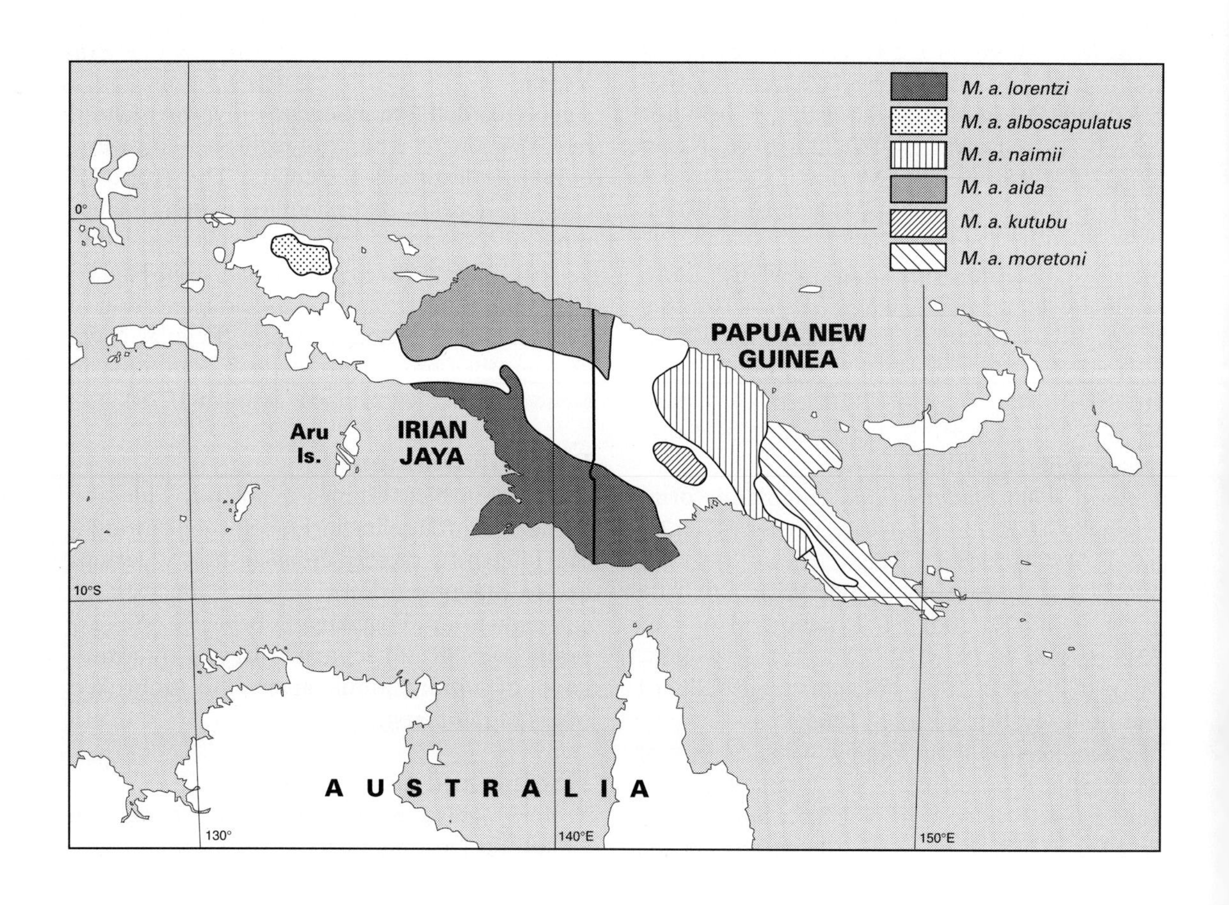

Carries tail erect; flies reluctantly and spends most of time foraging in dense cover, gleaning from vegetation and ground. Food: spiders and insects including beetles, grasshoppers, moths, and cicadas (Bell 1971; Nicholson and Coates 1975). Inquisitive, readily respond to squeaking. Frequently several birds found foraging together keeping contact by brief calls. Co-operative breeding species, generally seen in pairs or small groups of 2–8. No colour-banding studies, but appears to be resident in territories of 0.1–2 ha. Allopreening by group members between bursts of foraging.

Displays and breeding behaviour

Copulation brief, without any display preliminaries (Nicholson and Coates 1975). No data on EPCs, Petal-carrying, or Rodent-run displays.

Breeding and life cycle

Only one group has been followed intensively (Nicholson and Coates 1975), and they nested over several months, with several broods produced by a pair and helpers that were the progeny of earlier nestings. Sightings and the collection of juveniles with partly grown tails suggest eggs may be laid in any month, in both wet and dry seasons. Nest built by female; oval, domed (150 × 80 mm), with a 40-mm-diameter side entrance near the top; mainly of grass and dried leaves, twigs, rootlets, and fibre, lined with finer grass and sometimes feathers; 200–1500 mm above the ground in a small shrub or tall grass. Clutch 2–4, eggs (15.5–18 × 11–14 mm, Schodde 1982*b*) tapered oval, pinkish-white, speckled with red-brown spotting in zone around blunter end, laid on successive days. Parasitized by Brush Cuckoo (*Cacomantis variolosus*) (Diamond 1972). Incubation by female for 11–12 days, nestlings in nest 12–13 days, fed and faecal sacs removed by all group members. Fledglings cryptic 7–10 days, fed by group for 1 month, remain in family after independence (Coates 1990). The group studied by Nicholson and Coates (1975) already included two immatures presumably from a recent unrecorded nesting when they nested twice (first egg laid 1 Oct and 9 Feb), the first nest fledging one grey individual (28 Oct) which was in immature plumage by Jan, when the group consisted of breeding males and females, two young adults (immatures in October), and the immature from the first nesting.

Broad-billed Fairy-wren *Malurus grayi* (Wallace, 1862)

Todopsis grayi Wallace, 1862. *Proceedings of the Zoological Society of London,* **1862**, 166.

PLATE 4

Polytypic. Two subspecies. *Malurus grayi grayi* (Wallace, 1862); *Malurus grayi campbelli* Schodde and Weatherley, 1982.

Description

M. g. grayi

ADULT MALE: crown, mottled, the feathers basally charcoal with pale sky-blue tips; upper back and scapulars, smoky-blue grading to royal blue lower back; wings, brown; tail, brownish-grey, tipped white; black malar band from lores around eye to nape collar; sky-blue brow extending from bill to back of head, above black band; long, lanceolate, sky-blue ear tufts extend back from lower eye rim; ventrally pale sky-blue from chin to vent; bill, black, long, broad, and flattened; eye, dark brown; feet, brown.

ADULT FEMALE: like male above, but crown solid, unflecked charcoal; black lores and malar stripe separated from crown by blue brows and ear tufts, as male; blue throat and breast grade into white belly; buff around vent; eyes, bill, and feet as male.

IMMATURE: cinnamon-brown above, tawny breast, and white belly; face and malar stripe dusky and ear tufts tawny; gradually become blue, first on brow, then lower back and breast.

MOULT: both sexes once a year; no eclipse plumage recorded; with age, an increase in blue on male crown, female flanks.

History and subspecies

Two subspecies currently recognized (LeCroy and Diamond 1995). *Malurus grayi grayi* (Wallace, 1862), from N of central cordillera and Vogelkop; *Malurus grayi campbelli* Schodde and Weatherley, 1982, S of central cordillera at Mt. Bosavi; significantly smaller; both sexes have solid black crowns, more tawny scapulars, more distinct malar stripe, otherwise similar to *M. g. grayi*.

Malurus grayi was first described as *Todopsis grayi* by Wallace (1862), from a specimen collected by his assistant Charles Allen. Its broad bill prompted the French ornithologist Oustalet in 1878 to place it in a new monotypic genus *Chenorhamphus* (= 'goose-bill'; see Fig. 4.6) and this was followed by many subsequent authors (Salvadori 1881; Reichenow 1920; Mayr 1941; Mayr and Amadon 1951; Rand and Gilliard 1967). In general body form and colour, *M. grayi* matches the Australian members of the genus, with the black eye-stripe and contrasting, tapered blue ear tufts. Schodde and Weatherley in Schodde (1982*b*) named a population of Broad-billed Fairy-wrens from Mt Bosavi, south of the central highlands, as a separate species *M. campbelli*, smaller than *M. grayi*, with slight plumage differences (see also Schodde 1982*b*, 1984, Schodde and Weatherly 1983). Le Croy and Diamond (1995) compared 32 of the 37 known specimens of *M. grayi* and *M. campbelli*, and considered that while *campbelli* is smaller, it shows only minor colour differences from other populations and falls within the range of plumage variation of *M. grayi*; 'Schodde was unaware that the solid dark crown of *campbelli* is shared with northern females' (J. M. Diamond, *in litt.* 26 Aug 1994). LeCroy and Diamond (1995) treat *campbelli* as a subspecies of *grayi*, as did Beehler *et al.* (1986) and Vuilleumier *et al.* (1992). LeCroy and Diamond consider a third subspecies *M. g. pileatus* (Reichenow, 1920) from N central New Guinea as not separable from *M. g. grayi* on the few existing specimens.

Measurements and weights

Malurus grayi grayi: wing, M (n = 7) 60.2 ± 2.2, F (n = 10) 59.3 ± 1.5; tail, M (n = 7) 55.4 ± 2.2, F (n = 10) 53.1 ± 2.0; bill, M (n = 7) 14.8 ± 0.4, F (n = 10) 14.7 ± 0.4; tarsus, M (n = 7) 24.8 ± 0.3, F (n = 10) 24.8 ± 0.6; weight: M (n = 6) 14–17 (15.9 ± 1.1), F (n = 8) 13–16 (14.5 ± 1.0). (Weights from LeCroy and Diamond 1995.)

Malurus grayi campbelli: wing, M (n = 4) 54–57 (56), F (n = 3) 49–53 (51); tail, M (n = 2) 49–50, F (n = 3) 44–49; bill, M (n = 4) 13–16 (14.5), F (n = 3) 13–16 (14.3); tarsus, M (n = 2) 21–22, F (n = 3) 21–22; weight: M (n = 3) 10–12 (11.0 ± 1.0), F (n = 3) 8.0–11.0 (9.7 ± 1.5). (Weights from LeCroy and Diamond 1995.)

Field characters

Holds tail only slightly cocked; hops through vegetation and makes short weak flights. Occurs in similar places to Emperor Fairy-wren (*M. cyanocephalus*) but never together; not in secondary growth, preferring natural clearings caused by tree falls, land slides, rivers, or tracks. No overlap with Orange-crowned Wren (*Clytomyias insignis*) or White-shouldered Fairy-wren (*M. alboscapulatus*). Distinguished from other malurids by black eye-line, pale blue brow and ear tufts, and heavy, flattened bill (see Fig. 4.6(d)). Wallace's Wren (*Sipodotus wallacii*) has white underparts and white in wing. (See photograph of *M. g. campbelli* by R. Mackay, in Coates 1990, p. 89.)

Voice

Diamond (1981, p. 98) described three calls heard in Gauthier Mts: 'high-pitched slightly piercing notes 'ssss'—the usual call of foraging parties'; 'a very high-pitched sibilant, slightly prolonged up-slur'; and 'a very rapidly repeated short spitted note, 'ts-ts-ts', similar to the song of *M. alboscapulatus* and *Todopsis* (= *Malurus*) *cyanocephalus*' (see sonogram).

Range and status

Uncommon wherever found; probably permanently resident, to 1040 m (Diamond 1981) in the lowland and foothills rainforest of NW New Guinea from Vogelkop to Torricelli Mts and to Maprik on Sepik R.; also on N slopes of central mountain range from Weyland Mts to May and Frieda R. on Sepik drainage. Subspecies *M. g. campbelli* based on 10 birds from Mt. Bosavi, on the S side of the central range (Schodde and Weatherly 1983; Schodde 1984*a*).

Habitat and general habits

Frequents understorey of tall primary rainforest, particularly where tree-fall, creek, or track permits a tangle of vines and shrubs to develop; not in openings in secondary forest and areas disturbed by people (Diamond 1981). Ecologically and morphologically similar to *Clytomyias* but no overlap in altitude. *Sipodotus* occupies same altitude range of rainforest (to *c.* 1000 m) but forages to higher levels, moves more quickly, and frequently joins mixed species flocks, which *M. grayi* never does. Despite these similarities, only twice recorded at same collecting localities (Arfak Mts and Maanderberg Mt.) Usually found in groups of 2–4, apparently in well separated territories (three groups on 8 km transect); 'make short hops through dense vegetation, scurry through ground litter and make short weak flights' (Diamond 1981). Carries tail partly cocked, as *Clytomyias*.

Display and breeding behaviour

No information on displays or breeding behaviour of either subspecies.

Breeding and life cycle

Only one nest found; it was of *M. g. grayi* at 1040 m in the Gauttier Mts and contained two naked nestlings, 15 and 17 Oct, 1979 (J. M. Diamond, *in litt.* 26 Aug 1993); the

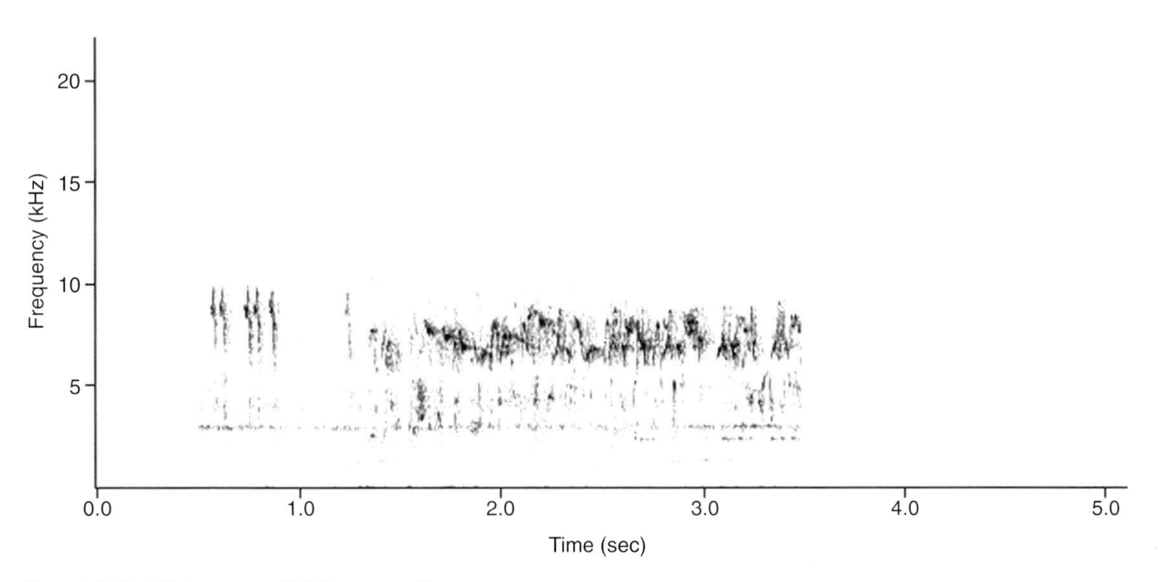

Broad-billed Fairy-wren (*Malurus grayi*)
M. g. grayi, song. K. D. Bishop and Jared Diamond, Western Van Rees Mountains, Irian Jaya.

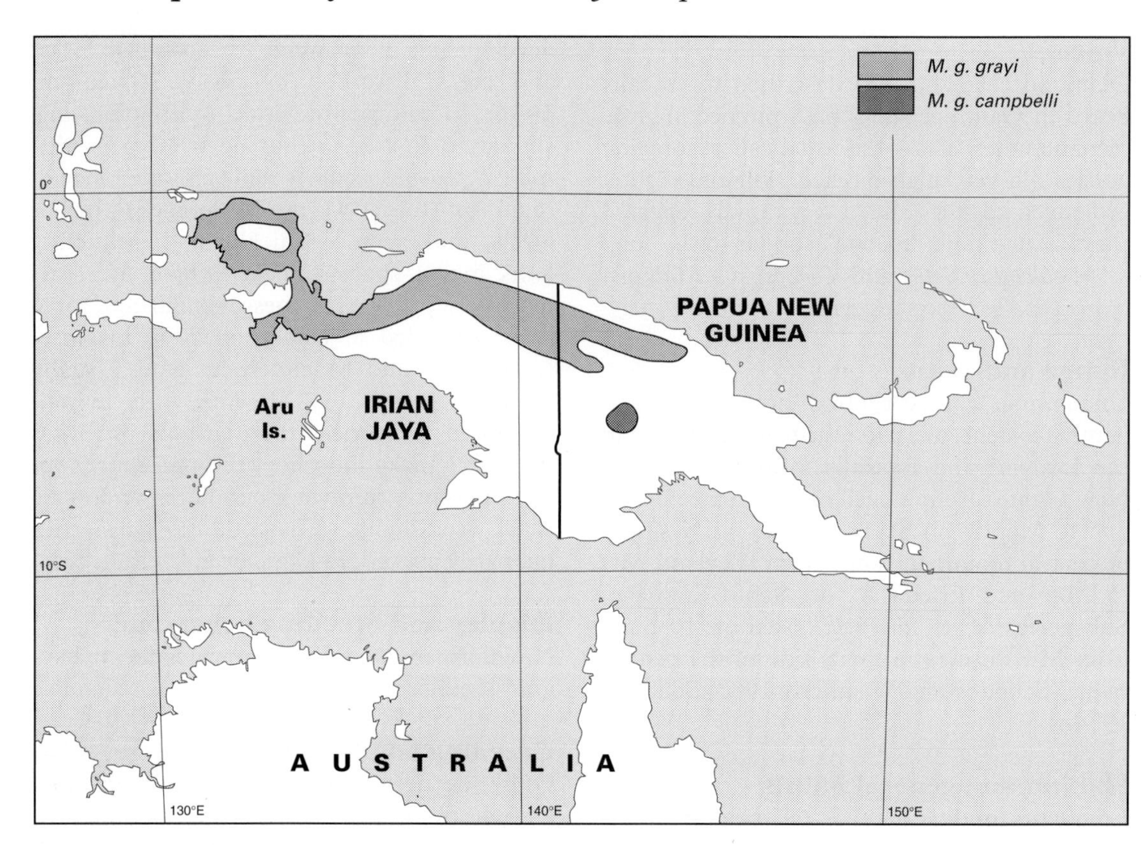

nestlings remained silent and the sitting bird flew off silently when the nest was approached. Nest unusual for a malurid; 'near a fallen log was a sapling, four centimetres in diameter and hung with moss. At a height of 0.5 m in the moss on the sapling's vertical trunk, *Chenoramphus* (*M. grayi*) had made a cavity eight centimetres deep and five wide with a side entrance, and lined it with dry needles and strips of bark' (Diamond 1981, p. 98.). Fledglings found also in Feb (Salvadori 1881; Mayr and Meyer de Schauensee 1939). Two fledglings of *M. g. campbelli*, which we estimate at 3–4 weeks old from the description given by Schodde (1984*a*, p. 250), were collected on 3 Nov 1982 at Mt. Bosavi, indicating laying in late Sept. Nothing more is known about this species.

Emperor Fairy-wren *Malurus cyanocephalus* (Quoy and Gaimard, 1830)

Todus cyanocephalus Quoy and Gaimard, 1830. In J. S. C. Dumont d'Urville, *Voyage de la corvette l'Astrolabe, execute par ordre du Roi pendant les annees 1826, 1827, 1828, 1829, etc.* Vol. 1, Zoologie, p. 227, pl. 5, fig. 4. PLATE 4

Polytypic. Three subspecies. *Malurus cyanocephalus cyanocephalus* (Quoy and Gaimard, 1830); *Malurus cyanocephalus mysorensis* (A. B. Meyer, 1874); *Malurus cyanocephalus bonapartii* (G. R. Gray, 1859).

Description

M. c. cyanocephalus

ADULT MALE: crown, light metallic royal blue; upper back, scapulars, and upper tail coverts, deep turquoise-blue; lower back, blue-black;

wings and tail, dusky black with blue tinges; forehead, lores, and sides of head to narrow collar, black; no coloured ear tufts; throat and belly, deep navy blue, breast darker; eyes, dark brown; bill, black; feet, dark grey-brown.

ADULT FEMALE: pattern of crown, face, and sides of head as male, blue of crown duller and deeper; back, chestnut; tail, black with bluish tinge, broad white tips; throat, deep blue, sharply distinct from white breast and belly; flight feathers, dusky grey, chestnut edges; flanks, thighs, and round vent, light chestnut; eyes and bill as male; feet, mid-brown.

IMMATURE: juvenile basically like female, but back rusty brown and head, including face, dusky-black; entire under-surface, dull white. Young females gradually acquire ad female plumage. Sub-ad males in transitional plumage are a mixture of ad male and female plumage, as young males pass through female plumage before attaining ad male.

MOULT: little known; moult recorded in all months, including breeding birds (Mayr and Rand 1937). No eclipse plumages recorded. Juvenile females moult directly into adult plumage; Juvenile males first acquire female-like plumage and then the blue back and belly of ad male. The time taken for these changes is not known.

History and subspecies

Three subspecies recognized (Schodde 1982*b*): *Malurus cyanocephalus cyanocephalus* (Quoy and Gaimard, 1830): crown light royal blue, size large; from Vogelkop and along N coast to Ramu R.; *M. c. mysorensis* (A. B. Meyer, 1874): crown light cobalt, size smaller; only on Biak Is.; *M. c. bonapartii* (G. R. Gray, 1859): crown light cobalt, size large (separated from *mysorensis* by nominate race); Aru Is. and S. New Guinea.

The first known malurids from New Guinea were specimens of *M. cyanocephalus* collected by Quoy and Gaimard in NW New Guinea on the voyage of *l'Astrolabe* in 1827. Although they had earlier collected fairy-wrens in Australia (*M. leucopterus* in 1818 on Dirk Hartog Is. and *M. splendens* at King George Sound, in 1826), they did not recognize the new bird as a fairy-wren, but identified it as one of the South American group of todies, and in 1830 named it *Todus cyanocephalus*. This is surprising in view of its broad similarity with *Malurus splendens*, the brilliant cobalt blue of the male, with black face and cocked tail. Bonaparte (1854) recognized that it was not a tody, and erected a new genus *Todopsis*. Subsequently, two other, similar species were named, *T. bonapartii* from Aru Is. (Gray 1859) and *T. mysorensis* from the small island of Biak (Meyer 1874). A further two species described by Rosenberg (1863) were later regarded as synonyms for *mysorensis* and *bonapartii*. All these species differed mainly in colour, and Rothschild and Hartert (1903) treated them as subspecies of *T. cyanocephalus*, adding a fourth, *dohertyi*, from the N coast of Irian Jaya, with darker, more chestnut females. This subspecies has not been upheld because of female variability (Mayr 1941; Ripley 1964) and is included in *M. c. cyanocephalus* by Schodde, who points out that much of the debate about species and subspecies of the Emperor Fairy-wren arose from inadequate, poorly documented collections (1982*b*).

Measurements and weights

Malurus cyanocephalus cyanocephalus: wing, M (n = 15) 59.7 ± 1.5, F (n = 13) 56.9 ± 1.3; tail, M (n =15) 58.1 ± 2.1, F (n = 13) 56.5 ± 2.4; bill, M (n =15) 12.9 ± 0.6, F (n = 13) 12.5 ± 0.6; tarsus, M (n = 15) 24.8 ± 1.2, F (n = 13) 23.1 ± 1.2; weight, M (n = 8) 10.0–16.0 (13.4 ± 1.5), F (n = 7) 10.0–14.0 (11.3 ± 1.0). (From Vogelkop; mainland N. New Guinea similar.)

Malurus cyanocephalus mysorensis: wing, M (n = 8) 57.6 ± 0.5, F (n = 2) 54.5 ± 0.7; tail, M (n =8) 53.7 ± 1.5, F (n = 2) 54.5 ± 0.7; bill, M (n =8) 12.8 ± 0.3, F (n = 2) 12.6 ± 0.6; tarsus, M (n = 8) 24.0 ± 0.3, F (n = 2) 22.4 ± 0.1 (From Biak Is., NW New Guinea; no weights recorded for this subspecies.)

Malurus cyanocephalus bonapartii: wing, M (n = 31) 59.4 ± 1.8, F (n = 26) 55.9 ± 1.6; tail, M (n =31) 57.4 ± 2.1, F (n = 26) 54.9 ± 2.4;

bill, M (n = 31) 13.1 ± 0.7, F (n = 26) 12.7 ± 0.7; tarsus, M (n = 31) 24.2 ± 0.8, F (n = 26) 22.9 ± 0.9; weight, M (n = 14) 10–17.0 (13.3 ± 1.5), F (n = 11) 9.5–14.0 (11.6 ± 1.0). (From southern New Guinea; Aru Is. similar; weights from ANWC, Ripley 1964, Hartert 1930.)

Field characters

No other male fairy-wren in New Guinea blue and black. Female recognized by blue head, chestnut back, deep blue throat sharply distinct from white underparts; females of other malurids have white throats. Larger than other malurids except Broad-billed Fairy-wren (*M. grayi*), cocks tail even beyond vertical and may fan it (as Lovely Fairy-wren *M. amabilis*). Occurs in same locations as White-shouldered Fairy-wren (*M. alboscapulatus*) that 'inhabits grassland and tolerates some bush' while *M. cyanocephalus* 'inhabits bushes and tolerates some grassland' (Bell 1969, p. 204). (See colour photographs by W. S. Peckover and B. Coates, p. 91 in Coates 1990.)

Voice

Tape-recorded (see sonogram). Amongst the first birds to call before dawn. Song 'a slow drawn-out nasal slurring ending in a series of sets of three notes. It also has a more typical *Malurus*-like twittering and readily reacts to being called up' (Finch 1982). The groups of three notes reminded Ripley (1964) of whiplash, start with sharp up-slurred whistle;

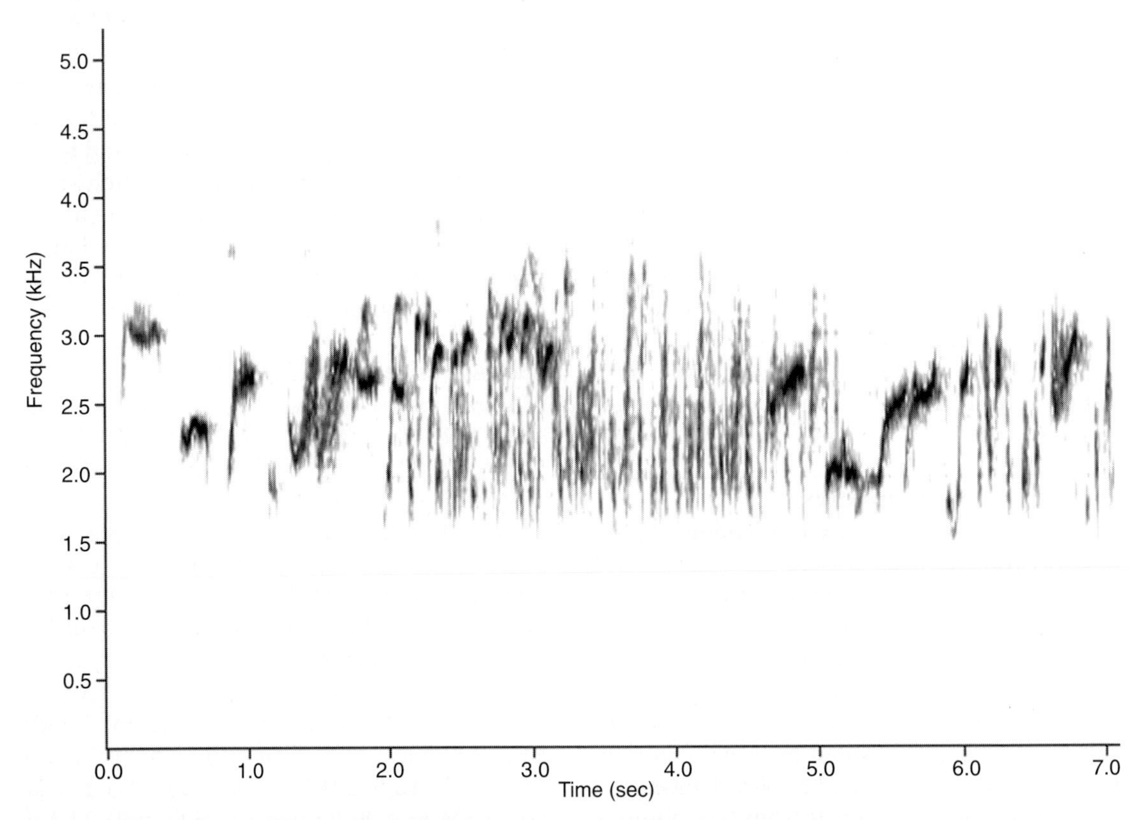

Emperor Fairy-wren (*Malurus cyanocephalus*)
K. D. Bishop and Jared Diamond, Western Van Rees Mountains, Irian Jaya.
M. c. cyanocephalus, song. Because the song is longer, sonogram shows whole song (7 sec) with expanded frequency scale and compressed time scale. It is possible that two birds are singing. Blackbird-like syllables from 0–2 sec are repeated at 5–7 sec; in between there is a section of *Malurus*-like reel.

'pause frequently in course of... feeding to gather into huddles and burst into song together' (Schodde 1982*b*). Beehler *et al.* (1986, p. 161) list two different songs, one a typical *Malurus* reel similar to that of the White-shouldered Fairy-wren, the second 'a hoarse, short, medium pitched thrush-like warble, suggesting the song of a European Blackbird'. Contact calls between group members repeated soft chirps 'tst-tst-tst-tst' etc.; also 'tschik' alarm note.

Range and status

Resident throughout lowland New Guinea except for NE corner (Madang to Milne Bay). Usually below 100 m, but at Moro, 750 m (Schodde and Hitchcock 1968). Central cordillera separates two main subspecies (*M. c. cyanocephalus* and *M. c. bonapartii*); *M. c. mysorensis* on Biak Is. resembles *M. c. bonapartii* but separated by *M. c. cyanocephalus* and the mountains. Widespread, not rare; likes disturbed situations; not threatened.

Habitat and general habits

Favours dense secondary growth at edges of forest and forest openings, riversides, roadside, and overgrown gardens. In suitable habitat, may be quite common: three groups located in 25 ha at Nomad R. (Bell 1970). Tamer and more inquisitive than other New Guinea malurids. Forages noisily in family groups usually within 1 m of ground; insectivorous; eats beetles, bugs, moths, spiders, grasshoppers (Schodde 1982*b*); gleans from leaves, palm fronds, branches, and debris (Bell 1984). Keeps tail well cocked over back. Not known to join mixed species flocks. Co-operative breeding species; generally seen in pairs or small groups, some with more than one adult male. Group of four males and one female netted together attending two

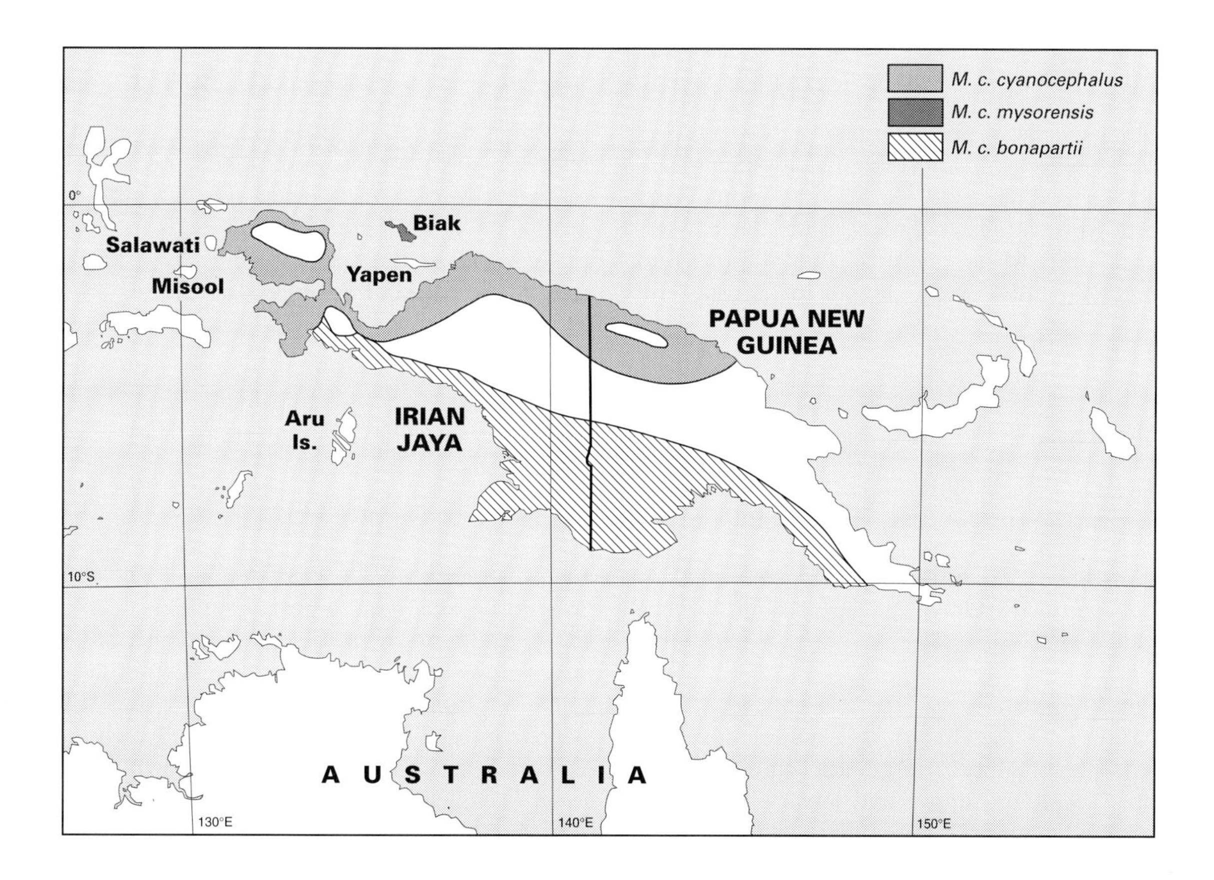

recently fledged young (Filewood 1970). Probably territorial: banding over 8 years at Brown R. near Port Moresby showed some individuals remained in the same area for several years (Filewood 1971); others move when secondary vegetation ceases to suit them (Bell 1982*a*). One adult male survived for 8 years, and was caught 12 times within 100 m radius, with three different females in succession (L. W. Filewood in *litt.* 10 Sept 1993).

Displays and breeding behaviour

Schodde (1982*b*) describes male flaring feathers on head and neck in reaction to territorial intruders. Rodent-run distraction display by all group members when dependent young netted (L. W. Filewood *in litt.* 10 Sept 1993). No other displays or breeding behaviour recorded.

Breeding and life cycle

Limited data suggest nesting at any time of year; juveniles recorded from Mar, Jun, Oct, Sept, and Dec (Schodde 1982*b*; Coates 1990). Only one nest found: 'A nest with four young in a bush about three feet from the ground was gourd-shaped, made of strips of fern, dead leaves in strips, and pieces of moss, all closely woven together. The entrance hole was on the side and the nest was closed above' (Ripley 1964). Many reports suggest young remain in group after independence. No detailed studies of biology.

Genus *Sipodotus* G. M. Mathews

Sipodotus G. M. Mathews, 1928. *Bulletin of the British Ornithologists' Club*, **48**, 83.

(type by original designation, *Todopsis wallacii* Gray, 1862).

Monospecific genus, confined to New Guinea where it lives in rainforest bordering the mountain ranges from W to E; some variation occurs over this extensive range. Blue cap, russet back, and white breast; sexual dimorphism slight, mainly in the colour of throat and breast. Bill long and broad; five pairs of tail feathers, short, graded, rounded, and white-tipped; tail is not held cocked. Usually seen in groups of 4–8, mainly foraging 2–10 m from the ground in the tangled climbers draping trees at the edge of openings in the forest.

Wallace's Wren *Sipodotus wallacii* (G. R. Gray, 1862)

Todopsis wallacii G. R. Gray, 1862. *Proceedings of the Zoological Society of London*, **1862**, 429.

PLATE 4

Polytypic. Two subspecies. *Sipodotus wallacii wallacii* (G. R. Gray, 1862); *Sipodotus wallacii coronatus* (Gould, 1878).

Description

ADULT MALE: Crown and nape, speckled black, the feathers sooty-black with light blue tips; scapulars, upper and lower back, rusty brown; wings, brownish grey; white tips to brownish-grey upper wing-coverts form two broad, broken bars across shoulder of the wing; tail, short, rounded, brownish-grey, white-tipped; face, black with incomplete white eye ring extended back into elongated white ear tufts; throat, breast, and belly, white; feet and legs, brown; eye, red-brown; bill, long, broad, black with white tip; rictal bristles, very long (see Fig 4.6(c)).

ADULT FEMALE: as male, with crown slightly duller, throat and breast yellowish.

IMMATURE: like adults but generally duller; crown speckled buff, not blue; ear tufts short; bill shorter and browner; probably acquire adult plumage in first year.

MOULT: once a year, Nov to Mar; no distinctive nuptial or eclipse plumage.

History and subspecies

Two subspecies recognized (Mayr and Cottrell 1986; Coates 1990): *Sipodotus wallacii wallacii* (G. R. Gray, 1862) in N and W; both sexes have white flanks and bellies; throat and breast white in male, yellowish buff in female; Misool and Yapen Is., Vogelkop and N coast of New Guinea, E from Geelvink Bay; *Sipodotus wallacii coronatus* Gould, 1878 in S and E; both sexes creamy buff on flanks and belly, but throat and breast of female less buff than in W birds; Aru Is., S New Guinea from the Setekwa R. to Milne Bay, and in the N to the Hydrographer Mts.

Wallace's Wren was collected in 1860 by Wallace's assistant, Charles Allen, on Misool Is., and described as *Todopsis wallacii* by Gray in 1862. Specimens from Aru Is. were named as *Todopsis coronata* by Gould (1878). In one of his rare excursions into the New Guinea avifauna, Mathews (1928) recognized the significantly different shape of the bill and the almost identical plumage of male and female, and established a new genus *Sipodotus* for Wallace's Wren. Although this genus was not recognized by many later workers (Mayr 1941; Rand 1942; Ripley 1964; Rand and Gilliard 1967; Gilliard and Le Croy 1970; Mayr and Cottrell 1986), Schodde (1982*b*) included *T. cyanocephalus* and *T. grayi* in *Malurus*, but recognized Wallace's Wren as a monospecific genus, for which Mathews' name *Sipodotus* was available, and to which he gave the vernacular name tree-wrens. Although Schodde (1982*b*) describes the variation between eastern and western forms, he then considered that subspecific differentiation could not be confirmed from the material available, since the material is patchy and specimens were often mis-sexed. Coates (1990) lists two subspecies, a north-western *S. w. wallacii* and a southern *S. w. capillatus*, following Mayr and Cottrell (1986), and Schodde (personal communication) now agrees with this arrangement. However, Mayr retained Wallace's Wren in the genus *Malurus* and when he needed to name the southern and south-eastern subspecies, he found Gould's (1878) name for Aru Is. specimens *Todopsis coronata* to be preoccupied by *Malurus coronatus*. He therefore coined a new subspecific name, *capillatus* nom. nov. Mayr, 1986. However, Schodde, by placing Wallace's Wren in a monospecific genus *Sipodotus* renders Gould's 1878 name still valid, and therefore *S. w. coronatus* is the southern race.

Measurements and weights

Sipodotus wallacii wallacii; wing, M (n = 49) 47.8 ± 1.5, F (n = 31) 45.9 ± 1.9; tail, M (n = 49) 46.5 ± 2.5, F (n = 31) 45.4 ± 2.6; bill, M (n = 49) 14.8 ± 0.6, F (n = 31) 14.3 ± 0.9; tarsus, M (n = 49) 19.1 ± 0.8, F (n = 31) 18.8 ± 0.7; weight: M (n = 4) 7.0–8.5 (7.9 ± 0.6), F (n = 2) 6.5–8.0. (Weights ANWC and Hartert 1930.)

Field characters

The only malurid that does not cock its tail. White throat and wing bars obvious. Long bill distinguishes from small *Gerygone* spp. with white underparts. Emperor Fairy-wren *M. cyanocephalus* has no white ear tufts or wing bars; Broad-billed Fairy-wren (*M. grayi*) has pale blue ear tufts but lacks wing bars and rufous back. Largely arboreal in forest midstage between understorey and canopy; joins mixed species flocks.

Voice

Most obvious feature is incessant soft, very high pitched, sibilant contact calls of foraging parties: 'see-see-see-see'. 'The song is a series of "sees" running into a high-pitched sibilant twitter' (Finch 1982), with none of the warbles and trills of other malurids. Readily attracted by squeaking. No tape-recordings located.

Range and status

Resident, in foothill rainforest throughout all New Guinea between 100 and 800 m, occasionally to 1200 m and to lowland plains where vegetation suitable, including isolated populations in rainforest at mouths of Fly and Digul R. Also Misool, Aru, Yapen (=Japen) Is.; doubtful record from Goodenough Is. Common in suitable habitat; main threat to species, large-scale clear-felling of forest.

Habitat and general habits

More in trees than undergrowth, mostly 2–10 m from ground, but to ground level or to canopy at 40 m. In trees with tangles of vines and climbing bamboo at edges of forest. Continuous foraging (as *Gerygone* spp.) by gleaning, probing, hovering, and flying sorties; may hang upside down to probe stems; moves about outermost branches, usually just beneath the outer leaves. Flight swift and undulating from tree to tree. Joins mixed species flocks foraging throughout the forest; usually in groups of 4–8 that appear to be family parties (Bell *et al.* 1979; Schodde 1982*b*). No marked birds studied, social organization unknown.

Displays and breeding behaviour

No breeding behaviour or displays recorded.

Breeding and life cycle

Gonads of collected specimens suggest breeding throughout year with slight peak Sept–Dec (Schodde 1982*b*). Only two nests found, both at Brown R.; eggs laid Nov and Jun (Bell *et al.* 1979). Nests suspended in tangle of vines, 5–10 m above ground; domed, 150–155 ×

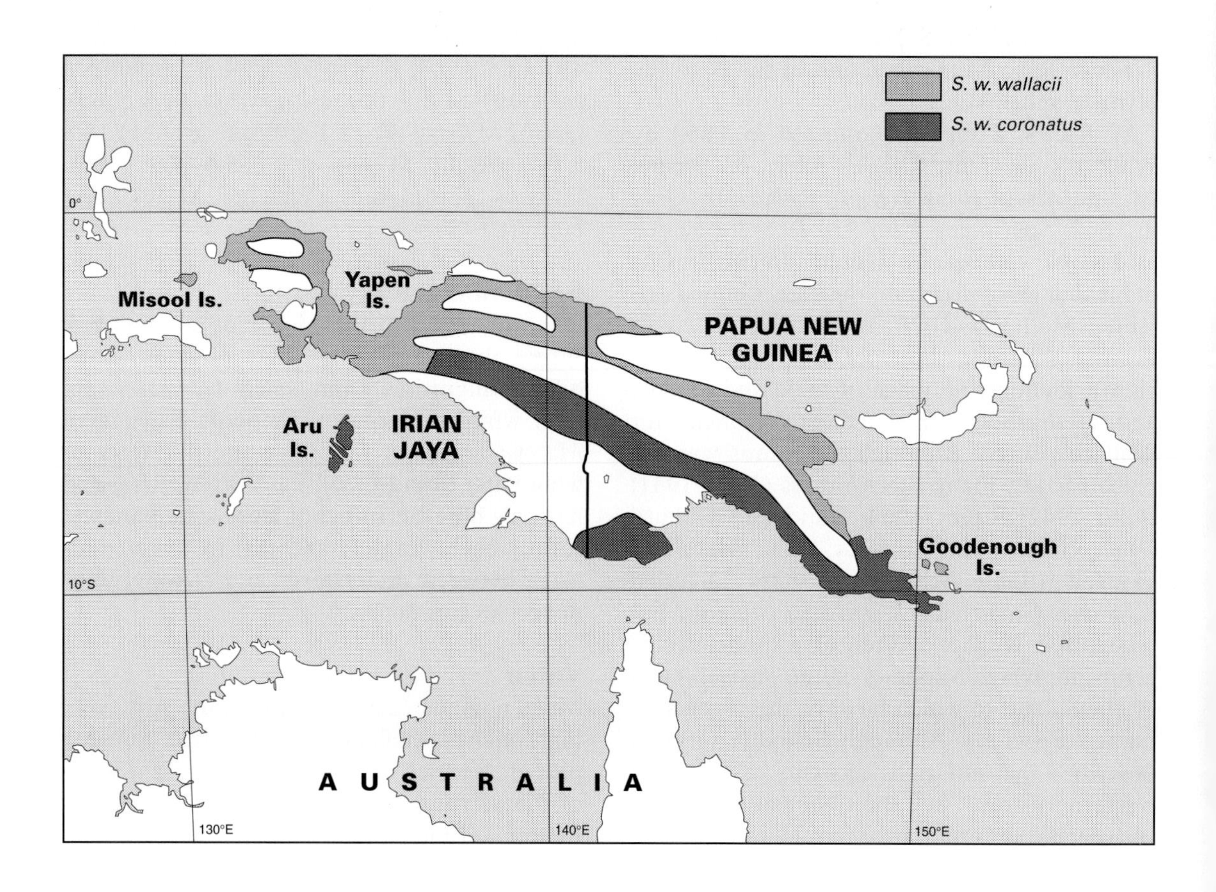

55–85 mm, with slightly hooded entrance near top; constructed of closely woven fine grasses, fibres, strips of palm fronds, and cobwebs, lined with thin, interwoven fibre and strips of leaf. One clutch known; two eggs (13.5 × 9.5 mm, n = 1), tapered oval, white, lightly marked with small rufous blotches towards blunter end. Schodde (1982*b*) considered five clutches labelled *Todopsis wallacii* in Berlin Museum to be misidentified. Little known of incubation or rearing of young; both known nests attended by three (unbanded) birds. At first nest, uncertain if all three birds fed nestlings. At second nest, two birds (adult and immature) carried nesting material (Coates 1973; Bell *et al.* 1979). Usually encountered in groups which may be family parties, suggesting that immatures may remain with family group.

Genus *Clytomyias* R. B. Sharpe

Clytomyias R. B. Sharpe, 1879. *Notes Leyden Museum*, **1**, 31.

Monospecific genus confined to montane New Guinea; least known of any malurid. Unstreaked, back olive-brown, underparts white or buff; sexual dimorphism slight and confined to under surface of one subspecies.

Russet-coloured cap distinctive. Long slender, wedge-shaped tail with four pairs of graded feathers frayed at edges, 'with flimsy decomposed vanes, resembling those of Rufous-crowned Emu-wrens: many of the web-holding hooks on the barbules are missing' (Schodde 1982*b*, p. 116); generally half-cocked and slightly spread; the fifth pair of rectrices are miniscule and generally missed.

Orange-crowned Wren *Clytomyias insignis* Sharpe, 1879

Clytomyias insignis Sharpe, 1879. *Notes Leyden Museum*, **1**, 31. PLATE 4

Polytypic. Two subspecies. *Clytomyias insignis insignis* Sharpe, 1879; *Clytomyias insignis oorti* Rothschild and Hartert, 1907.

Description

C. i. insignis

ADULT MALE: crown, face, ear coverts, and nape, orange-rufous; back, olive-brown; throat, breast, and upper belly, creamy-white; rufous thighs; tail, light rufous; under-tail coverts, pale ochreish-grey; wings, brown; bill, broad, black; eyes, dark brown; legs, pink-brown.

ADULT FEMALE: indistinguishable from male.

IMMATURE: fledgling duller, darker, uniformly ochreish-brown; crown, brownish-rufous; tail, brown; eyes and bill, as adult, except lower mandible pale brown.

MOULT: no defined season of moult; probably moults once in year (Schodde 1982*b*); no eclipse plumage recorded.

History and subspecies

Two subspecies recognized; *Clytomyias insignis insignis* Sharpe, 1879: confined to Vogelkop, with throat, breast, and upper belly creamy-white, under-tail coverts, pale ochreish-grey. *C. i. oorti* Rothschild and Hartert, 1907: on central cordillera; throat, breast, and belly, ochreish buff in males, creamier buff in females, under-tail coverts, dull ochreish.

Clytomyias insignis was described from one specimen collected in the Arfak Mountains in the Vogelkop of W New Guinea (Sharpe 1879*a*). Rothschild and Hartert (1907) described a subspecies *C. i. oorti* from the high montane forests of the central cordillera, and recognized its affinity with other malurids because of its reduced number of rectrices. The genus has always been recognized as distinct, and was usually placed with *Todopsis* and *Malurus* in lists of species. Mayr (1941) placed it within the subfamily Malurinae of the Muscicapidae. The presence of an interscapular gap in the spinal feather tract reinforced its affinities (Harrison 1969). Schodde (1982*b*) records geographical variation as slight, even within the subspecies *oorti* which extends the length of the central cordillera.

Measurements and weights

Clytomyias insignis insignis: wing, M (n = 1) 55, F (n = 1) 57; tail, M (n =1) 63, F (n = 1) 67; bill, F (n = 1) 13.3; tarsus, M (n = 1) 23.7, F (n = 1) 23.0; weight: none recorded.
Clytomyias insignis oorti: wing, M (n = 14) 56.2 ± 2.2, F (n = 15) 56.6 ± 2.4; tail, M (n = 14) 65.8 ± 3.4, F (n = 15) 67.9 ± 2.9; bill, M (n = 14) 13.1 ± 0.6, F (n = 15) 13.3 ± 0.8; tarsus, M (n = 14) 24.7 ± 0.8, F (n = 15) 25.2 ± 1.1; weight: M (n = 12) 10.0–14.0 (12.3 ± 1.3), F (n = 3) 12.0–14.0 (13.0 ± 1.0). (Weights from museum specimens and Hartert 1930.)

Field characters

Orange crown, half-cocked tail, and ceaseless activity diagnostic; usually in groups of 4–12; the only malurid in mountain forest, and only mountain bird with rufous crown (see colour photograph by W. S. Peckover in Coates 1990, p. 97).

Voice

Song a brief, high-pitched chattering reel, with a few introductory chirps; contact calls sharp, high pitched, sibilant chirps, 'trrt, trrrt, tst, tsst' (sonogram), similar to most fairy-wrens, but even more frequent (Schodde 1982*b*). Mouse-like squeaks: 'jib-jib'; also some high-pitched chattery notes like those of *M. alboscapulatus* (Beehler 1978; Beehler *et al.* 1986). Subdued twitter and an alarm note like that of *M. cyaneus* (Bell 1969). Tape-recordings of song not located.

Range and status

Between 2000 and 3000 m (rarely to 1200 m) along both N and S sides of the central cordillera of New Guinea from Owen Stanley Ra. in SE to Weyland Mts in W; birds on Vogel-

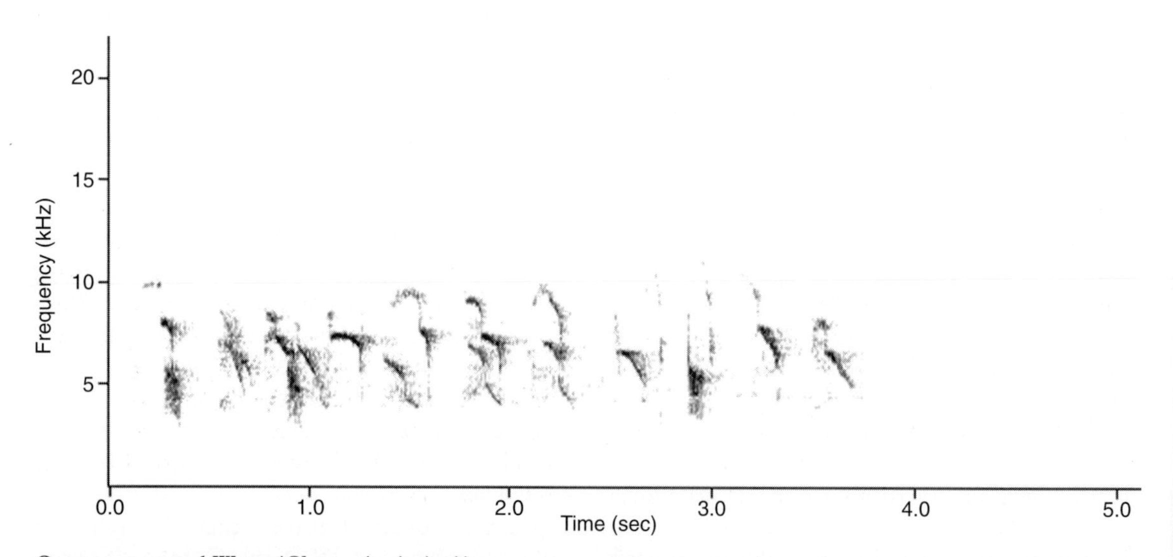

Orange-crowned Wren (*Clytomyias insignis*)
K. D. Bishop and Jared Diamond, at ca. 2000 m in Eastern Jayaywijaya Ranges of Irian Jaya, New Guinea. *C. i. oorti*, short segment of contact calls or travelling party; the three central elements are 'seep' sounds, with flanking 'tu' elements sounding lower.

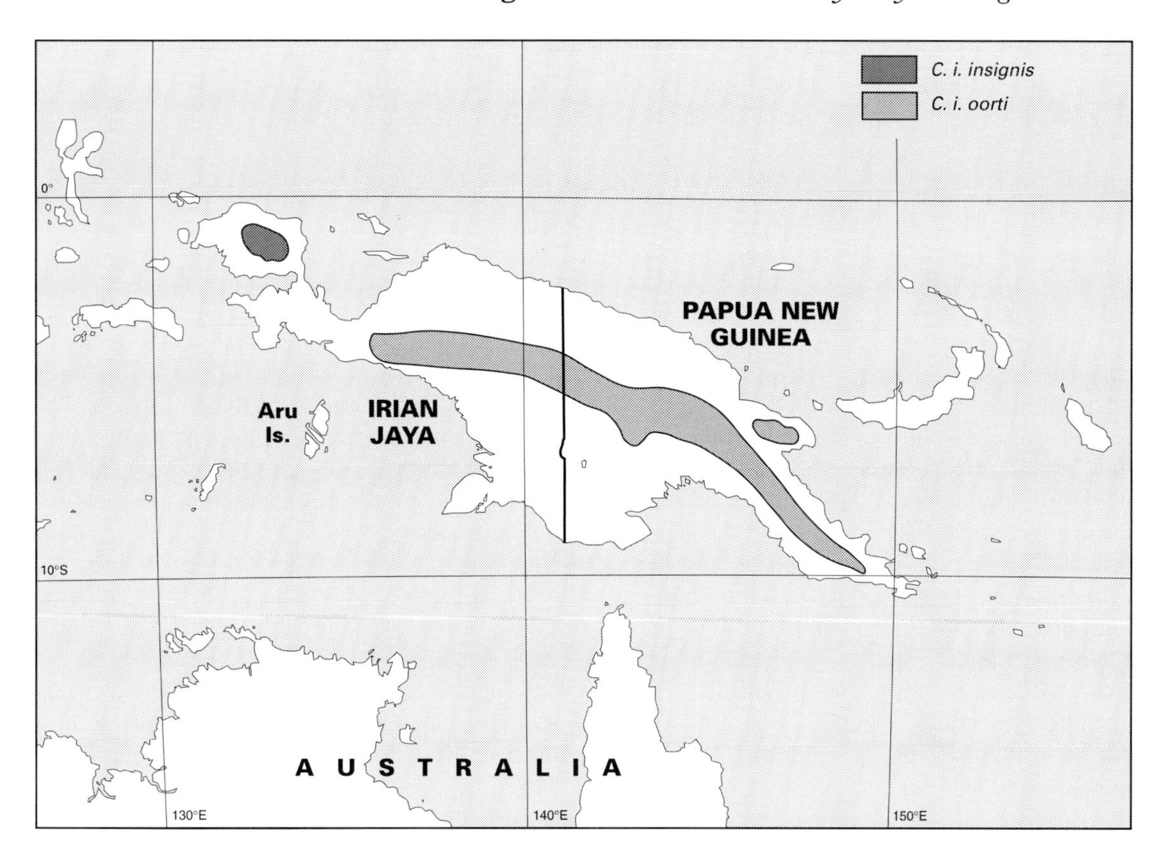

kop (nominate subspecies) and Saruwaged-Finisterre Ra. on Huon Peninsula isolated from main population; the latter are not separable as subspecies. Not recorded from Van Rees-Gauttier or Torricelli Mts in N. Rarely seen but locally common in suitable habitat. Little threatened, since high mountain forest inaccessible to clear-felling.

Habitat and general habits

Usually in thick cover provided by climbing bamboo, ferns, and vines in the sub-canopy of high mountain rainforest between 2000 and 3000 m. Suitable dense tangles occur in openings caused by tree fall or tracks. At Tari Gap, mostly associated with the scrambling bamboo *Nastus* sp. that grows on the edge of the forest, but also in dense understorey right into the forest (Frith and Frith 1992). Rarely flies, moves very rapidly through thick vegetation, gleaning from underside of leaves; carries tail half-cocked, slightly spread. Do not join mixed species flocks that tend to follow them (B. W. Finch, personal communication in Schodde 1982*b*). Groups appear to remain in the same patch of habitat throughout the year, working it over regularly and quickly searching for insects (unspecified) deep in dense foliage. Rarely seen on ground. Always in pairs or small groups, commonly 6–8 birds, all year round. Four adults, 'obviously fellow flock members', caught in same mist-net, Tari Gap (Frith and Frith 1993)

Displays and breeding behaviour

Nothing known about displays or breeding behaviour. No nest ever found.

Breeding and life cycle.

Little information available; female containing two eggs collected by Diamond at Awande, mid-June (1972); male and female with active gonads collected in Snow Mts, Aug (Ripley 1964); five males and four females collected in July at Mt. Hagen had small gonads, one male had gonads beginning to enlarge; eight of these birds were moulting, stage variable (Mayr and Gilliard 1937).

Genus *Stipiturus* Lesson

Stipiturus R. P. Lesson, 1831. *Traite d'Ornithologie, Livraison*, **8**, 414.

Streaked brown plumage with characteristically only six very long, graded filamentous tail feathers, the barbules of which scarcely interlock at all. Wings small and rounded as in *Malurus*; flight weak, legs long. Sexually dimorphic, males with pale blue throats and females tawny, as for rest of underparts; this difference is visible even at the time of fledging. Habitat, dense heath, spinifex, or swamp through which they forage energetically, seldom showing themselves and rarely cocking their tails.

The three currently recognized species of emu-wrens (*Stipiturus*) differ in range and the extent to which they vary across it; all are very habitat specific. The Southern Emu-wren (*S. malachurus*) occurs across S Australia and Tasmania wherever habitat is suitable. Populations tend to be isolated from each other, and have diverged considerably. The Mallee Emu-wren (*S. mallee*) has a very local distribution, with no described variation. The Rufous-crowned Emu-wren (*S. ruficeps*) has a wide distribution and varies little. Attempts to name these forms in a way that reflected relationships have produced a variety of schemes in the last 50 years.

The only form known for a century was the Southern Emu-wren, first named as *Muscicapa malachura* from near Port Jackson (Shaw 1798). Shaw's formal description was an appendix to an 'Account of a new species of *Muscicapa* from New South Wales' to the Linnean Society in London by Major-General Thomas Davies. Many later accounts proposed new names: *Motacilla fimbriata* Wilkes, 1817; *Malurus palustris* Vieillot, 1818; *Malurus gularis* Stephens, 1826; *Malurus emitis* Ewing, 1841; Hindwood (1931) gives a detailed account of early records of the emu-wren. The French ornithologist René Lesson visited Port Jackson during the voyage of the *Coquille* (1823–5) and collected specimens, from which he separated the emu-wrens in a new genus *Stipiturus* (1831).

By the time of Gould's accounts (1841, 1865) the name had stabilized as the Emu-wren (*Stipiturus malachurus*), which he placed among the Insessores, following *Malurus* and *Amytis* (=*Amytornis*). Campbell (1901*a*) and North (1904) both placed *Malurus* among the Muscicapidae, but *Stipiturus* was included by Campbell with the Timaliidae, the Babbling Thrushes, along with *Amytis* and *Acanthiza*, while North placed *Stipiturus* and *Amytis* in the family Sylviidae.

In 1899, Campbell published a brief, provisional description of a new emu-wren *Stipiturus ruficeps*, the Rufous-crowned Emu-wren (1899*a*). He had been sent a female specimen collected at North West Cape in WA by Tom Carter, who, not recognizing it as different from *S. malachurus*, had sent the male of the pair to his father. Carter procured another male for Campbell, who published a full description (1899*b*). In 1908, Campbell was alerted by a report in the Melbourne *Argus*, which mentioned a Rufous-crowned Emu-wren in the Victorian Mallee. Campbell knew the Rufous-crowned Emu-wren only from North West Cape and inland north-western deserts and acquired a Victorian skin for examination, from which he described the Mallee Emu-wren (*Stipiturus mallee*). Campbell noted that emu-wrens from WA were greyer, lighter in colour, with narrower tail feathers, and proposed a separate species *Stipiturus westernensis* (1912).

Thus initially, four separate species were described and the 1913 Checklist maintained this. Mathews placed *mallee* and *westernensis* as subspecies of *malachurus*, along with another three described by him (Mathews 1912*a,b*). He maintained this treatment of *mallee* as one of many subspecies of *malachurus* (1922–3, 1930). As many as 11 subspecies of *malachurus* have been described at various times (see species accounts for details).

Contrary to Mathews, the 1926 Checklist maintained three separate species, *malachurus* (including *westernensis*), *mallee*, and *ruficeps*, as did Serventy and Whittell in the various editions of *The Birds of Western Australia*. Condon treated *mallee* and *ruficeps* as subspecies of *malachurus*, but in 1962 he treated them as separate species, quoting Keast (1957) as 'presenting a fairly good case for maintaining the three forms as distinct species' (p. 129). Keast actually said 'Most workers would have serious misgivings about listing *mallee* as a species and it should be placed with *malachurus*.' (p. 52). Ford argued that *mallee* is derived from *ruficeps*, on the basis of the colour pattern of the head, and the blue face, and treated *mallee* as a subspecies of *ruficeps* (1970). Schodde (1982*b*) considered that while the rufous crown and blue face of *mallee* resemble *ruficeps*, the rest of the plumage (breast, belly, back, and wings) are more similar to *malachurus* and the fledglings are quite distinct; accordingly, he treated it as a distinct species, confirmed by biochemical data (Christidis and Schodde, in press).

Southern Emu-wren *Stipiturus malachurus* (Shaw, 1798)

Muscicapa malachura Shaw, 1798. *Transactions of the Linnean Society of London*, **4**, 242.

PLATE 5

Polytypic. Seven subspecies. *Stipiturus malachurus malachurus* (Shaw, 1798); *Stipiturus malachurus littleri* Mathews, 1912; *Stipiturus malachurus westernensis* Campbell, 1912; *Stipiturus malachurus hartogi* Carter, 1916; *Stipiturus malachurus intermedius* Ashby, 1920; *Stipiturus malachurus halmaturinus* Parsons, 1920; *Stipiturus malachurus parimeda* Schodde and Weatherly, 1981.

Description

S. m. malachurus

ADULT MALE: upper parts, rusty-brown, streaked black from crown to rump; front of crown, more rufous, unstreaked; wings, grey-brown; flight feathers edged rufous; tail, about twice as long as body, dark grey, of only six relatively broad, filamentous, loosely webbed feathers, central two much longer (Fig. 4.2); ear coverts, light brown, may have paler shaft-streaks; lower half of eye rimmed white; throat, upper breast, and brief eye-brow, mid sky-blue; rest of underparts, rufous-tawny, whiter on belly; eye, brown; bill, black; feet, brown.

ADULT FEMALE: as male, but crown and upper surface more darkly streaked; lacks rufous fore-crown; no blue coloration, throat and breast same reddish-brown as rest of underparts; bill dark brown with light grey base to lower mandible.

IMMATURE: as female, but duller, more diffuse streaking on upper parts, face unstreaked, bill brown. Sex differences apparent at 5 days: male with pale blue-grey throat and breast, female pale reddish-brown.

MOULT: once a year after breeding, no eclipse plumage; tail feathers replaced throughout year as they wear.

History and subspecies

Owing to considerable variation between isolated populations over wide geographical range, 11 subspecies of *Stipiturus malachurus* have been described; six were recognized by Schodde (1982*b*); we recognize a seventh, *S. m. hartogi*. In the past, *S. mallee*, or both *S. mallee* and *S. ruficeps*, have been included as subspecies of *S. m. malachurus* (Mathews 1912*a*, 1922–3, 1930; Condon 1951; Keast 1957), but are now recognized as separate species (RAOU Checklist 1926; Serventy and Whittell 1948, 1976; Schodde 1982*b*; Cracraft 1986; Christidis and Boles 1994).

Stipiturus malachurus malachurus (Shaw, 1798) (includes *S. m. tregallasi* Mathews, 1912 and *S. m. richmondi* Mathews, 1923). E and S of Dividing Ra., from Noosa, Qld to mouth of Murray R. in SA. In drier areas in E of SA, less red, paler blue, and less streaked.

Stipiturus malachurus littleri Mathews, 1912. Tas. Similar to *S. m. malachurus*, but more tawny and dorsal streaks finer; belly all rufous; smaller, with shorter tail, broader rectrices than *S. m. malachurus*.

Stipiturus malachurus intermedius Ashby, 1920. Small localized population in Mt. Lofty Ra., SA. Size, tail, and blue areas similar to *S. m. malachurus*. Much darker and more heavily streaked olive-grey on back. Ear coverts plain to white-streaked. Adult male usually with black-streaked, olive-brown fore-crown.

Stipiturus malachurus halmaturinus Parsons, 1920. Kangaroo Is., SA. Largest form. Tail and blue areas as *S. m. malachurus*, but paler, greyer back with fine black shaft-streaks, white-streaked ear coverts, and plain dull rufous fore-crown.

Stipiturus malachurus parimeda Schodde and Weatherly, 1981. Tip of Eyre Peninsula, SA. Back pale olive-grey with fine black streaks, pallid tawny flanks, belly white. Plain rufous fore-crown and broad tail feathers (as E subspecies), longer blue eye-brow, and pale sky-blue throat and breast.

Stipiturus malachurus westernensis Campbell, 1912 (includes *S. m. rothschildi* Mathews, 1912 and *S. m. media* Mathews 1919). SW WA. White-streaked ear coverts, slender rectrices, olive-grey back with black shaft-streaks. Rufous fore-crown, with black shaft-streaks over head. Extensive blue eye-brow, extending to lores. In far SW, rufous and blue tones similar to SE Australia, in N and E, paler and smaller (Ford 1970).

Stipiturus malachurus hartogi Carter, 1916. Confined to Dirk Hartog Is., at least 300 km N of nearest (doubtful) record of *S. m. westernensis*. Smaller, paler and greyer; streaks on head and back very narrow, almost absent.

Measurements and weights

Stipiturus malachurus malachurus: wing, M (n = 57) 43.1 ± 1.1, F (n = 31) 41.6 ± 1.7; tail, M (n = 57) 113.3 ± 6.5, F (n = 31) 107.6 ± 5.8; bill, M (n = 57) 8.3 ± 0.4, F (n = 31) 8.1 ± 0.6; tarsus, M (n = 57) 19.4 ± 0.8, F (n = 31) 18.8 ± 0.8; weight, M (n = 15) 5.5–9.0 (7.7 ± 0.9), F (n = 7) 7.4–8.3 (7.8 ± 0.3).
Stipiturus malachurus littleri: wing, M (n = 14) 42.1 ± 1.2, F (n = 4) 39.8 ± 1.5; tail, M (n = 14) 100.2 ± 4.6, F (n = 4) 91.8 ± 4.5; bill, M (n = 14) 8.3 ± 0.6, F (n = 4) 8.3 ± 0.7; tarsus, M (n = 14) 18.7 ± 1.1, F (n = 4) 18.7 ± 1.1.
Stipiturus malachurus intermedius: wing, M (n = 13) 43.3 ± 1.3, F (n = 11) 41.8 ± 0.9; tail, M (n = 13) 110.5 ± 5.2, F (n = 11) 107.4 ± 4.4; bill, M (n = 13) 8.8 ± 0.6, F (n = 11) 8.3 ± 2.4; tarsus, M (n = 13) 19.3 ± 0.7, F (n = 11) 18.7 ± 1.0.
Stipiturus malachurus halmaturinus: wing, M (n = 4) 44.5 ± 0.6, F (n = 6) 44.5 ± 1.1; tail, M (n = 4) 115.3 ± 8.4, F (n = 6) 119.0 ± 5.6; bill, M (n = 4) 9.6 ± 0.7, F (n = 6) 8.9 ± 0.3; tarsus, M (n = 4) 20.5 ± 0.6, F (n = 6) 20.1 ± 0.7; weight, M (n = 4) 7.0–8.3 (7.4), F (n = 2) 7.9–8.3 (8.1).
Stipiturus malachurus parimeda: wing, M (n = 3) 42.7 ± 0.6, F (n = 2) 41.5 ± 2.1; tail, M (n = 3) 108.0 ± 13.1, F (n = 2) 108.0 ± 5.7; bill, M (n = 3) 8.6 ± 0.2, F (n = 2) 8.5 ± 0.7; tarsus, M (n = 3) 19.2 ± 0.7, F (n = 2) 18.6 ± 1.5.
Stipiturus malachurus westernensis: wing, M (n = 21) 44.0 ± 1.8, F (n = 16) 42.4 ± 1.7; tail,

M (n = 21) 117.9 ± 6.2, F (n = 16) 109.6 ± 6.7; bill, M (n = 21) 8.7 ± 0.5, F (n =16) 8.3 ± 0.4; tarsus, M (n = 21) 20.0 ± 0.7, F (n = 16) 19.3 ± 0.7; weight, M (n =2) 7.0, 7.0.
Stipiturus malachurus hartogi: wing, M (n = 5) 40.2 ± 1.8, F (n = 7) 38.3 ± 0.9; tail, M (n = 5) 118.0 ± 6.1, F (n = 7) 109.5 ± 7.7; bill, M (n = 5) 8.5 ± 0.7, F (n =7) 7.9 ± 0.6; tarsus, M (n = 5) 19.9 ± 0.7, F (n = 7) 19.5 ± 1.7.
(Weights from museum specimens and ABBBS.)

Field characters

Small size, long filamentous tail, and streaked ear coverts distinguish them from any female fairy-wren that may overlap; easily distinguished from male fairy-wren in breeding plumage. No overlap with Rufous-crowned Emu-wren (*Stipiturus ruficeps*), 300 km NE of Dirk Hartog Is., so locality and lack of spinifex assist identification. Mallee Emu-wren (*S. mallee*) has more rufous, less streaked crown, blue on ear coverts, and shorter tail.

Voice

Song a soft short descending trill 'deedle-deedle-deedle', 4–5 short notes; heard all year, more frequently in breeding season, possibly only by male (see sonograms); similar to fairy-wrens, but much softer. Contact call high pitched, very soft 'pree-pree', answered by single 'pree', continuous when foraging. Alarm, shrill scream 'steet steet steet'. Nest call soft 'tweet' by sitting bird to approaching partner (Hutton 1991). Tape-recordings available.

Range and status

Resident in coastal heaths, swamps, and dune thickets. Coastal E Australia from Noosa in Qld, S to Wilson's Promontory, and along S coast to mouth of Murray R.; inland in Vic. to Grampians and Ninety Mile Desert; isolated subspecies: southern Mt. Lofty Ra., Kangaroo Is., and tip of Eyre Peninsula. In SW Australia, coastal from Israelite Bay in E to Geraldton in N, extending patchily inland, with isolated population on Dirk Hartog Is. Tas., except central highlands and E coast. Locally abundant in suitable habitat, although rarely seen. Very wide range. Restricted subspecies becoming rare: *S. m. intermedius* classified as endangered, *S. m. parimeda* classified as vulnerable by Garnett (1992*a*) and *S. m. hartogi* requires more data before its status can be assessed. Main threats to survival: habitat clearing and drainage of wetlands in densely populated coastal areas.

Habitat and general habits

Favoured habitat low heathland vegetation, often on coastal sandplains, sand-dunes, and swampy areas bordering wetlands; populations in inland areas occupy drier heathlands. In Tas., dense wet heaths on poorly drained flats, with thick cover of Cutting Rushes (*Gahnia* spp.) and Button Grass (*Gymnoschoenus sphaerocephalus*). Hard to observe in dense cover; little known from field studies, except for observations of Fletcher (1913, 1915) and observations of birds in captivity (Hutton 1991). Flight weak, slow and just above vegetation; difficult to flush from cover, and then fly only short distance, undulating, tail trailing behind. On still days, show themselves briefly on top of bush, on windy days, remain invisible low in cover. Forage mostly by hopping on ground (tail erect), probing in vegetation, capture resting insects by flying sorties to 1 m, also forage among leaves and twigs of shrub cover. Food: wide variety of small insects and spiders (Barker and Vestjens 1990); beat moths to remove wings; split stems of reeds (*Juncus* sp.) to take insects and larvae (Fletcher 1915). No studies of marked birds, so social organization unknown. Encountered singly, in pairs, or, in autumn and winter, small groups, probably family parties (North 1904; Fletcher 1915; Chisholm and Cayley 1929). Pair or family perch closely huddled together and allopreen (Hutton 1991).

Displays and breeding behaviour

Most behaviour known only from aviary observations (Hutton 1991). First indication of breeding: increased territorial calls by male

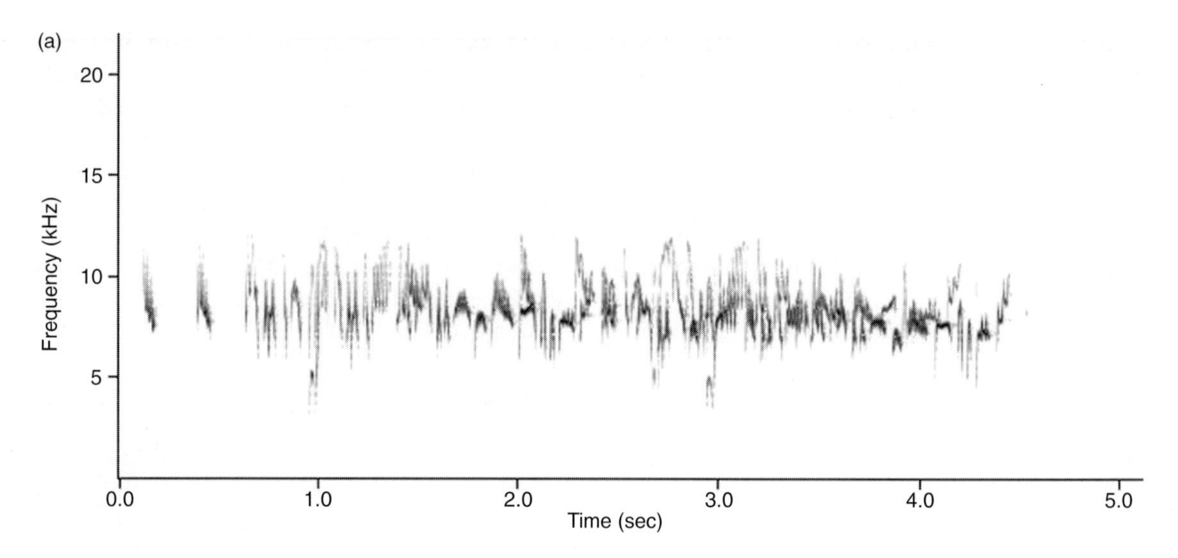

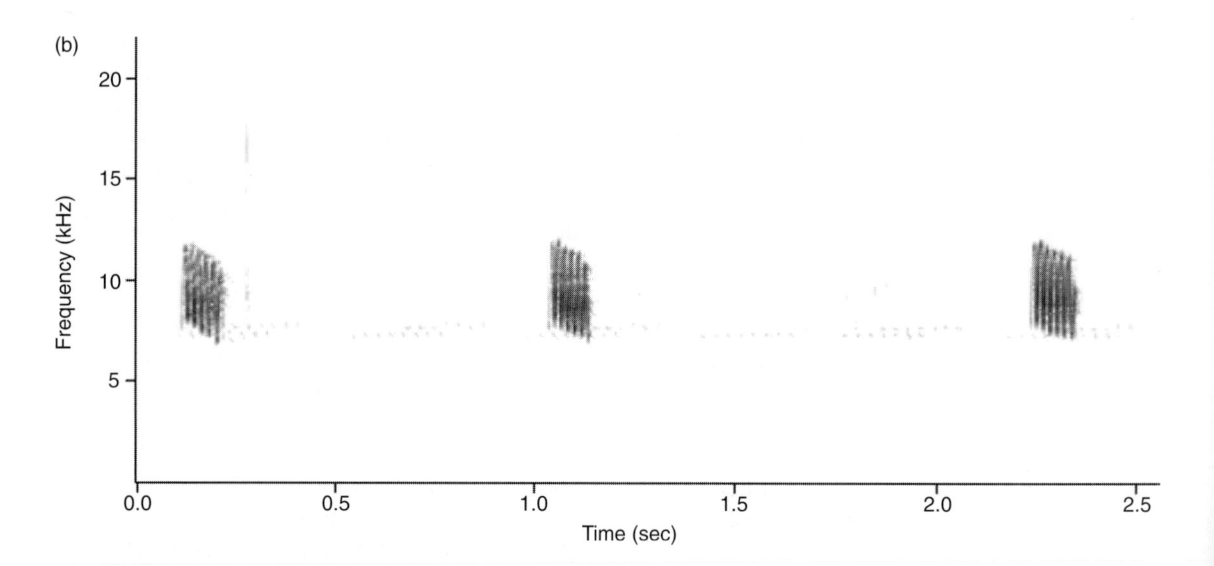

Southern Emu-wren (*Stipiturus malachurus*)
(a) *S. m. malachurus*, song. D. Stewart, Barrenjoey National Park, NSW. (b) *S. m. littleri*, alarm call. D. Stewart, Tasmania.

in morning and throughout day, followed by courtship feeding. Courtship and mating frequent (2–3 times a day in captivity) and obvious (rarely seen in *Malurus* spp.): after courtship feeding, male remains in front of female, bobs up and down with head forward, blue breast fluffed; moves to one side of female, then to other, still bobbing up and down, mounts and copulates; female remains still throughout. Rodent-run distraction display when nest or young threatened (K. A. Hindwood *in litt.* 22 Dec. 1960).

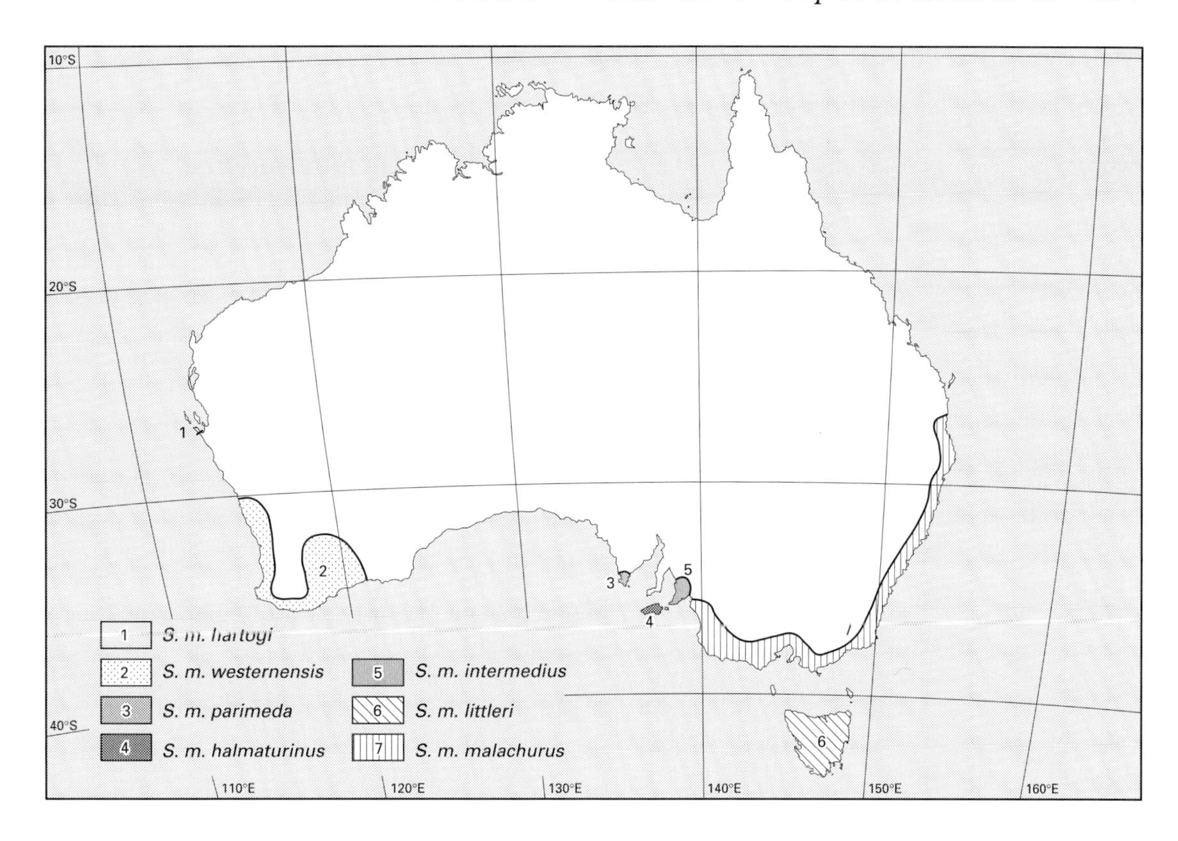

Breeding and life cycle

Breeding pairs dispersed in individual territories. Laying season in austral spring–summer Aug–Jan, perhaps earlier in N; most nests Oct–Nov, but Aug nests recorded even in Tas. (Fletcher 1913, 1915; RAOU NRS). No marked female studied, but female thought to re-nest after failure and to lay second clutch *c.* 8 weeks after first clutch fledges (Fletcher 1915). Female alone builds nest, fed by male during construction (*c.* 10 days). Nests small (80–140 mm high × 70–80 mm wide), with 30–50 mm side entrance, domed, usually concealed in dense shrubs, tussocks of Cutting Grass (*Gahnia* sp.) or reeds. More spherical than nest of *Malurus* spp., loosely and roughly woven of moss, with stems and leaves of dry and green grass, lined with finer grass, plant down, feathers, or fur. In Tas., moss may be main nest material (Fletcher 1915). Clutch 2–4, mainly 3; eggs large (16–17 × 12–13 mm, n = 48, Schodde 1982*b*), similar in size and appearance to those of *M. cyaneus*: tapered oval, white, lustreless, finely marked with reddish-brown spots and flecks, concentrated at larger end, laid daily in the morning. Occasionally parasitized by Horsfield's Bronze-Cuckoo (*Chrysococcyx basalis*), Shining Bronze-Cuckoo (*Ch. lucidus*), Fan-tailed Cuckoo (*Cacomantis flabelliformis*), Pallid Cuckoo (*Cuculus pallidus*), Brush Cuckoo (*Cac. variolosus*) (Brooker and Brooker 1989*a*). Most incubation (13–14 days) by female, relieved for short period by male (only female develops bent tail during incubation). Nestlings fed by both parents for 11–15 days (RAOU NRS). After fledging, young cryptic 1 week, independent at 2 months (Fletcher 1913, 1915; Hutton 1991). No records of helpers at nest; reports of groups of more than two suggest young stay with parents after fledging, at least until winter or start of next breeding season.

Mallee Emu-wren *Stipiturus mallee* Campbell, 1908

Stipiturus mallee Campbell, 1908. *Emu*, **8**, 14.

PLATE 5

Monotypic

Description

ADULT MALE: crown, unstreaked, pale rufous; back, mid olive-brown, streaked black; wings and tail, grey-brown; tail, shorter than in Southern Emu-wren, six feathers with few barbules, web of vanes largely unmeshed; central feathers one and a half times as long as outer feathers; lores, eyebrow, ear coverts, throat, and breast, mid sky-blue, ear coverts streaked black; lower half of eye rimmed white; undersurface rufous on flanks, white at centre of belly; eye, dark brown; bill, black; feet, pinkish brown.

ADULT FEMALE: as male, but rufous confined to forehead; breast and flanks, tawny; belly, white; lores, white; ear coverts, grey, streaked white; eyes and bill, dark brown; feet, pinkish brown.

IMMATURES: sexable before fledging from colour of throat (female tawny, male greyish white); uniform olive-grey with dusky streaking from fore-crown to rump; breast greyish-white, extending further onto lower breast and belly than in fledglings of other emu-wrens; flanks, tawny; lores and cheeks, white; ear coverts, streaked white; eyes, bill, and feet, brown.

MOULT: once a year after breeding; no eclipse plumage.

History and subspecies

No subspecies recognized. Since original description by Campbell in 1908 as *S. mallee*, considered as subspecies of Southern Emu-wren (*S. malachurus*) (Mathews 1912*a*, 1922–3; Condon 1951; Keast 1957), as subspecies of Rufous-crowned Emu-wren (*S. ruficeps*) (Ford 1970; Parker 1975; Blakers *et al.* 1984), and as separate species (RAOU Checklist 1926; Serventy and Whittell 1948, 1976; Schodde 1982*b*; Christidis and Boles 1994).

Measurements and weights

Wing, M (n = 16) 39.8 ± 1.0, F (n = 6) 38.7 ± 1.4; tail, M (n = 16) 85.1 ± 5.7, F (n = 6) 79.5 ± 5.9; bill, M (n = 16) 8.6 ± 0.5, F (n = 6) 8.3 ± 0.2; tarsus, M (n = 16) 15.5 ± 0.6, F (n = 6) 15.1 ± 0.6); weight, M (n = 1) 5.5. (Weight from specimen in South Australian Museum.)

Field characters

Only possible overlap is with *S. malachurus*, separated on habitat: *S. mallee*, spinifex in well-grown mallee; *S. malachurus*, swampy heath on sandplain. Distinguished by paler, unstreaked, more rufous crown, more extensive blue on face of male (including ear coverts) and shorter tail. Small size, long filamentous tail, and streaked ear coverts distinguish them from any female fairy-wren that may overlap; easily distinguishable from any male fairy-wren in breeding plumage.

Voice

Song a high-pitched twittering reel of 1–2 sec, similar to reel of fairy-wrens but higher and weaker (see sonogram). Contact calls, a continuously repeated, almost inaudible high-pitched 'treee', hard to distinguish from cicada or grasshopper. In alarm, louder 'trrrt'.

Range and status

Resident with restricted range confined to uncleared low eucalypt woodland (mallee) with understorey of spinifex; on both sides of border between SA and Vic., S of Murray R. In SA, in narrow strip W of Vic. border, Nadda, Peebinga, Pinnaroo in N, S to Comet Bore (35° 44′ S, 140° 48′ E; Eckert 1977). More extensive in NW Vic. S to Yanac (36° 07′ S, 141° 27′ E) and E through Big Desert and Sunset Country to Hattah-Kulkyne National Park (Chisholm 1946; Schodde 1982*b*; Blakers *et al.* 1984; Emison *et al.* 1987). S limit probably determined by S limit of spinifex. Only in suitable habitat; localized and rare in SA, rare in most parts of NW Vic, more common in

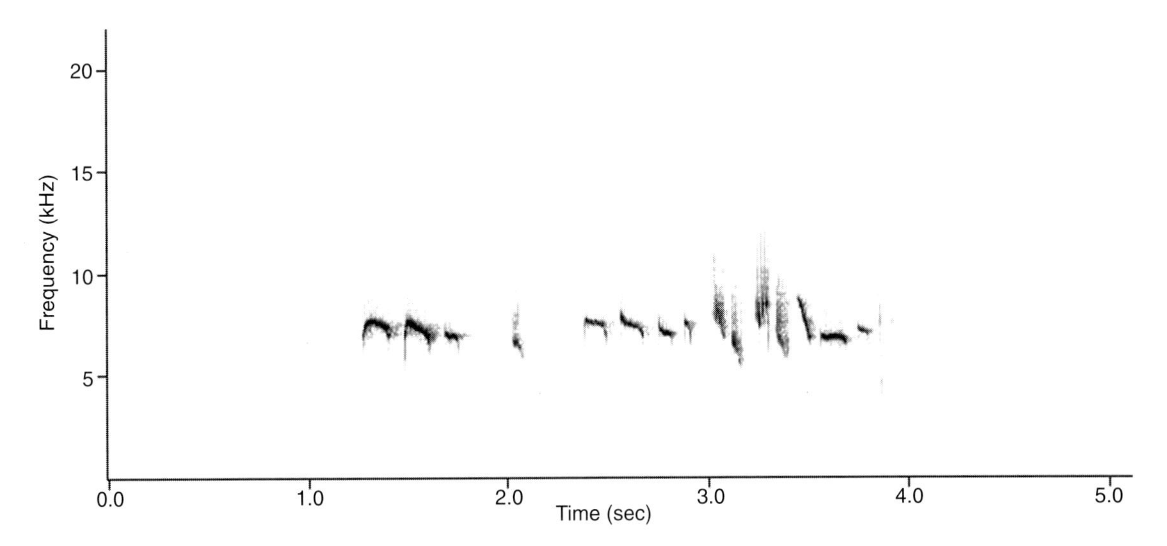

Mallee Emu-wren (*Stipiturus mallee*)
Song. R. Buckingham. Pink Lakes State Park, Victoria (from *A Field Guide to Australian Bird Song*, cassette 9; see p. 20 of Plan)

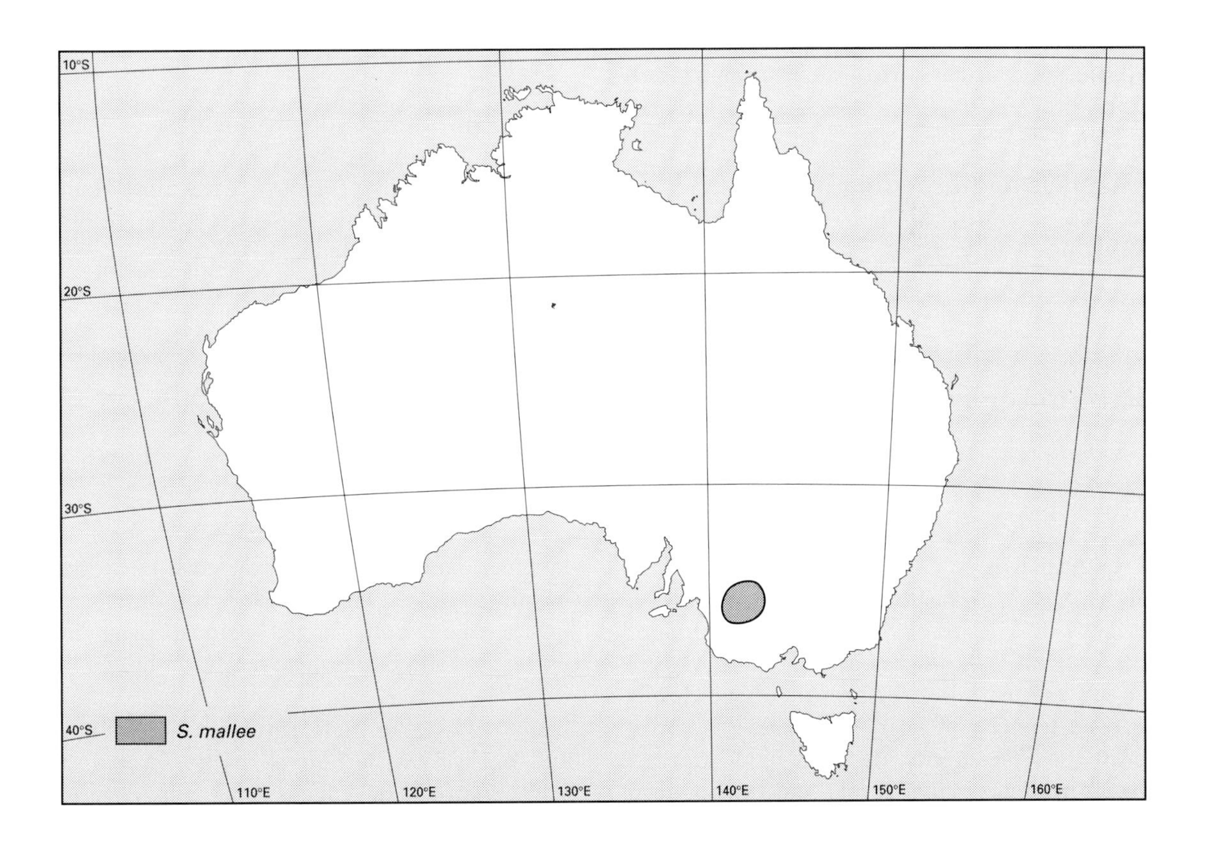

Hattah, nowhere abundant. Conservation status: vulnerable (Garnett 1992*a*).

Habitat and general habits

Lives only in hummock grassland (*Triodia*), usually on low dune systems as understorey below low woodland dominated by Cypress Pine (*Callitris*) and mallee forms of *Eucalyptus* spp. Also in Mallee Teatree (*Leptospermum laevigata*)–*Triodia* heathlands, with *Allocasuarina*, *Banksia* spp., and other shrubs (Emison *et al.* 1987). Little known, since hard to observe in clumps of spinifex. Forages rapidly through shrubs and spinifex, generally within 0.5 m of ground. In open, carries tail half-cocked; in cover, tail trails behind. In calm weather, may venture onto open ground to forage or call from top of bush; in windy weather, almost invisible. Food mainly insects, but includes some seeds (Schodde 1982*b*). Encountered singly, in pairs, or in small groups: a group of five early in spring before any young fledged (Garnett 1992*a*); seven in Apr (McLennan 1906 in Howe 1933). No studies of marked birds; social organization unknown.

Displays and breeding behaviour

Typical malurid Rodent-run display when fledglings threatened (Schodde 1982*b*). Nothing else recorded of displays or breeding behaviour.

Breeding and life cycle

Most nests recorded Sept–Nov (McGilp 1921, 1942; Howe 1933; RAOU NRS). No marked birds studied, so number of broods raised per year not known. Nest built by female, escorted by male (Howe 1933). Nests small (80–120 mm high × 80–90 mm wide), domed, usually placed in centre of spinifex hummock, 150–450 mm above ground; 30–40 mm wide side entrance, slightly hooded, usually faces into centre of clump. Woven of thin grass stems, spinifex spines, and other fibrous material, lined with plant down, feathers, and fur. Clutch 3, rarely 2; eggs (13.5–16 × 10–12 mm, n = 12, Schodde 1982*b*); tapered oval; white with little gloss, covered with reddish-brown freckles and blotches, concentrated at larger end (more freckled all over than other emu-wren eggs). Only female incubates and broods; incubation and nestling periods not known, probably not very different from *S. malachurus*. Young reared by both parents, no details known. Party of six brown birds and one male in April probably a family party (Howe 1933); this and other records of groups with more than two birds suggest young may stay with family group after independence.

Rufous-crowned Emu-wren *Stipiturus ruficeps* Campbell, 1899

Stipiturus ruficeps Campbell, 1899. *Victorian Naturalist*, **15**, p. 116. PLATE 5

Monotypic

Description

ADULT MALE: crown and nape, unstreaked, bright rufous; back cinnamon-brown with faint dusky streaks; wings grey-brown, flight feathers edged rufous; tail, grey-brown, of six closely barbed feathers of nearly equal length; lores, brow, and ear coverts, deep sky-blue, streaked black; white eye ring; throat and breast, deep sky-blue, dark streaks on lower edge of breast showing as broken blackish band; rest of ventral surface plain buff, tinged rufous, paler in centre; eyes and bill, brown; feet, light brown.

ADULT FEMALE: as male, throat and breast not blue but yellowish buff as belly; lores and brow unstreaked; cheeks and ear coverts streaked light and dark, with pale sky-blue tinge; bill slightly paler than male.

IMMATURE: greyish-brown above, less streaked than adult; lores and brow dull whitish, ear coverts with faint pale streaks; eyes, bills and feet, as adult. Sexes distinct before fledging; male: creamy white throat and breast soon changing to faint blue; female: throat, breast, and belly, yellowish-buff.

MOULT: once a year after breeding, unknown when juvenile male acquires blue breast of adult; no eclipse plumage.

History and subspecies

No subspecies recognized; no significant geographical variation over their large range. Birds from WA may be redder and female bluer on lores and face (Schodde 1982*b*). Since its original description by Campbell (1899*a*), *S. ruficeps* has been considered as one of three species of emu-wren (RAOU Checklist 1926; Serventy and Whittell 1948, 1976; Schodde 1982*b*; Cracraft 1986; Christidis and Boles 1994), as conspecific with Mallee Emu-wren in *S. ruficeps*, with two subspecies *S. r. ruficeps* and *S. r. mallee* (Ford 1970; Ford and Parker 1974; Blakers *et al.* 1984), as one of two species, *S. ruficeps* and *S. malachurus* (including *S. m. malachurus* and *S. m. mallee*, Mathews 1912*a*, 1922–3, 1930; Keast 1957), and as one of three subspecies of *S. malachurus* (Condon 1951).

Measurements and weights

Wing, M (n = 31) 38.9 ± 1.2, F (n = 14) 37.9 ± 1.6; tail, M (n = 31) 70. 9 ± 5.5, F (n = 14) 68.9 ± 7.0; bill, M (n = 31) 7.8 ± 0.5, F (n = 14) 7.3 ± 0.5; tarsus, M (n = 31) 14.9 ± 0.8, F (n = 14) 14.8 ± 0.7; weight, M (n = 9) 4.9–6.5 (5.4 ± 0.7), F (n = 4) 4.0–5.0 (4.7 ± 0.5). (Weights from museum specimens and field measurements by authors.)

Field characters

Does not overlap with ranges of Mallee Emu-wren (*S. mallee*) and Southern Emu-wren (*S. malachurus*); locality and spinifex habitat assist identification. Unstriated rufous crown diagnostic, male has more blue on face and throat than other *Stipiturus*. Tail about same length as head and body, shorter than tail of *S. mallee*. Small size, filamentous tail, and streaked ear coverts distinguish them from any female fairy-wren that overlaps. Easily distinguished from male fairy-wren in breeding plumage.

Voice

Tape-recorded (ANWC). Song a high-pitched twittering reel of 1–2 sec, similar to reel of fairy-wrens but higher and weaker, audible over shorter distance (see sonogram). Contact calls continuously repeated, almost inaudible high-pitched 'tseee', hard to distinguish from cicada or grasshopper. In alarm, louder 'trrrrt'.

Range and status

Resident in arid regions of Australia where spinifex occurs (*Triodia* and *Plectrachne*). In WA, from North West Cape through Pilbara region to Cape Keraudren (S edge of Great Sandy Desert); NE of line from North West Cape–Kennedy Range (Ford 1978*a*)–Wiluna (Lake Way)–Naretha (on NW fringe of Nullarbor Plain; Collins 1934); E through Gibson Desert, Great Victoria Desert, and Warburton Range to Tanami Desert and Macdonnell Ra. in NT; E to SW fringe of Barkly Tableland (Frewena—Ford and Parker 1974), and in the S Simpson Desert (May 1977*b*; Blakers *et al.* 1984); isolated population in patches of mallee–spinifex in an area bounded by Dajarra, Winton, Fermoy, and Boulia in SW Qld (Officer 1970; Vernon 1972; Ford and Parker 1974). Probably locally common in patches of suitable habitat; hard to locate.

Habitat and general habits

Optimum habitat unburnt, old-growth spinifex, with large hummocks interspersed with mallee eucalypts or shrubs of *Acacia*, *Cassia*, *Melaleuca*, or *Thryptomene* (Whitlock 1924; Mees 1961; Ford and Parker 1974; Schodde 1982*b*). Usually on sand-plain or stony hills,

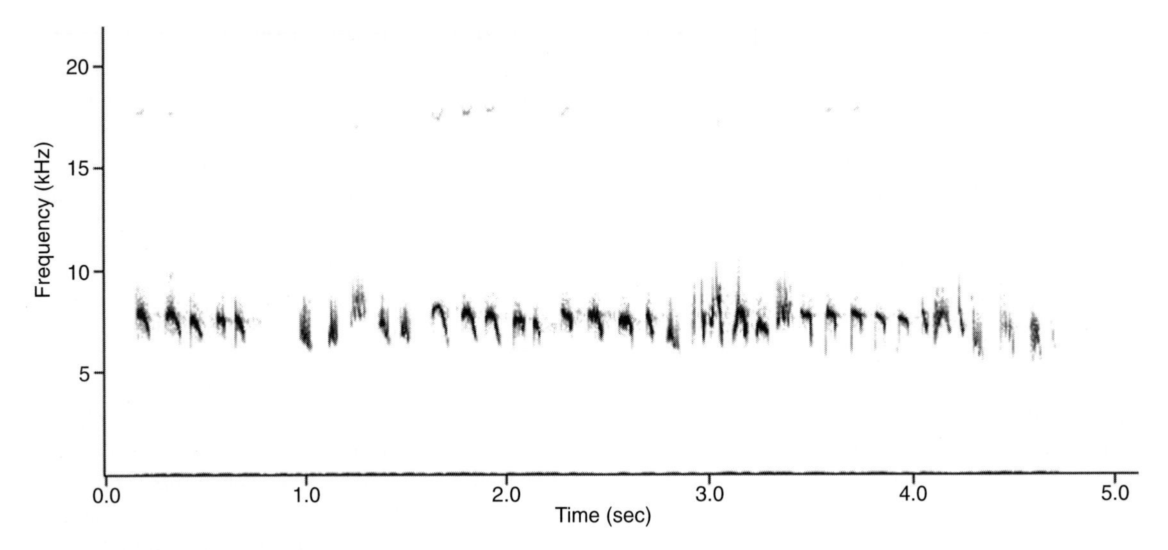

Rufous-crowned Emu-wren (*Stipiturus ruficeps*)
Song. G. S. Chapman. Opalton, Queensland.

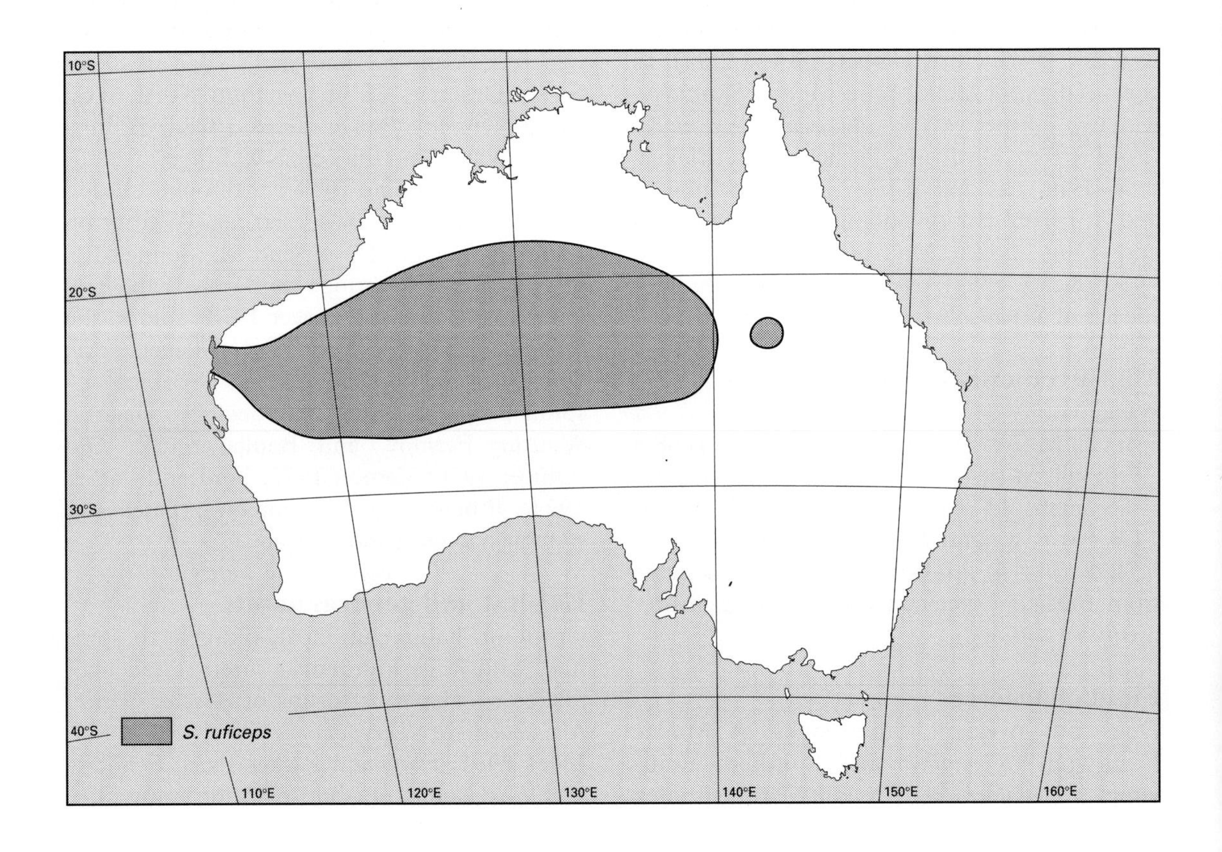

but also in cane-grass (*Zygochloa*) on sand dunes (in Simpson Desert, May 1977*b*). Little known, since hard to observe in clumps of spinifex; no detailed studies. Rarely flies, moves from bush to bush hopping along ground; forages rapidly through shrubs and spinifex, generally within 0.5 m of ground; no records of diet from specimens, but observations of foraging suggest it is similar to other emu-wrens: predominantly small insects and spiders. In open, carries tail half-cocked, but in cover, tail trails behind. In calm weather, may venture onto open ground to forage or call from top of bush, but in windy weather, almost invisible. Encountered singly, in pairs, or small groups; Hitchcock and Jarman (1944) recorded 'perhaps six' in July near Alice Springs. No studies of marked birds, so social organization unknown.

Displays and breeding behaviour

Nothing recorded of displays. Only records of breeding behaviour suggest that when female nest building, calling more frequent, louder than usual and birds more visible; female carries nesting material obviously, outside spinifex clumps and bushes, with male in attendance (Hitchcock and Jarman 1944).

Breeding and life cycle

Nests and other evidence of breeding Aug–Oct (Carter 1903; Whitlock 1910, 1924; White 1917; Hitchcock and Jarman 1944; Storr 1984; RAOU NRS). No marked birds studied, so number of broods raised per year not known; perhaps only one; seasonal conditions may dictate timing and length of breeding season (Beruldsen 1980; Schodde 1982*b*). Nests small (90–110 mm high × 60–80 mm wide), domed, usually placed in top or centre of spinifex tussock 200–350 mm above ground, side entrance 30–40 mm wide. Woven of thin grass stems and cobwebs, lined with plant down and feathers. Clutch 2–3, eggs (14.5–15.5 × 10–12 mm, *n* = 6, Schodde 1982*b*) tapered oval, white, slightly glossy, well marked all over, particularly at larger end, with reddish-brown freckles and blotches. Only female builds; nothing known of time of laying, incubation, or nestling periods; probably not very different from *S. malachurus*. Parasitized by Horsfield's Bronze-Cuckoo (*Chrysococcyx basalis*) (Brooker and Brooker 1989*a*). Young reared by both parents, but no details known. Records of groups with more than two birds suggest that young may stay with family groups after independence.

Genus *Amytornis* Stejneger

Amytornis L. Stejneger, 1885. p. 499 in Vol. 4 of *The Standard Natural History*, edited by J. S. Kingsley.

Synonyms *Amytis* Lesson, 1831; *Diaphorillas* Oberholser, 1899; *Magnamytis* Mathews, 1912; *Eyramytis* Mathews, 1912; *Mytisa* Mathews, 1913.

The largest of the malurids, grasswrens are well camouflaged for survival in sparsely vegetated habitats; brown, with strong black and white streaking of the upper surfaces and sometimes below as well. Individual feathers are strongly streaked, with an area close to the central shaft contrasting strongly with the rest of the vane; the shaft itself is frequently white and may be edged with black, while the vane is brown or rufous (see Plate 6). Grasswrens spend most of their time on the ground, running and hopping in search of seeds and insects. Wings short and rounded; they fly rarely. Tarsus short compared with *Malurus*. Tail long, with five pairs of feathers that grade evenly in size, the smallest being quite large compared with other malurid genera where the outer rectrices are small to miniscule. Most species of grasswren cock their tails. Generally found in arid habitats from rocky sandstone escarpments to spinifex, cane-grass, lignum, and saltbush thickets. Frequently in family parties; probably co-operatively breeding.

Grasswrens were assigned to several genera, chiefly by Mathews (1912*a*, 1922–3, 1930), before it was accepted that the species were best combined in one genus, *Amytornis*. Occurring in remote areas, difficult of access, several species remained unseen for decades after their initial discovery by pioneering collectors at the beginning of the twentieth century or earlier. Five of the eight species (*A. barbatus*, *A. goyderi*, *A. woodwardi*, *A. dorotheae*, and *A. housei*) have very restricted distributions; only *A. barbatus* has populations isolated sufficiently to have diverged into subspecies. The other three species, *A. striatus*, *A. textilis*, and *A. purnelli*, are found over much larger ranges within which there is significant variation, and, over many years, much confusion arose in their naming. The position was finally clarified by Parker (1972).

An excellent general article on grasswrens with outstanding colour photographs of all eight species appears in *Wingspan* March 1996 (Chapman 1996). In the text that follows, we use the order in Christidis and Boles (1994); the grouping of species in the plates was dictated by the requirements of space and the number of subspecies to be illustrated.

Grey Grasswren *Amytornis barbatus* Favaloro and McEvey, 1968

Amytornis barbatus Favaloro and McEvey, 1968. *Memoirs of the National Museum of Victoria*, **28**, 1–9.

PLATE 8

Polytypic. Two subspecies. *Amytornis barbatus barbatus* Favaloro and McEvey, 1968; *Amytornis barbatus diamantina* Schodde and Christidis, 1987.

Description

A. b. barbatus

ADULT MALE: crown and forehead, blackish, grading to gingery-brown hind-crown, back, and shoulders; greyer lower back and rump; on crown, distinct white feather shafts give appearance of black streaked with white; hind neck and mantle striated, white shafts narrowly edged black, with gingery-rufous feather margins; streaks sparser on lower back and rump; primaries grey-brown, wing-coverts lighter, both with pale shafts and edges; tail very long and tapering, dusky with white shafts and creamy edges; face white, with broad white eyebrows meeting in front; a thick black line from base of bill through eye to side of neck; on each side of face below the white ear coverts, a black V extends ventrally from the black eye-stripe across the lower throat; a thin, indistinct, black malar stripe runs between front of black V and base of bill; ventral surface, white, except for pale buff flanks and faint black streaks on breast; eye, dark brown; bill, black, short, deeper than wide, finch-like; feet, black.

ADULT FEMALE: generally similar to male; a little smaller and duller, chest markings less distinct; lacks rufous flank patches of other female grasswrens.

IMMATURE: like adult in general markings, but duller and fluffier; face markings indistinct; eyes paler, bills and feet grey.

MOULT: adults moult once a year after breeding, generally Oct–Dec; no eclipse plumage; tail feathers replaced as they wear (Schodde 1982*b*).

History and subspecies

Two subspecies recognized. *A. b. barbatus* Favaloro and McEvey, 1968 from lower Bulloo R. drainage. *A. b. diamantina* Schodde and Christidis, 1987, five discrete populations on Cooper–Diamantina drainage.

Amytornis barbatus was described by Favaloro and McEvey (1968) from floodplains of the Bulloo Overflow in NW NSW. Subsequently, other localized populations were reported on the Cooper–Diamantina and Eyre Creek systems in SA and SW Qld (Cox 1974; Joseph 1982; May 1982), recognized by Schodde and Christidis (1987) as a new subspecies. They found no differentiation in tissue isozymes, but *A. b. diamantina* from the Cooper–Diamantina system is larger, more cinnamon dorsally, and less densely streaked than *A. b. barbatus*.

Measurements and weights

Amytornis barbatus barbatus (Bulloo drainage).: wing, M (n = 8) 58.5 ± 1.1, F (n = 7) 56.3 ± 1.5; tail, M (n = 7) 101.8 ± 7.3, F (n = 7) 94.3 ± 4.5; bill, M (n = 8) 9.7 ± 0.7, F (n = 3) 9.0 ± 0.8; tarsus, M (n = 5) 24.7 ± 0.6, F (n = 3) 23.4 ± 0.4; weight, M (n = 7) 17.5–20.0 (18.5 ± 0.9), F (n = 7) 16.5–19.0 (17.8 ± 0.8).
Amytornis barbatus diamantina (Cooper–Daimantina drainage): wing, M (n = 7) 62.7 ± 1.4, F (n = 5) 58.6 ± 1.5; tail, M (n = 7) 109.9 ± 2.9, F (n = 5) 103.6 ± 5.0; bill, M (n = 7) 11.1 ± 0.7, F (n = 5) 10.1 ± 0.3; tarsus, M (n = 7) 25.2 ± 1.5, F (n = 5) 26.3 ± 1.6; weight, M (n = 1) 21.0 g. (Weights from museum specimens and ABBBS.)

Field characters

Similar in size to other central Australian grasswrens, but distinguished by pale appearance, black-and-white facial markings, and very long tails. Only grasswren in Lignum-covered swamplands and floodplains of inland river systems. Hard to see in dense habitat; when undisturbed may sit and preen at top of a Lignum bush. At Goyder's Lagoon SA, in Lignum and sedge of floodplain, Eyrean Grasswren (*A. goyderi*) on sand ridges nearby.

Voice

Tape-recorded (ANWC). More vocal than other grasswrens, but appear to lack their elaborate songs. Most frequent call, a series of high-pitched, metallic ringing notes, 'trip-ip-ip' (Cox 1976) or 'tsit-tsit-tsit' (Robinson 1973); this was the only call heard between male and female even at height of breeding season (Favaloro and McEvey 1968; Robinson 1973); also used as contact call. A group of birds foraging together produces a continuous series, individual birds give two or three (see sonogram). Other contact calls are a high soft trill. Alarm call single, high-pitched piercing 'eep'.

Range and status

Resident. First collected by N. J. Favaloro in July 1967 in swamp on floodplain of Bulloo R in NW NSW; since then, species identified at several other locations. Present known range: *A. b. barbatus*: floodplain at mouth of Bulloo R. ('Bulloorine'), including swampy areas Jerrira Swamp, Bulloo R. Overflow, and Caryapundy Swamp which extends across border into Qld (Favaloro and McEvey 1968). *A. b. diamantina*: five apparently discrete populations: 1. floodplain at mouth of Diamantina R., known as Goyder's Lagoon (Cox 1976); 2. between Lake Eyre and Goyder's Lagoon, near junction of Kallakoopah Creek and Warburton R.; 3. Embarka Waterhole on Cooper Creek N of Innamincka (May 1982); 4. floodout of Eyre Creek, Lakes Mippia, Machattie, and Koolivoo on Glengyle Station, N of Birdsville (Joseph 1982); 5. Lake Cuddapan and Farrar's Creek, between Diamantina R. and Cooper Creek (personal communication by H. Rabig and H. B. Gill to Schodde 1982*b*). Known populations appear isolated; much potentially suitable, but impenetrable, habitat on floodplains of Lake Eyre drainage and Bulloo R. not yet fully investigated. Locally common in suitable habitat; populations probably fluctuate owing to alternation between severe drought and good seasons after heavy rain. Greatest threat probably restricted distribution in single climatic region, also destruction of habitat by fire; grazing by introduced mammals, especially feral goats, pigs, and horses; reduction of flow in inland rivers to provide water for irrigation. No known major populations at present in any conservation reserve.

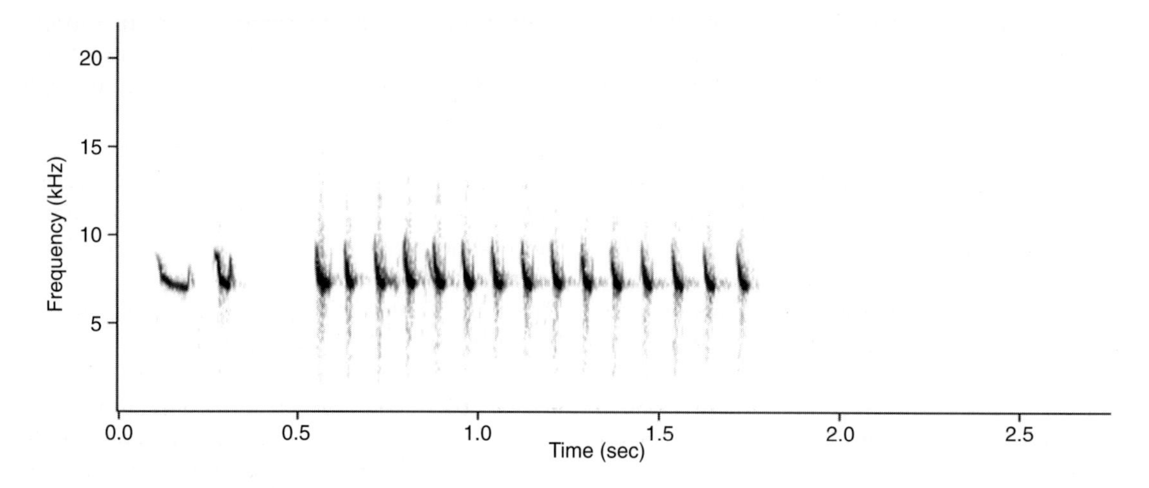

Grey Grasswren (*Amytornis barbatus*)
G. S. Chapman, Teurika, New South Wales.

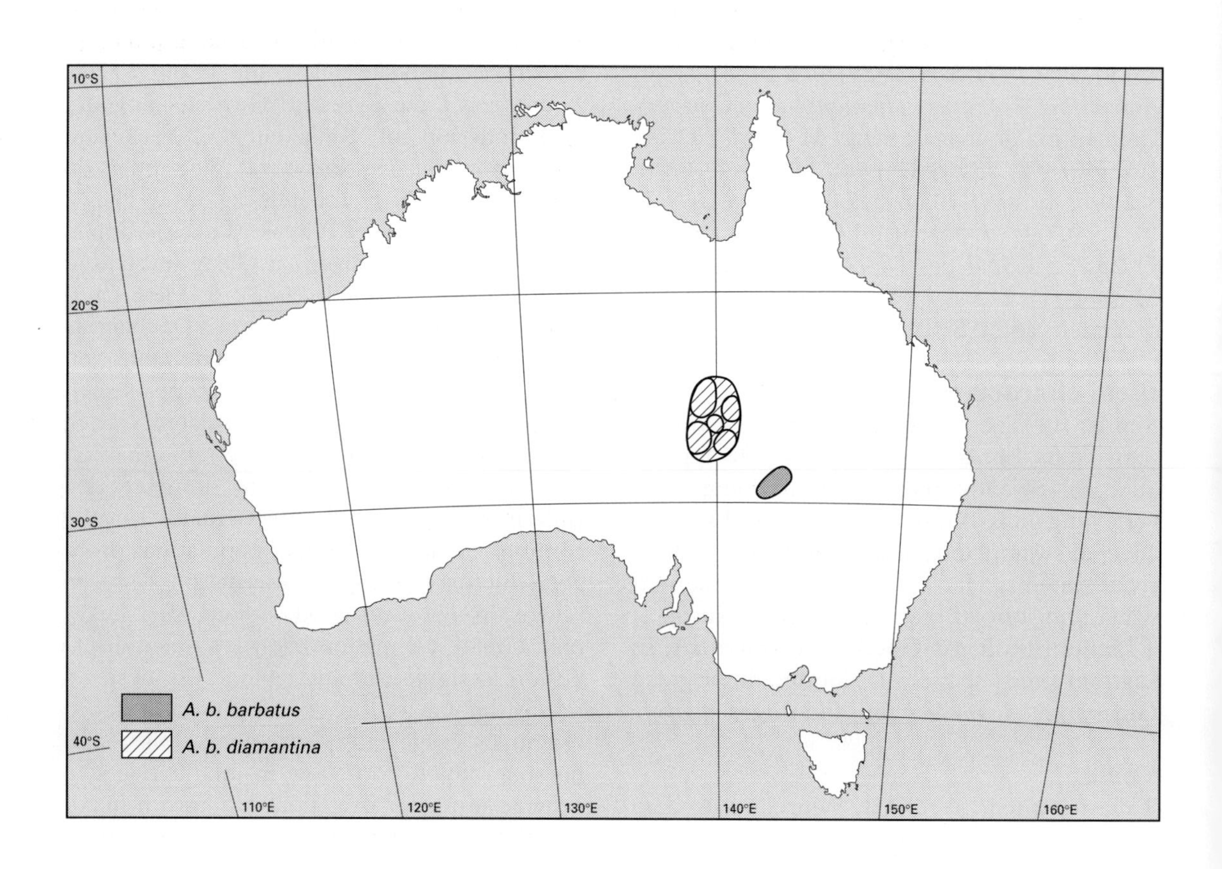

Habitat and general habits

Occurs on floodplains that are periodically inundated when normally dry rivers flow and flood out over an extensive area. Most of this water dries up, but some places remain which carry water except in extreme drought. Major flooding follows extensive rainfall that may occur far away, months earlier, but swamps also receive run-on after local rain in normal seasons. The essential feature of habitat in all known localities is Lignum (*Muehlenbeckia cunninghamii*), a many-stemmed, intricately branching, woody shrub with reduced leaves; some wetter areas carry a more or less continuous cover of Lignum *c.* 2 m tall in normal and good seasons. In Bulloo R. swamps, occurs in thickets of Lignum 1–2 m high, one to several m across, interspersed with 1–2 m-high clumps of Swamp Cane-grass (*Eragrostis australasica*) (Plate 4 in Favaloro and McEvey 1968). At Goyder's Lagoon, habitat denser (Cox 1976): thickets of Lignum interspersed with dense swards of sedge (*Eleocharis pallens*) 0.5 m high. Height of shrub cover may be important during floods when water may be 1 m deep; vegetation taller than this is essential for survival; in drought, dense cover much reduced. Difficult to flush from cover; frequently fly downwards, from high in one clump to base of next, tail trailing as *Malurus* spp. Run and bounce on ground with tail cocked. Forage in open and through shrubs; diet: seeds, insects, and occasionally water snails (Favaloro and McEvey 1968; Parker 1980; Robinson 1973). Encountered singly, in pairs, or small groups: the first recorded sighting was of five birds in July (Favaloro and McEvey 1968). No detailed studies of marked birds; social organization not known.

Displays and breeding behaviour

Even at height of breeding season, nothing similar to song of other grasswrens heard. One record of copulation, with no indication of courtship display: '....early in the morning while two birds were observed sunning themselves on top of a Lignum bush, they were joined by a third. All three were calling intermittently as the latest arrival approached and eventually took up a position close to the bird near the end of the branch. After a short interval, but without further ceremony or display, mating took place. The female remained to preen herself as the male returned to the centre of the Lignum' (Favaloro and McEvey 1968, p. 8). No other records of mating behaviour.

Breeding and life cycle

Lives in the most arid and variable climatic region of Australia; significant rainfall or none at all may occur in any month. Pairs or groups dispersed in individual territories of 1–2 ha (Cox 1976; Schodde 1982*b*). Few nests known, laying recorded winter and spring (early July, mid-Aug, Sept) after a flow of water through swamp (Favaloro and McEvey 1968; Robinson 1973; Schodde 1982*b*). Seasonal conditions may dictate timing and length of breeding season; birds may not breed in dry years. No marked birds studied, so number of broods raised per year not known. Nests semi-domed, bulky (100–120 mm high × 80–100 mm wide), not concealed but blending well with surroundings in the centre of a Lignum or cane-grass clump. Large opening at side and sometimes a short platform at entrance; loosely woven of grass stems, lined with finer, softer grass, some plant down, and a few feathers. Clutches 2–3, eggs (18.5–21.0 × 14.5–15.5 mm, n = 22, Schodde 1982*b*) rounded oval, creamy-white, dull, densely marked all over with reddish-brown spots and blotches, some concentration at blunter end; laid in mornings at 24-hr intervals (Schodde 1982*b*). Only female incubates and broods; incubation and nestling periods unknown, probably similar to *A. striatus* (similar size; incubation period 14–15 days, nestling period 12–14 days, Hutton 1991). Both parents feed young and remove faecal sacs; may be a co-operative breeder, with one record of three adults at nest (Robinson 1973); no detailed studies. Adults very cautious approaching nest, with much calling.

Black Grasswren *Amytornis housei* (Milligan, 1902)

Amytis housei Milligan, 1902. *Report on exploration of the north-west Kimberley, 1901*, p. 52.

PLATE 7

Monotypic

Description

ADULT MALE: head and upper back, black, each feather with a white shaft-streak; lower back, rump, and upper tail coverts, unstreaked, chestnut; tail and wings, black except for chestnut edges and shafts of innermost secondaries; tail, moderately long; rectrices broad and rounded at tips, with six of the 10 nearly equal in length, giving a bushy appearance; face, black, streaked white with traces of white lower eye-rim; no brow streak; entire ventral surface black, streaked white on throat, less streaked on breast, abdomen unstreaked; eyes, brown; bill, grey-black, relatively long and slender; feet, grey-black.

ADULT FEMALE: like male, except ventral surface. Lower breast, belly, and underwing-coverts, rufous-tan; throat and breast as male; chin and lower eye-rim, white.

IMMATURE: initially uniformly dusky, sexes alike, fluffier than adults; faint whitish streaks over head and breast; mouth bright yellow, white gape. After post-juvenile moult, sex differences begin to appear.

MOULT: adults moult once a year, in summer (Jan–Mar), at same time as breeding (Schodde 1982*b*); no sign of moult in July (Harrison 1974); no eclipse plumage.

History and subspecies

Amytornis housei was first collected in 1901 by F. M. House on the Brockman Expedition through the central Kimberleys of NW Australia, and was described from two males and one female as *Amytis housei* (Milligan 1902). Black Grasswrens did not escape Gregory Mathews' attention, and he described the tawny-bellied female as a new species *Magnamytis kimberleyi* in his *Birds of Australia* (1922–3). *A. housei* was not seen again in the field until 1968, when the Harold Hall expedition of the British Museum made a special effort to find it, retracing the steps of the Brockman expedition, and locating the birds close to where House first collected them (Freeman 1970); since then, they have been found at other locations in the Kimberleys. No subspecies are recognized.

Measurements and weights

Wing, M (n = 11) 74.7 ± 1.3, F (n = 11) 72.7 ± 1.6; tail, M (n = 11) 92.4 ± 3.2, F (n = 11) 87.1 ± 3.9; bill, M (n = 11) 14.2 ± 0.8, F (n = 11) 13.4 ± 0.8; tarsus, M (n = 11) 25.9 ± 0.6; F (n = 11) 25.3 ± 0.5; weight, M (n = 5) 29.0–31.0 (29.9 ± 0.8), F (n = 5) 23.5–27.9 (25.5 ± 0.2). (Weights from museum specimens and Harrison 1974.)

Field characters

Large stocky dark grasswren, with long, dark, broad tail, generally held low. No other grasswren within 500 km.

Voice

Tape-recorded (ANWC). Probably breeds in the summer wet season, when habitat inaccessible except by helicopter, which means that we do not know if the songs recorded are all they sing, or if more varied songs are heard in the breeding season. Songs of other grasswrens are heard, less often, all year round, so it is likely that recordings in dry season are representative. Song relatively low-pitched (3–6.5 kHz) and short, with introductory notes followed by a buzzing vibrato and a brief trill with occasional up or down-slurred whistle, more variable than *Malurus* reel; Schodde (1982*b*) refers to metallic quality of their chattering trills. Male more vocal than female. Contact calls are sharp chips and purrs, loud ticking, and rattle in alarm (see sonogram).

Range and status

Resident; tropical NW Australia, restricted to W Kimberley region, N to mouth of Mitchell R., S almost to King Leopold Ra., W to coast, E to *c*. longitude 126° E. Moderately abundant in appropriate very specific habitat. Main threat is habitat alteration caused by too frequent fire.

Habitat and general habits

Found only in tumbled sandstone escarpments, gorges, ravines, and outcrops, sparsely

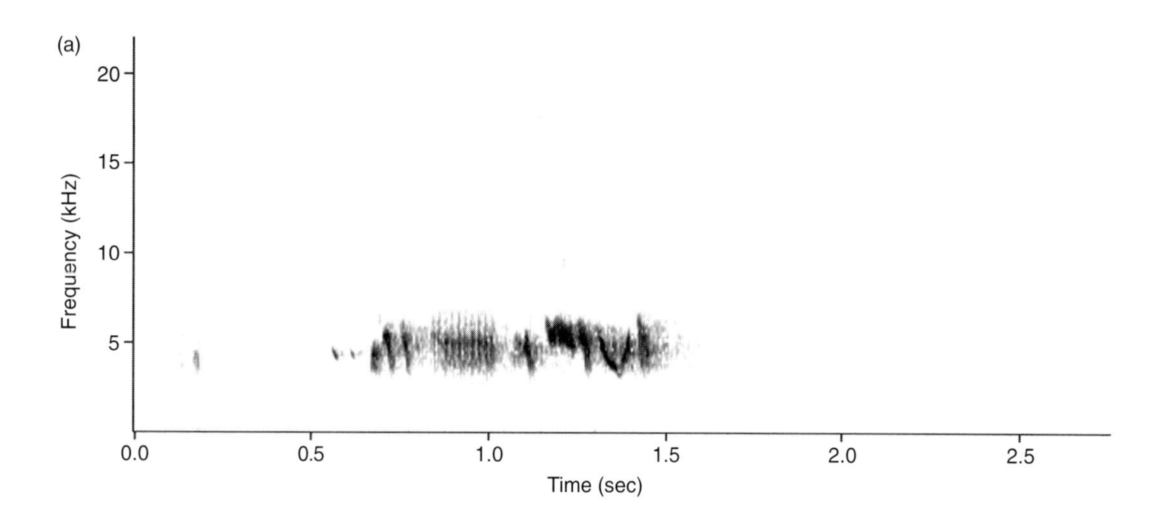

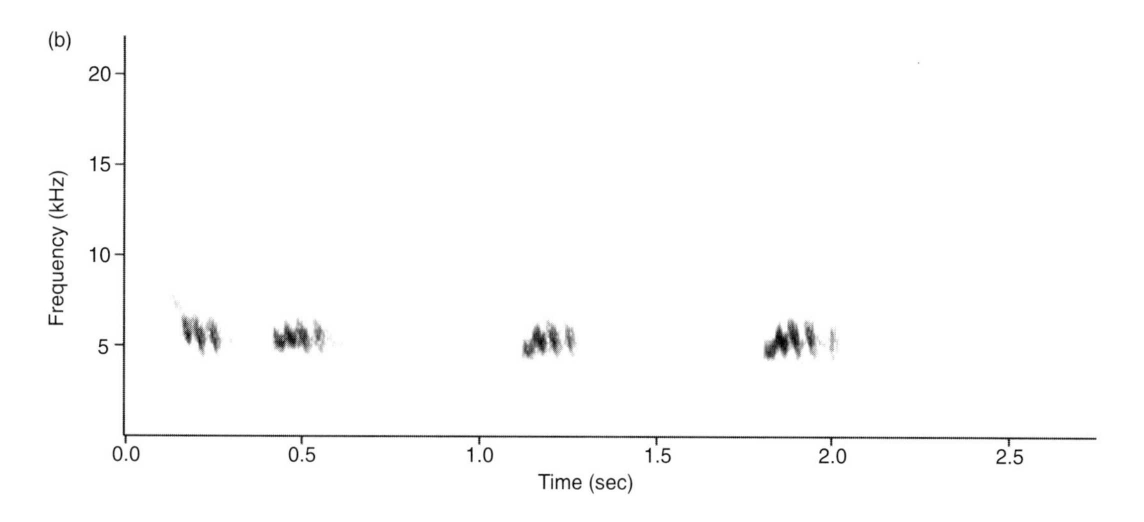

Black Grasswren (*Amytornis housei*)
(a) song. (b) contact 'seep'. G. S. Chapman. Mitchell Plateau, Western Australia.

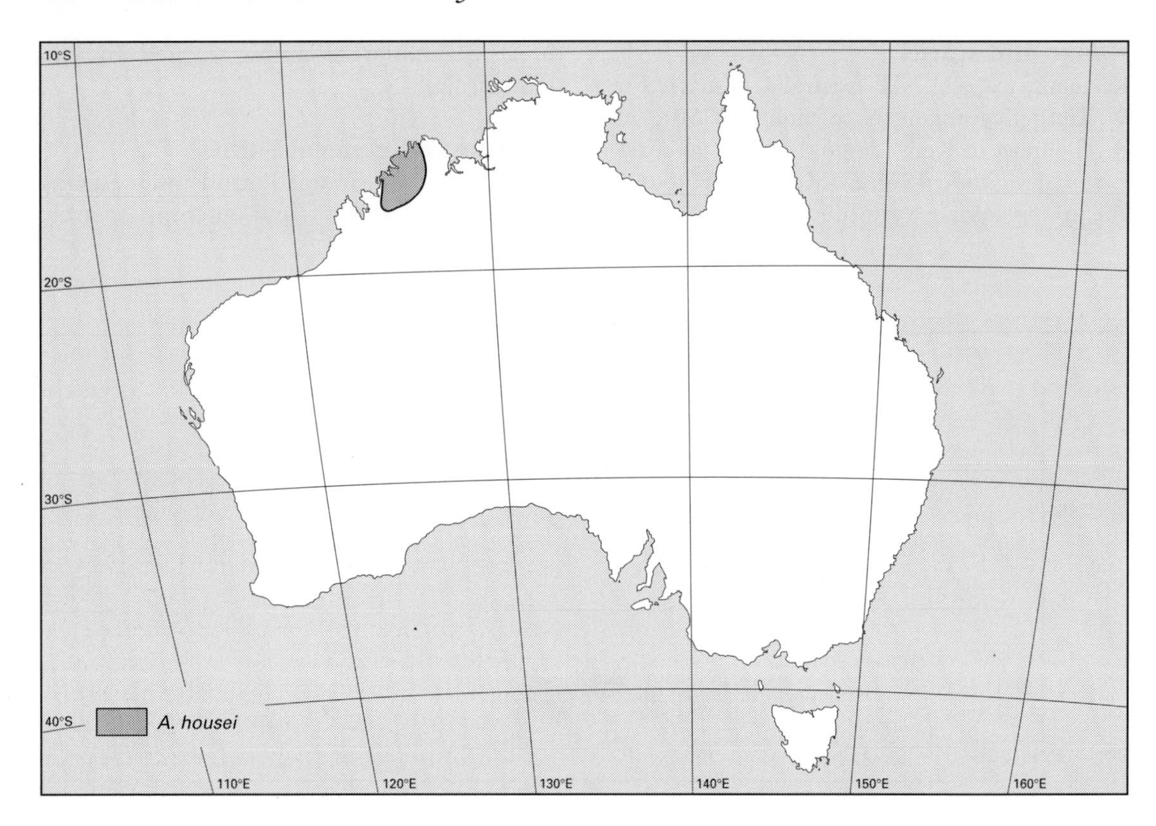

treed and covered with tussocks of spinifex (*Triodia*, *Plectrachne*) and dense shrubby growth. Well camouflaged in light and shade of red sandstone habitat. Forages on ground in dense cover among rock crevices and around edges of spinifex tussocks. Eats insects and seeds, particularly grasses (*Sorghum*, *Setaria*, and *Panicum*; Freeman 1970; Harrison 1974). Flight rare; foraging birds occasionally glide low for short distances across rocky slopes or downhill. Most movement by hopping or running. Foraging groups keep in contact by intermittent calls (Freeman 1970; Harrison 1974). 'At the first note of alarm the whole group scatters, diving headlong in different directions into dense shrubberies and crevices under boulders' (Schodde 1982*b*, p. 172).

Displays and breeding behaviour

Very little is known. In a striking display of unknown context during breeding season, male sang briefly from on top of a rock, then charged down at female, with his head forward, tail partly cocked and fanned, wings half-spread with drooping tip (Schodde 1982*b*). Typical malurid Rodent-run displays recorded when one of group threatened (Freeman 1970; Schodde 1982*b*).

Breeding and life cycle

No detailed studies of biology. Pairs or groups dispersed in territories of 1–3 ha (Schodde 1982*b*). Extent of laying season unknown. New nest found and female with egg in oviduct collected Jan, recent fledgling caught Feb (Johnstone and Smith 1981); dependent young in April from nest late Jan–early Feb (Nevill 1994); dependent juvenile with partly grown flight feathers collected early July, probably fledged late June (Freeman 1970). Only one nest described: bulky (200 × 150 mm), domed, with central side entrance, walls of woven spinifex lined with softer grass, at top of spinifex tussock (Johnstone and Smith 1981). Eggs and other details of breeding unknown. Probably co-operative breeding; all accounts refer to birds seen in pairs or groups of up to six (Freeman 1970; Harrison 1974; Schodde 1982*b*). One observation of juvenile attended and fed by three adults (1M, 2F; Freeman 1970).

White-throated Grasswren *Amytornis woodwardi* Hartert, 1905

Amytornis woodwardi Hartert, 1905. *Bulletin of the British Ornithologists' Club*, **16**, 30.

PLATE 8

Monotypic

Description

ADULT MALE: Head and upper back, black with distinct white shaft-streaks, grading to dark chestnut lower back and rump; wings, dusky black, edged chestnut; tail, long, black, rectrices broad and rounded at tips, with six of the 10 nearly equal in length, giving a bushy appearance; face, black, streaked white; lower eye-rim, white; black malar band from base of bill down sides of neck; throat and breast, white, separated from tawny belly and flanks by band of white feathers broadly edged black; thighs and undertail, dark chestnut; eye, brown; bill, relatively long and slender, grey-black, paler at base; feet, grey-brown.

ADULT FEMALE: like male except for uniformly deeper chestnut belly, flanks, and thighs.

IMMATURES: patterned as adults; upper surface duller, streaks less distinct; belly paler, tawny in both sexes; breast band less developed; bill pale.

MOULT: adults appear to moult over an extended period; individuals found in partial moult most months; no eclipse plumage (Schodde 1982*b*).

History and subspecies

Amytornis woodwardi was first collected by John Tunney on his expedition to NW Australia financed by Rothschild and sponsored by the WA Museum. He spent nearly 3 months in W Arnhem Land, and collected a large grasswren on its sandstone bluffs and escarpments. It was originally identified by Hartert at Tring as *A. housei* (1905*a*), which he had not seen. Later, he recognized his error, and described it as a new species *A. woodwardi* (Hartert 1905*b*). He clearly recognized the sexual dimorphism, but Mathews (1922–3) did not, naming the female *Magnamytis alligator. A. woodwardi* was not seen in the field for nearly 50 years, but as access to W Arnhem Land improved, it was found at a number of sites on the Arnhem Land escarpment (Schodde 1982*b*; Noske 1992).

Measurements and weights

Wing, M (n = 9) 75.6 ± 1.4, F (n = 10) 75.3 ± 1.5; tail, M (n = 9) 104.0 ± 3.2, F (n = 10) 101.2 ± 4.0; bill, M (n = 9) 14.0 ±0.3, F (n = 10) 13.9 ± 0.3; tarsus, M (n = 9) 28.4 ± 0.9, F (n = 10) 28.0 ± 1.2; weight, M (n = 8) 33.0–40.0 (36.8 ± 2.9), F (n = 8) 30.0–37.0 (33.4 ± 2.4). (Weights from museum specimens.)

Field characters

Largest and most strikingly plumaged of all grasswrens; from a distance, gives impression of large pied grasswren; white throat contrasts strongly with dark back and belly. Distinguished from Carpentarian Grasswren (*A. dorotheae*) by larger size, darker head and shoulders, and streaked breast band; ranges do not overlap.

Voice

Tape-recorded (ANWC). Song a sweet-sounding extended (4–5 sec) variable sequence of syllables including trills, buzzes, up- and down-slurred notes; includes frequencies above 8 kHz. Song heard in all months (Noske 1992); quite unlike typical stereotyped *Malurus* reel (see sonograms). Contact calls a high-pitched 'peep' (Harrison 1974) or a reedy cricket-like chirp (Schodde 1982*b*). Alarm calls single or repeated strong, sharp 'tsit'.

Range and status

Resident; range restricted to NW half of Arnhem Land sandstone plateau in Wet–Dry tropics region of NT, in an area between W escarpment of Arnhem Land Plateau and headwaters of Mann and Katherine R. Status moderately common; estimated 48 000–59 000 birds in 1 400 000 ha of suitable habitat (Noske 1992). Greatest threat to species is altered fire regimes leading to changes in sandstone plant communities

Habitat and general habits

Two major types of habitat identified by Noske (1992). 1. Terraced hillsides along broad valleys, including narrow side gullies, tors, and pinnacles. 2. Relatively bare flat plateaux and ridge-tops with few boulders. Significant features of both habitats are bare rock and presence of spinifex (*Triodia*). Optimum habitat, flat sparsely vegetated unbroken plateaux: crevices under boulders and in cliffs provide shelter from sun, and refuge from predators and fire. Forage by gleaning and hop-search on bare rock, among litter, in spinifex clumps, and on mats of grass (*Micraira*). Flight rare, usually short and low; tail cocked infrequently. Runs rapidly with tail and head lowered; when disturbed, darts into crevices and remains hidden (Schodde and

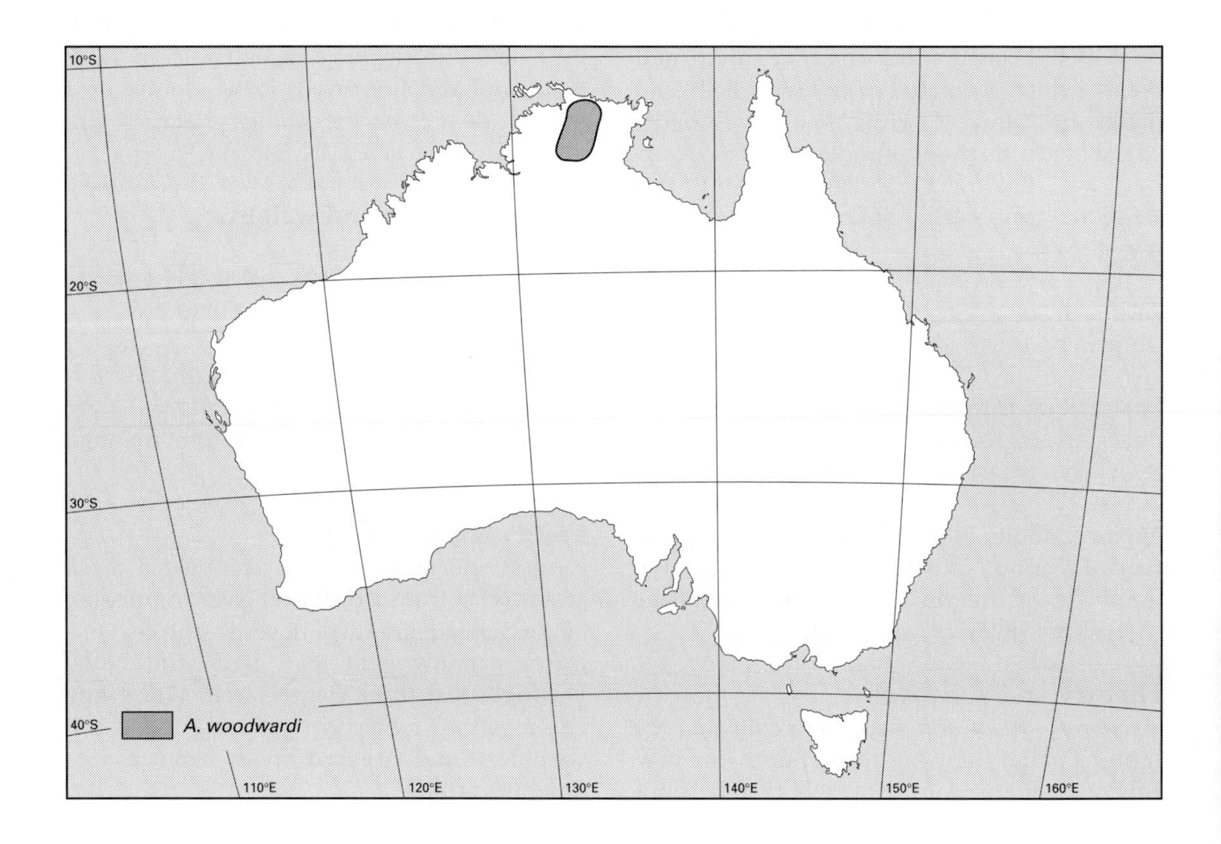

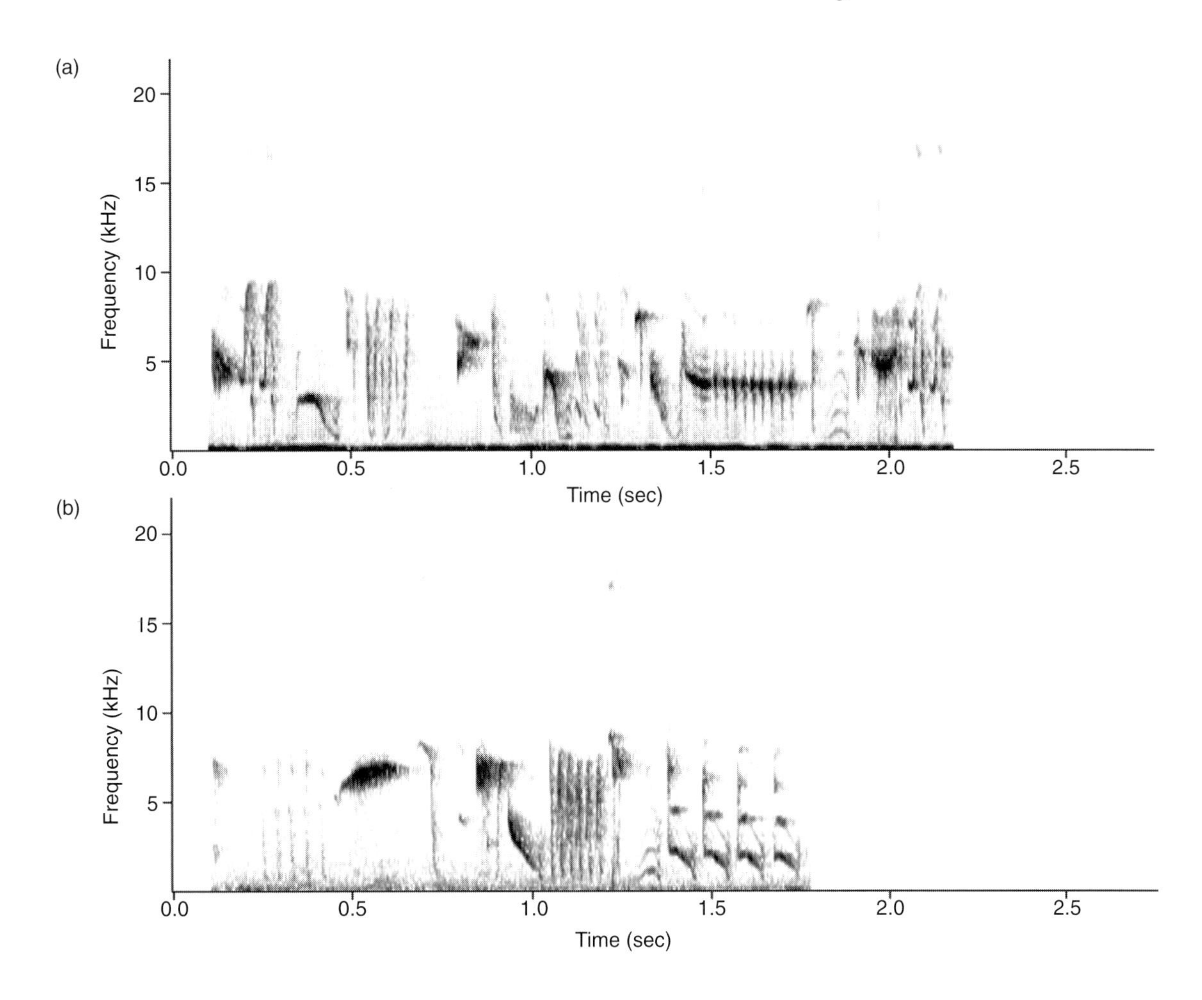

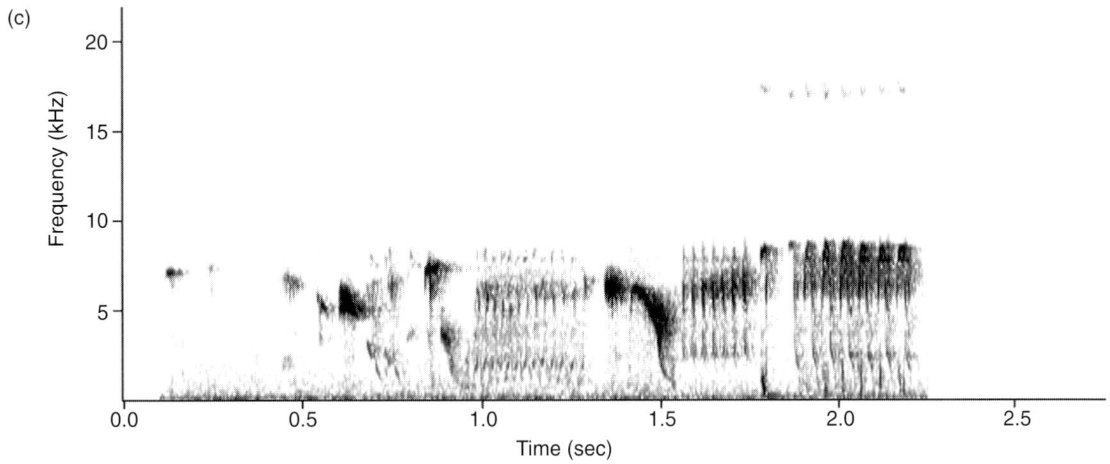

White-throated Grasswren (*Amytornis woodwardi*)
(a), (b), and (c) are examples of different phrases of song. G. S. Chapman. Gunlom Falls, Northern Territory.

Mason 1975). Diet of plant stems and seeds (including *Triodia*), ants, beetles, and small moths. Grit present in gizzard (Harrison 1974; Noske 1992). Many accounts of groups of up to seven birds in non-breeding season and up to four in breeding season (Harrison 1974; Schodde 1982*b*; Noske 1992).

Displays and breeding behaviour

Nothing recorded.

Breeding and life cycle

No detailed studies of biology. Pairs or groups dispersed in territories of *c.* 10 ha. Laying season at least Dec–May or June; female may produce two broods (Noske 1992). Few nests known; nest very bulky, domed (160–180 mm × 120–140 mm) with thick roof and 40–50 mm wide side entrance; constructed mainly of interwoven grass stems and fine leaves, with base of grass, broader leaves, and papery *Melaleuca* bark, lined with finer grass; well concealed towards top of spinifex clump, 200–400 mm above ground. Only recorded clutch of two eggs. Eggs tapered oval, 22 × 16 mm: off-white, dull, with fine speckles and blotches of reddish-brown and purplish-grey, concentrated in zone at larger end (Schodde and Mason 1975). Incubation, hatching, growth, and development undescribed. After fledging, young birds dependent on adults for at least 1 month, and remain with group after independence. Co-operative breeder (see photograph of group by G. S. Chapman in Noske 1991*b*); groups of three and four birds attending nestling and fledgling reported (Noske 1992).

Carpentarian Grasswren *Amytornis dorotheae* (Mathews, 1914)

Magnamytis woodwardi dorotheae Mathews, 1914. *Australasian avian Record*, **2**, 99.

PLATE 8

Monotypic

Description

ADULT MALE: ground colour of crown dusky black, grading to chestnut-russet upper and lower back and rump; from crown to mid-back and shoulders, feathers streaked with narrow black line on each side of white shaft; wings, dusky brown, flight feathers and coverts edged pale chestnut; tail, long, slender, dusky feathers with russet shafts and fringes; forehead and face, black with thin white streaks; brow, chestnut; lores and lower eye rim, white; black malar stripe runs from below lores to side of neck, contrasting with white chin; throat, breast, and upper belly, white, grading to tawny lower belly, flanks, and under tail; eye, brown; bill, relatively long and slender, dark grey; feet, dark grey.

ADULT FEMALE: like adult male, except flanks and lower belly deep chestnut, separated from white breast by an incomplete band of blackish-edged feathers encroaching inwards from sides of breast.

IMMATURES: patterned like adults; upper surface duller, dusky brown with faint streaks; below, dull white, flanks cinnamon-rufous. No brow or black malar stripe; bill pale. Sexes initially indistinguishable.

MOULT: adults moult once a year, probably after breeding, generally Feb–June; tail feathers replaced as they wear; no eclipse plumage (Schodde 1982*b*).

History and subspecies

Amytornis dorotheae is another of the long-lost grasswrens. In his ornithological notes of the Barclay Expedition to the NT, G. F. Hill (1913) mentioned a grasswren amongst the spinifex and rocks in the ranges near Borroloola, and tentatively identified it as *A. woodwardi,* but he collected no specimens. Almost immediately,

H. L. White despatched H. G. Barnard in search of the birds; several males were collected (Barnard 1914). Since there were no specimens in Australia of *A. woodwardi* from the Alligator R. for comparison, these McArthur R. birds were identified as *A. woodwardi.* Mathews (1914) separated them as a subspecies *dorotheae* of *woodwardi,* on account of their smaller size. He did not mention their most obvious difference, their lack of a streaked band across the lower breast. In 1917, he recognized them as a distinct species *Magnamytis dorotheae*; he was not misled by females in this case, since none was collected until 1974.

Measurements and weights

Wing, M (n = 6) 63.5 ± 2.0, F (n = 4) 61.5 ± 1.3; tail, M (n = 6) 86.8 ± 3.4, F (n = 4) 80.5 ± 2.4; bill, M (n = 6) 11.7 ± 0.7, F (n = 4) 11.7 ± 0.3; tarsus, M (n = 6) 23.8 ± 1.1, F (n = 4) 22.7 ± 0.8; weight: M 21–25, F 21–23 (from Schodde 1982*b*, no n given).

Field characters

Medium sized, much smaller than White-throated Grasswren (*A. woodwardi*). Distinguished from Striated Grasswren (*A. striatus*) or Dusky Grasswren (*A. purnelli ballarae*) by contrast between blackish head with black malar band and gleaming white throat and breast. Head and shoulders less black than *A. woodwardi*; male *A. dorotheae* lacks streaked breast band entirely; in female, breast band rudimentary, evident only at sides of breast.

Voice

Tape-recorded (ANWC). Song a sweet-sounding extended (4–5 sec) variable sequence, including trill, buzzes, and down-slurred whistles, with some frequencies above 8 kHz (see sonograms). Song heard in all months, usually delivered from top of rock or bush, with tail cocked and head thrown back (Schodde 1982*b*; see photograph in Chapman 1996, p. 24). Quite unlike stereotyped *Malurus* reel. Contact calls high-pitched cricket-like chirp 'ssstzz', given singly; alarm call harsher, louder, buzzing 'tzzzt' (Schodde 1982*b*).

Range and status

Resident; restricted to several dissected sandstone ranges lying *c.* 100 km inland from Gulf of Carpentaria coast; from Tawallah Ra., just E of Limmen Bight R. to Buckulara Ra. and China Wall in NT. In NW Qld, Lagoon Creek Gorge and Hells Gate (McKean and Martin 1989); also 100 km NW of Mt. Isa, Qld, and in Lawn Hill National Park, 300 km from main population (Beruldsen 1992 and *in litt.*; Harris 1992). Status data deficient, may not be threatened, since found to be more widespread than previously thought (Garnett *in litt.*); uncommon in restricted habitat. Greatest threat to species is altered fire regime: frequent dry season fires used in management of pastoral land (McKean and Martin 1989).

Habitat and general habits

Large mature stands of spinifex (*Triodia* and *Plectrachne*) on dissected plateaux and slopes in north of range; southern sites N of Mt. Isa less dissected siltstone ranges, with spinifex cover. Tumbled rocks and surface cracks provide recesses and clefts for shelter. Trees (*Eucalyptus* spp.) and shrubs (*Acacia* spp.) generally present, but not essential (McKean and Martin 1989; Harris 1992). Flight rare, usually short and low, with tail trailing. Travels across open areas by running or fast bouncy hop, with tail cocked. Forages mostly on ground, searching rock crevices and litter under shrubs and spinifex tussocks for insects and seeds (Schodde 1982*b*; Whitaker 1987). Forage as loose group, with occasional contact calls. When disturbed, dart into rock crevice or spinifex tussock and remain hidden for several minutes (Whitaker 1987).

Displays and breeding behaviour

Nothing recorded.

Breeding and life cycle

Little known. Nests recorded in Jan, Feb, Sept; newly fledged young in Jan and June (Hill 1913; White 1914; Schodde 1982*b*; Whitaker 1987; Fleming and Strong 1990). Laying probably in any month if seasonal conditions good. Nest

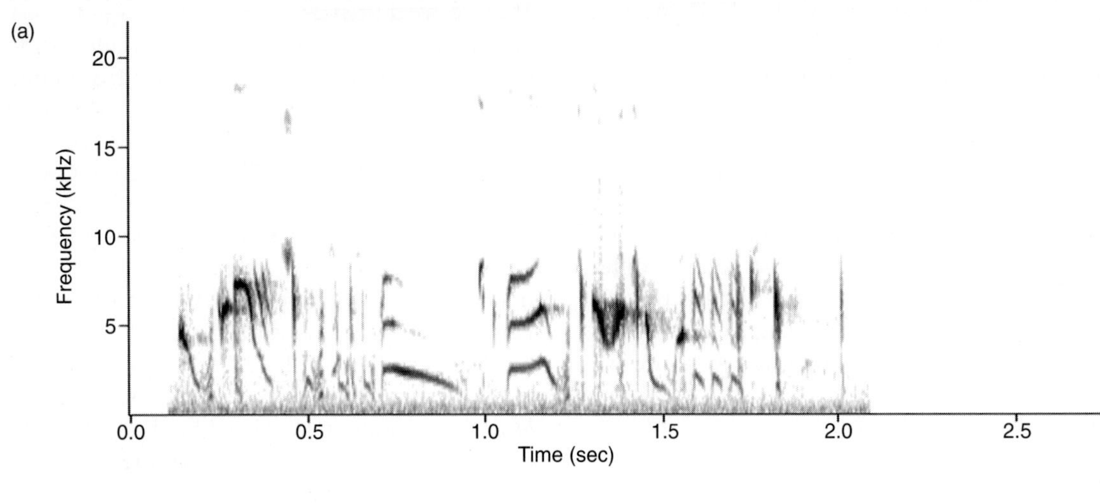

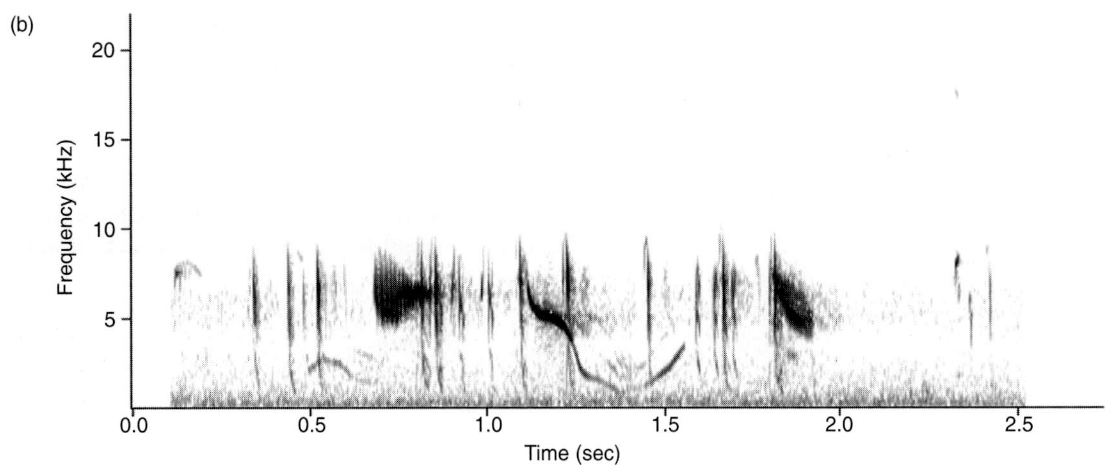

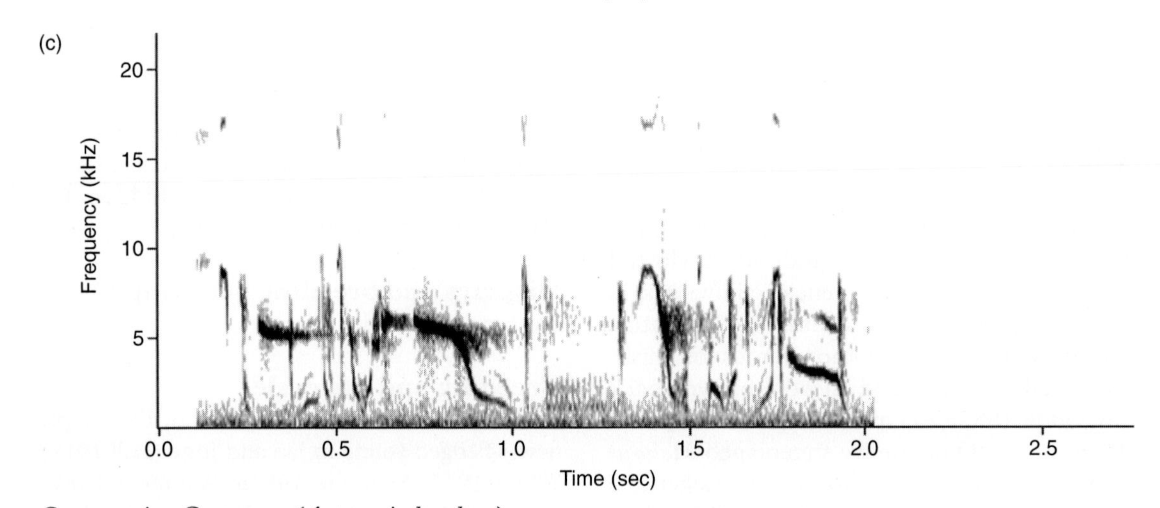

Carpentarian Grasswren (*Amytornis dorotheae*)
(a), (b) and (c) are examples of different phrases of song. G. S. Chapman. Mt. Isa, Queensland.

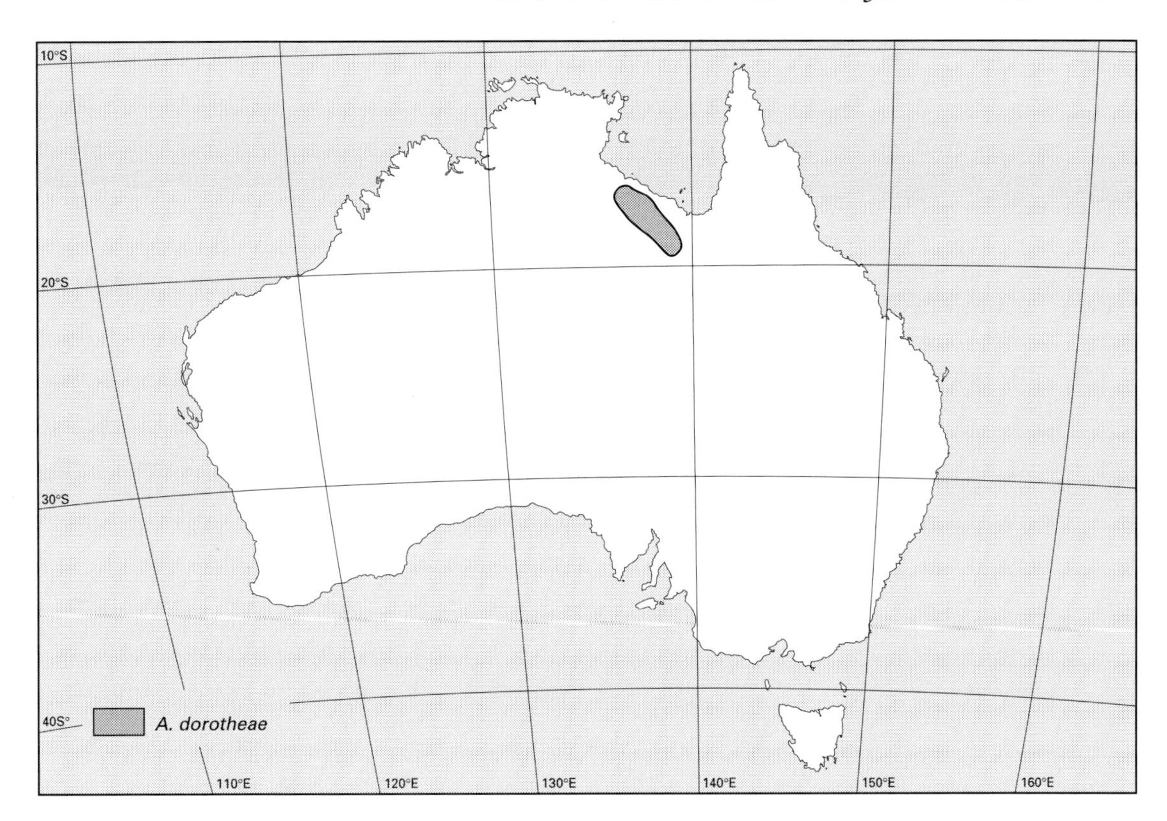

bulky, oval, domed (140–160 mm × 100–120 mm), with side entrance 40–50 mm wide; constructed of interwoven stems of spinifex and other grasses, with some leaves of *Eucalyptus* and *Acacia*; lined with softer grasses; embedded in top of spinifex tussock, *c.* 200–600 mm above ground. Clutch generally 2; eggs (19.5–20.5 × 15–16 mm, *n* = 9), broadly oval, pinkish white with fine, faint spots and blotches of brownish-red and mauve, more numerous at larger end; most sparsely marked of all grasswren eggs (Schodde 1982*b*). Incubation, hatching, growth, and development undescribed. One record of parasitism by Brush-Cuckoo (*Cacomantis variolosus*) (Brooker and Brooker 1989*a*). Young birds remain with group after independence; probably a co-operative breeding species; several accounts refer to groups of three and four adults (Schodde 1982*b*; Whitaker 1987; Beruldsen 1992; Harris 1992).

Striated Grasswren *Amytornis striatus* (Gould, 1840)

Dasyornis striatus Gould, 1840. *Proceedings of the Zoological Society of London*, **1839**, 143.

PLATE 6

Polytypic. Three subspecies *Amytornis striatus striatus* (Gould, 1840); *Amytornis striatus whitei* Mathews, 1910; *Amytornis striatus merrotsyi* Mellor, 1913.

Description

A. s. striatus

ADULT MALE: (in S and E of range): ground colour of crown, upper and lower back, and rump mid grey-brown; from crown to mid-back and shoulder, feathers with white shaft streaks edged black, less distinct on back and rump; wings, mid grey-brown with rufous patch at base of outer flight feathers and under wings; tail, slender, brown, with pale buff shafts and edges, long (tail:wing ratio 1.36); face with clear white streaks on dark grey-brown, bounded below by broad black malar stripe from lores to neck; bright rufous brow streak; lores and lower eye-rim, white; chin and upper throat, white, grading to buff on lower breast and belly, with grey streaks on upper breast; greyish flanks and thighs; eye, brown; bill, black, moderately slender, width similar to depth, short relative to other grass-wrens; feet, black. Birds of N and W, more rufous on back, wings, and tail, no black edges to white streaks; breast often plain white, ventral surface with more rufous tinge; bill and feet, leaden-grey.

ADULT FEMALE: as male, except for deep rufous patch on each side of lower breast just under wings; streaks on breast more marked.

IMMATURE: duller, streaks less distinct, no rufous brow or malar stripe; feet light brown; fledgling male and female, flanks similar dull rufous tawny, fades in male, brightens in female at 6–8 weeks (Hutton 1991).

MOULT: adults moult once a year, probably after breeding, from Jan in N to early May further S (Izzard *et al.* 1973; Schodde 1982*b*; Hutton 1991). Tail feathers replaced throughout the year, as they wear; no eclipse plumage.

History and subspecies

Three subspecies now recognized. *Amytornis striatus striatus* (Gould, 1840), originally widespread in central and eastern Australia. *Amytornis striatus whitei* Mathews, 1910, from mountain ranges of Pilbara region, WA; larger than NW form of *A. s. striatus*; generally redder, deep russet on back and margins of wings and tail; tail:wing ratio 1.31; white shaft-streaks of back feathers with fine black margins; large rufous brow stripe; lower breast and belly, tawny; bill, leaden-grey, long and slender. *Amytornis striatus merrotsyi* Mellor, 1913, from North Flinders Ra., South Australia: more rufous back and greyer underparts than *A. s. striatus*; tail shorter (tail:wing ratio 1.03), bill longer and deeper, tarsus longer than *A. s. striatus*; black malar stripe obscured, rufous brow much reduced.

Amytornis striatus was originally collected by Gould in central-western NSW, from a locality given as lower Namoi R., and described by him in 1840 as *Dasyornis striatus*. Over many years, speculation has failed to identify the exact location (Schodde 1982*b*; McAllan 1987). As exploration proceeded, specimens were found across inland Australia in spinifex grasslands; they generally differed somewhat in colour from the forms described in SE Australia, and were described as separate species: *A. whitei* from the Pilbara, WA (Mathews 1910); *A. merrotsyi* from the Flinders Ra., SA (Mellor 1913), and *A. rufa* from the Tanami Desert, NT (Campbell and Kershaw 1913). Mathews also distinguished two subspecies, *A. s. oweni* from the East Murchison, WA, and *A. s. howei* from the Murray Mallee of Vic. and SA (1911*a*,*b*). These various forms originally spanned most of the arid and semi-arid areas of Australia, and show considerable variation in size and colour. Various taxonomic treatments of this variation were proposed. Mathews (1930) included *merrotsyi* as a subspecies of *striatus*, together with *A. s. striatus* and *A. s. howei*. He maintained *A. whitei*, the Rufous Grasswren, as a separate species, including *A. w. whitei*, *A. w. oweni*, and *A. w. rufa*. Keast (1958) considered *whitei* distinct from the other two rufous forms *oweni* and *rufa* because of its longer bill; he treated *whitei* and *oweni* (including *rufa*) as subspecies of *striatus*, and included *merrotsyi* within the subspecies *striatus*. Mees (1961) maintained three separate subspecies, *whitei, oweni*, and *rufa*, for WA forms. In 1974, Ford and Parker examined a large series of specimens from the Great Sandy, Gibson, and Great Victoria Deserts, and united all three in *A. s. whitei*, since they considered the variation to be clinal. All authors (except Mathews) placed Mathews' *howei* within *A. s. striatus*. Schodde's 1982 revision maintained three subspecies on the basis that they form a sequence from large and grey-brown in the SE to small and russet in the N and W: *A. s. striatus*, including the sandplain spinifex forms *rufa, oweni*, and *howei*; *A. s. whitei*, the large, long-billed, reddish form from the spinifex-clad stony hills of the Pilbara; *A. s. merrotsyi*, the short-tailed form from the Flinders Ra. of SA.

Measurements and weights

Amytornis striatus striatus (south-east): wing, M (n = 42) 63.6 ± 1.9, F (n = 25) 60.6 ± 1.3; tail, M (n = 42) 86.4 ± 3.5, F (n = 25) 80.7 ± 3.4; bill, M (n = 40) 10.2 ± 0.6, F (n = 25) 9.9 ± 0.5; tarsus, M (n = 40) 24.5 ± 1.1, F (n = 25) 23.6 ± 1.1); weight, M (n = 5) 16.0–22.0 (18.5 ± 2.8). (Weights from museum specimens.)
Amytornis striatus striatus (north-west): wing, M (n = 25) 58.3 ± 1.8, F (n = 20) 56.9 ± 1.6; tail, M (n = 25) 80.0 ± 3.2, F (n = 20) 76.6 ± 3.8; bill, M (n = 25) 10.3 ± 0.5, F (n = 20) 10.1 ± 0.8; tarsus, M (n = 25) 22.9 ± 0.8, F (n = 20) 21.8 ± 0.8.
Amytornis striatus whitei: wing, M (n = 8) 62.8 ± 1.4, F (n = 4) 59.0 ± 1.2; tail, M (n = 8) 82.3 ± 4.1, F (n = 4) 76.3 ± 3.2; bill, M (n = 8) 12.2 ± 1.1, F (n = 4) 12.3 ± 0.2; tarsus, M (n = 8) 25.3 ± 0.8, F (n = 4) 23.9 ± 1.3.
Amytornis striatus merrotsyi: wing, M (n = 3) 63.7 ± 2.1, F (n = 4) 61.3 ± 0.6; tail, M (n = 3) 65.3 ± 7.4, F (n = 4) 65.6 ± 0.9; bill, M (n = 3) 11.5 ± 1.0, F (n = 4) 11.2 ± 0.8; tarsus, M (n = 3) 26.2 ± 0.2, F (n = 4) 24.4 ± 1.4.

Field characters

Medium sized grasswrens with black malar stripe, white throat. May overlap with or occur close to Thick-billed Grasswren (*A. textilis*), Dusky Grasswren (*A. purnelli*), also possibly Eyrean Grasswren (*A. goyderi*) and Grey Grasswren *A. barbatus* in central Australia. Distinguished from much darker *A. purnelli* and *A. textilis* by black malar stripe, white throat, and habitat—*A. purnelli* generally on rocky hillsides, *A. textilis* generally not where spinifex dominant. More slender bill, distinct malar stripe, streaked breast, and spinifex habitat distinguish *A. striatus* from *A. goyderi* in cane-grass on sand-dunes. The simple malar stripe, darker, browner appearance, and spinifex habitat distinguish *A. striatus* from long-tailed *A. barbatus* in Lignum.

Voice

Tape-recorded (ANWC). Song a sweet-sounding sequence of varied calls, generally introduced by staccato 'tew tew tew', followed by a melodic liquid ripple of song, including repeated trills and buzzes, finishing with more short, repeated notes; lasting up to several seconds. Given predominantly by male, also by female and immature. Sometimes occurs in bouts of sustained singing (Whitlock 1910). Song of subspecies *whitei* (Pilbara) and *striatus* (Yathong, NSW) are very similar (see sonograms). Significant feature of the recorded songs of *A. striatus* is the prominence of phrases with series of staccato notes. Contact calls a high, single, or repeated 'seep', or soft twittering trills; in breeding season, during nest-building and incubation, both sexes communicate with low, soft purring ripple. Alarm a louder, short, sharp 'tchirr'; more intense alarm a loud shrill call: a repeated scream, followed by a machine-gun rattle, 'eet eet eet rat tat tat tat' (Hutton 1991). More vocal on overcast than on sunny days (Howard and Howard 1984).

Range and status

Resident. Formerly widespread, ranging patchily in spinifex grasslands over most of Arid Zone; spinifex grasslands in pastoral areas are frequently burnt, destroying mature spinifex clumps important to *Amytornis*. Introduced

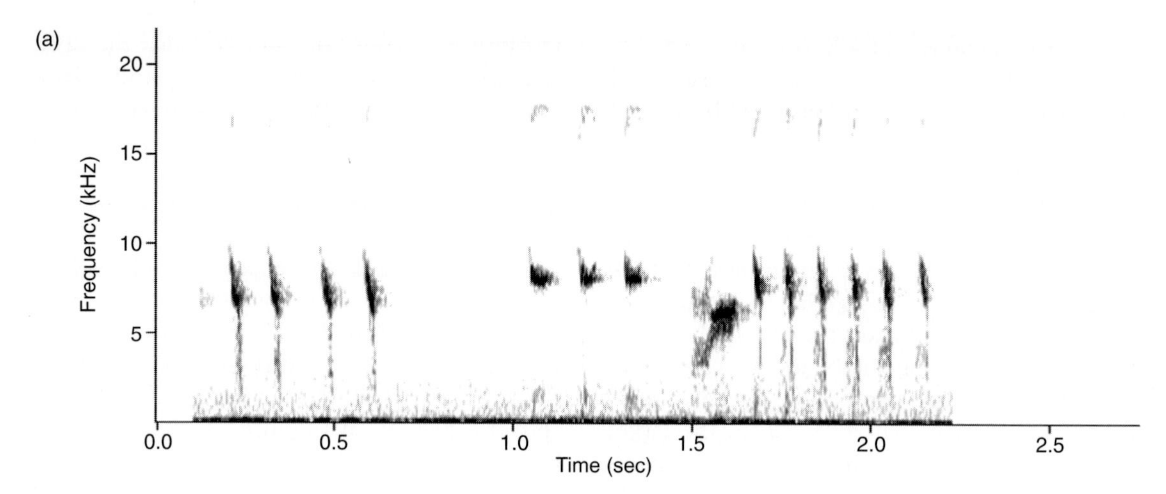

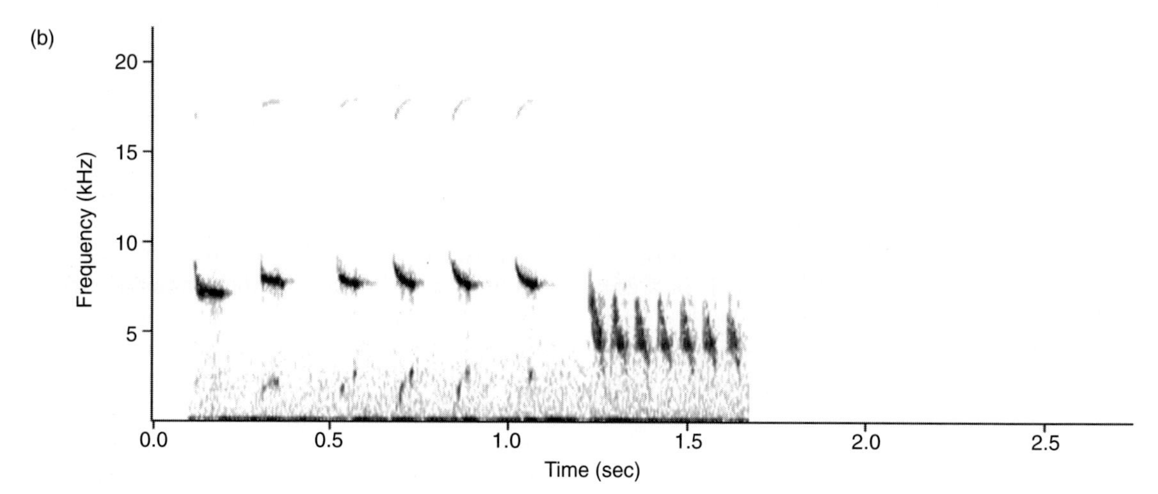

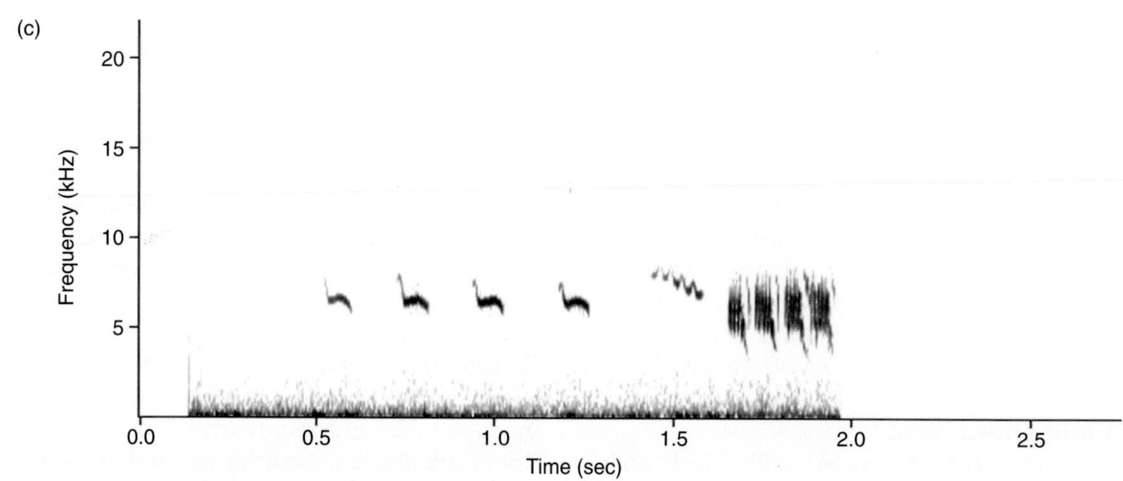

Striated Grasswren (*Amytornis striatus*)
(a) and (b) *A. s. striatus*, song. G. S. Chapman. Yathong Nature Reserve, New South Wales.
(c) *A. s. whitei*, song. G. S. Chapman. Croydon Station, Pilbara, Western Australia.

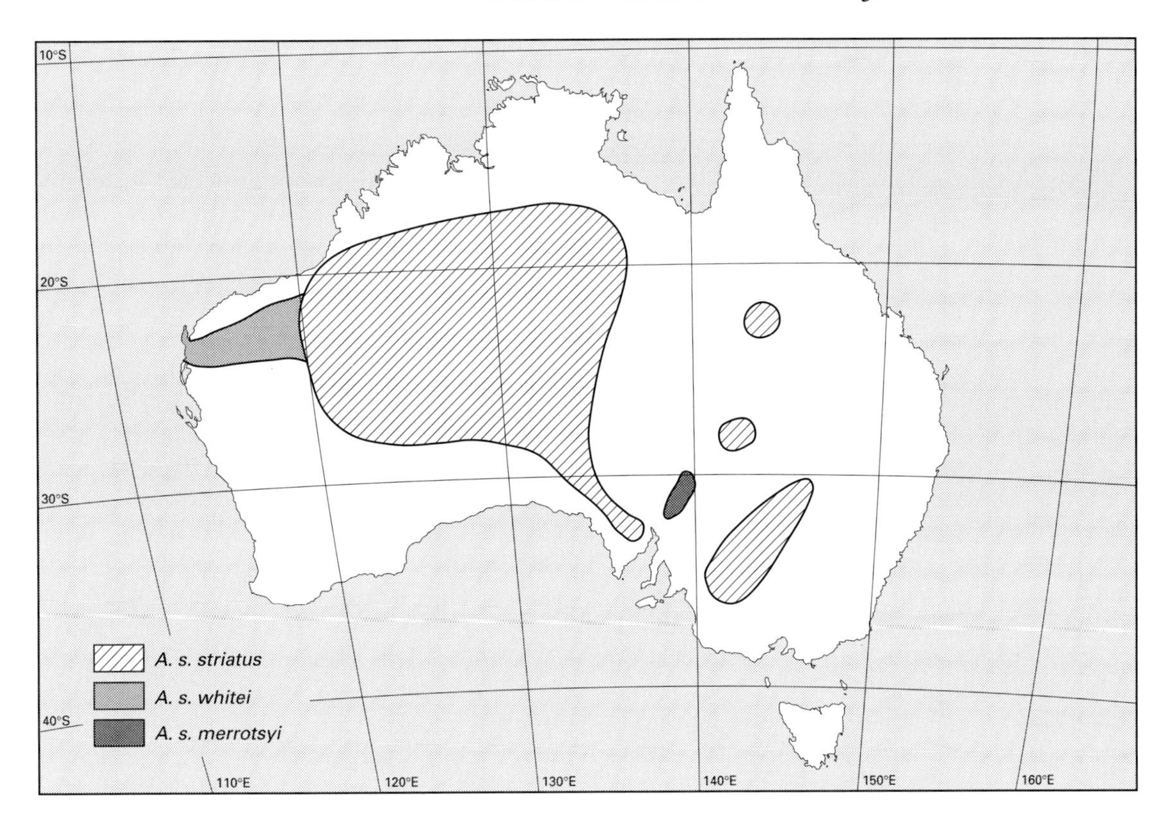

predators and excessive fire frequency may threaten populations of all subspecies.

A. s. striatus: throughout most of arid Australia, except Flinders Ra., SA, and Pilbara, WA. Replaced by *A, goyderi* in Strzlecki and Simpson Deserts. Patchily distributed in Great Sandy, Gibson, and Great Victoria Deserts in W, from S of Wiluna in SW, across N of Nullarbor Plain to upper Eyre Peninsula; in Tanami Desert, sandplain S of Petermann Ranges, near Ayer's Rock and Hermannsberg in NT. Populations in eastern half of range fragmented and much reduced. In isolated patches of spinifex near Opalton and Fermoy in central W, to Santos in SW corner of Qld. In SE, in Yathong Nature Reserve near Cobar, NSW (Morris *et al.* 1981), and Murray Mallee on Vic./SA border. Conservation status 'Insufficiently known' (Garnett 1992*a*); known populations in conservation reserves in all states except Qld.

A. s. whitei: hills of central and northern Pilbara, WA, from Cape Ra. (North West Cape) E to Coongan R. and Nullagine, S. to Barlee Ra. Present range not significantly reduced; habitat not subject to clearing; locally common. Conservation status 'secure' (Garnett 1992*a*); significant area included in Millstream–Chichester, Cape Ra., and Karijini (Hamersley Ra.) National Parks, WA.

A. s. merrotsyi: Still occurs in isolated patches in Flinders Ra., SA, from Freeling Heights in N to Mt Remarkable just S of Wilmington; excessive fire may reduce some populations, and fragmentation may prevent recolonization. Conservation status 'Insufficiently Known' (Garnett 1992*a*); in Gammon Ra. and Flinders Ra. National Parks.

Habitat and general habits

In spinifex (*Triodia*, *Plectrachne*) on sandplains (*A. s. striatus*) or stony hills (*A. s. merrotsyi* and

A. s. whitei). In sandplain habitat, spinifex often interspersed with shrubs or low trees (*Grevillea, Eremophila, Thryptomene, Acacia, Eucalyptus*). Prefer mature unburned spinifex with some tall hummocks and old clumps in the form of a ring with the centre long dead and now open, and with new growth at the edges. Forage on ground, moving rapidly through dense growth with tail horizontal; in open, hop and run with tail cocked. May sunbathe, preen, or allopreen at top of bush or sing from exposed perch (Whitlock 1910; photograph by G. S. Chapman in Pringle 1982, p. 96). To reach the top of bush, hop and flutter in zig-zag pattern through branches; glide or flutter downwards; reluctant to fly any distance in open. Detailed observation of pairs and groups in captivity recorded dust bathing, bathing in puddles of water, and apparent examples of play; also hunting for insects by moonlight (Hutton 1991). Diet: seeds, especially *Triodia*, and insects, especially ground-living beetles and ants (Barker and Vestjens 1990). Izzard *et al.* (1973) saw *A. s. striatus* at Yathong pulling off and eating red flowers of Cactus Pea (*Bossiaea walkeri*). In response to alarm call given by group member on sighting a predator, all birds instantly disappear into dense cover, then 'freeze', extending and flattening head and tail tight to the ground (Howard and Howard 1984). Encountered singly, in pairs, or small groups; no marked birds studied, so social organization in the wild not known.

Displays and breeding behaviour

Breeding behaviour known only from aviary observations. Courtship feeding begins *c.* 1 week before nesting. Mating occurs frequently during nest-building; male approaches female with insect, female lowers head and tail, crouches with wings extended and quivering; male hops up and down in front of female a few times, then swiftly approaches from side and mounts. Male feeds female insects during the 7–10 days of nest construction and 14–15 days of incubation. Collection and carrying of nest material all done on foot (Hutton 1991, from aviary studies).

Breeding and life cycle

Pairs or groups probably dispersed in territories of *c.* 1 ha (Howard 1986, for *A. s. whitei*). Laying season July–Nov if conditions suitable, may breed any time in response to sufficient rainfall (Whitlock 1910; Carter 1917; Beruldsen 1980; Parker 1982*a*; Schodde 1982*b*). Nesting frequency unknown. In captivity with abundant food, nested in Feb, Mar, Apr, May, Aug, and Nov; female re-nested 4–5 weeks after first brood fledged (Hutton 1991). Nest oval, domed (110–130 mm × 80–100 mm), side entrance near top, often with hood; woven of grass stems, *Triodia* spines, and strips of bark, lined with fine grass, rabbit fur, and vegetable down (Whitlock 1910; Mellor 1913; Sutton 1929; Parsons and McGilp 1934; Schodde 1982*b*; Howard and Howard 1984). Up to 250 mm above ground, almost always at edge or near top in spinifex clump. Old rings of spinifex preferred to tall dense young tussocks. Nest built by female attended by male; female incubates and broods young (Whitlock 1910; Chandler 1940). In captivity, male incubated for short periods, relieving female while she fed (Hutton 1991). Clutch 2–3, eggs (18–21 × 14–16.5 mm, n = 30, Schodde 1982*b*) rounded oval, white, with sparse purplish-red spots and blotches, mostly towards larger end; laid daily, in morning; incubation starts when clutch complete. Parasitized by Fantailed Cuckoo (*Cacomantis pyrrophanus*), Black-eared Cuckoo (*Chrysococcyx osculans*), and Horsfield's Bronze-Cuckoo (*Ch. basalis*) (Brooker and Brooker 1989*a*). Incubation period 13–15 days, nestling period 12–14 days (Schodde 1982*b*; Hutton 1991). Both parents feed young. After fledging, young cryptic 1–2 weeks, independent at 3–4 weeks (Hutton 1991). No records of helpers at nest; reports of more than two in groups suggest that young stay with parents well after they fledge (Izzard *et al.* 1973; Miller 1973).

Eyrean Grasswren *Amytornis goyderi* (Gould, 1875)

Amytis goyderi Gould, 1875. *The Annals and Magazine of Natural History*, **16**, 286.

PLATE 6

Monotypic

Description

ADULT MALE: ground colour of crown, upper and lower back, and rump, cinnamon brown; from crown to mid-back and shoulders, feathers with broad, whitish, shaft-streaks bordered dark-brown, less distinct on lower back and rump; wings, dark grey-brown, innermost primaries edged russet; tail, moderately long, dark brown, pale shafts and russet edges; face with uniform white streaks on dusky brown, no brow streak; lores and lower eye-rim, white, vestigial black malar stripe obscured by streaking; underside, dull white, light cinnamon on flanks; eye, brown; bill, leaden-grey, short, stout, and finch-like, deeper than wide; feet, brownish grey.

ADULT FEMALE: as male, except for chestnut flanks.

IMMATURE: much browner than adult dorsally, streaking indistinct, flanks of immature female less red than adult; feet paler brownish grey.

MOULT: adults moult once a year, probably after breeding; no eclipse plumage.

History and subspecies

The original six specimens of *A. goyderi* were collected somewhere NE of Lake Eyre near the Macumba R. by J. W. Andrews, naturalist of the 1874–5 Lake Eyre Expedition despatched by the SA Government to survey pastoral lands. Two specimens were sent to John Gould, who described a new species *Amytis goyderi* (1875); those two specimens in the British Museum were the only ones known until a third was unearthed in the Australian Museum (Hindwood 1945); the other three specimens have been lost. *A. goyderi* was not seen again in the field until 1961 (Morgan *et al.* 1961); May (1977*a*) later collected it. With easier access to the central Australian deserts, it is now known to be relatively widespread in the sand dunes of both the Simpson and Strzlecki Deserts; Schodde (1982*b*) considered that the two populations have not been separated to any significant effect. No subspecies are recognized.

Measurements and weights

Wing, M (n = 20) 59.0 ± 1.5, F (n = 9) 57.3 ± 1.4; tail, M (n = 20) 78.4 ± 6.3, F (n = 9) 77.4 ± 3.4; bill, M (n = 20) 10.8 ± 0.5, F (n = 9) 10.1 ± 0.4; tarsus, M (n = 20) 23.8 ± 0.7, F (n = 9) 22.3 ± 0.7; weight, M (n = 7) 15.0–19.0 (17.5 ± 1.6), F (n = 6) 15.0–18.0 (17.0 ± 1.3). (Weights from museum specimens.)

Field characters

May overlap or occur close to Grey Grasswren (*A. barbatus diamantina*), Striated Grasswren (*A. striatus striatus*), and Thick-billed Grasswren (*A. textilis modestus*); specific habitat is sand dunes covered with cane-grass (*Zygochloa paradoxa*); leave characteristic paired footprints on the sand (Parker *et al.* 1978). All four species generally similar in size, but blunt, deep heavy bill of *A. goyderi* diagnostic (Fig. 4.7(d)). Distinguished from *A. striatus* by unstreaked breast, less distinct malar stripe, lack of red brow. Distinguished from *A. barbatus* (they overlap at Goyder's Lagoon) by redder appearance, shorter tail, absence of distinct black face marks. Distinguished from *A. textilis* by redder appearance and white breast.

Voice

Tape-recorded (ANWC). Song sweet-sounding warble of varied short syllables, including trills, buzzes, up- and down-slurred whistles, and staccato pips (see sonograms), lasting up to 3–4 sec, and including some frequencies above 8 kHz; heard throughout year from both male and female, especially in breeding season; sometimes given from top of grass tussock. Contact calls a high, single, or repeated 'seep seep', also a high soft trill. Alarm a louder, sharp, high 'zeeet'.

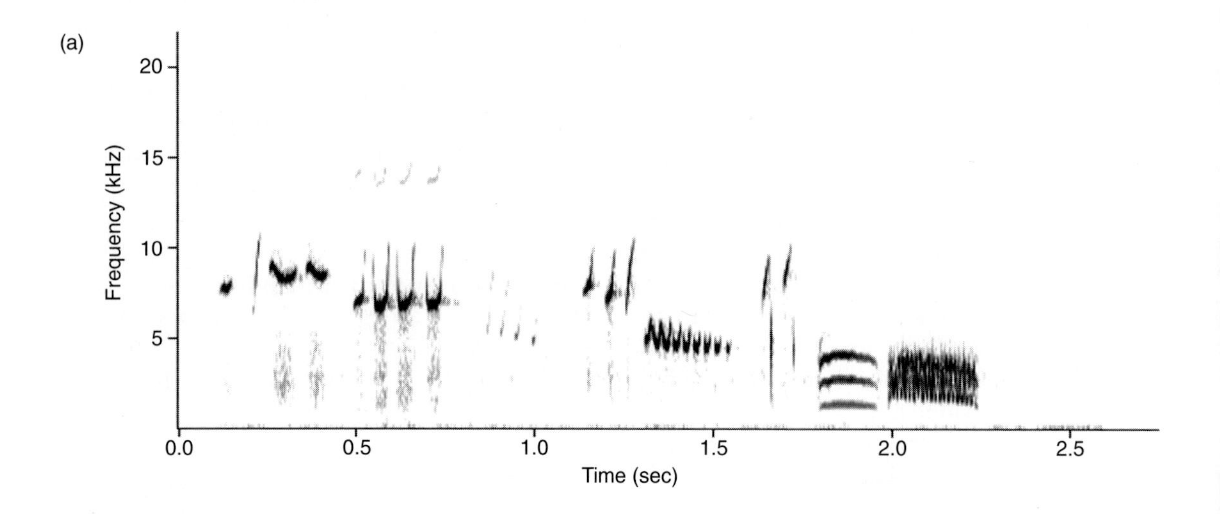

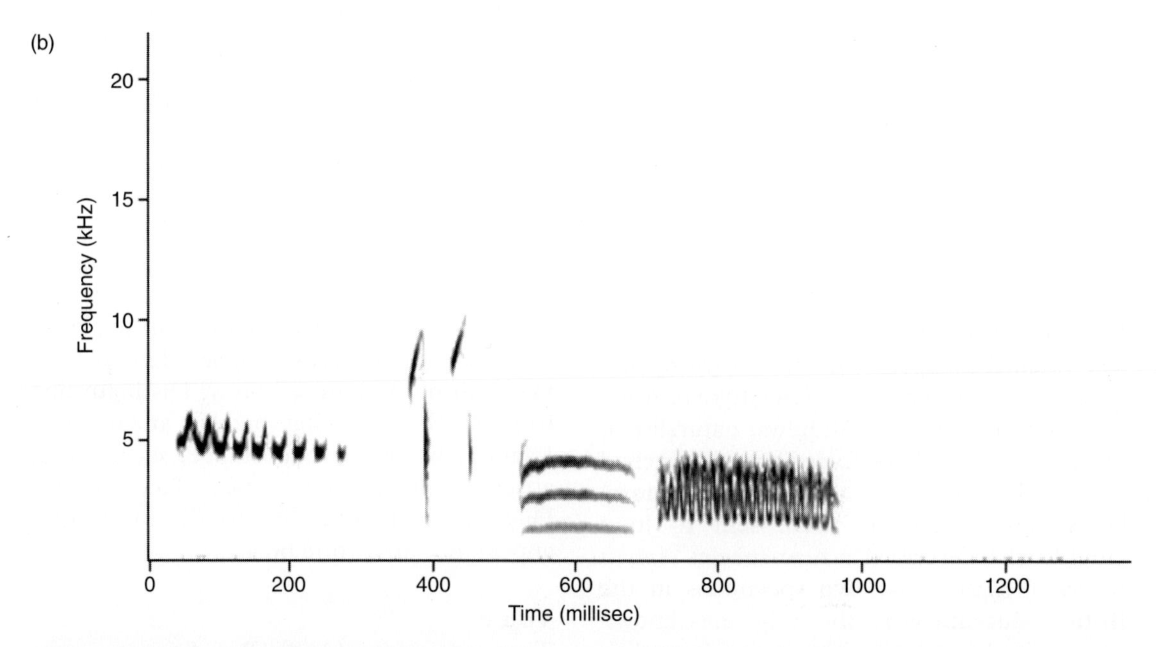

Eyrean Grasswren (*Amytornis goyderi*)
(a) Phrase of song at same scale as sonograms of other *Amytornis* spp.
(b) Detail of song (1.2 sec): the last four elements of song (a), with expanded time scale. G. S. Chapman, Moomba, South Australia.

Range and status

Resident, throughout sand dune areas of Simpson and Strzlecki deserts where SA, Qld, and NT meet. NW Simpson Desert in NT, where Hay, Plenty, and Hale R. and Illogowa Creek run to NW out of Simpson Desert; S to Andado Station and Mulligan R. (Cole and Gibson 1987). Widespread in area between Birdsville, Poeppel's Corner (Parker *et al.* 1978) and Goyder's Lagoon, S to Macumba R. just N of Lake Eyre (type locality) and SE to Strzlecki Creek in region of Moomba (Parker 1980; Blakers *et al.* 1984). First described in 1875 from Macumba R., then not seen until 1961 (Morgan *et al.* 1961). Subsequent sightings on Qld–SA border confirmed by two specimens (May 1977*a*). Recent sightings at edge of Simpson Desert suggest locally common in suitable habitat (Cole and Gibson 1987). Populations probably fluctuate due to alternation between severe drought and good seasons after heavy rain. Significant areas of habitat are included in Simpson Desert National Park in Qld and Simpson Desert Conservation Park in SA (Parker and Reid 1978). Greatest threat probably their restricted distribution in a single climatic region and damage to vegetation by introduced mammals (camels, rabbits, and horses; Cole and Gibson 1987).

Habitat and general habits

Most numerous among tussocks of Sandhill Cane-grass (*Zygochloa paradoxa*) growing in drifting sand on crests and sides of large sandhill ridges; tussocks *c.* 1.3 m high and 1–4 m across, 2–3 m apart. Sometimes seen among clumps of Dune Pea (*Swainsona rigida*) between cane-grass tussocks, rarely in spinifex (*Triodia basedowi*) and *Acacia* sp. on lower slopes of dunes and in swales (Parker *et al.* 1978). All movements quick and nervous, with tail cocked. Shy, hard to see, and difficult to flush. 'When they did flush, they half-flew

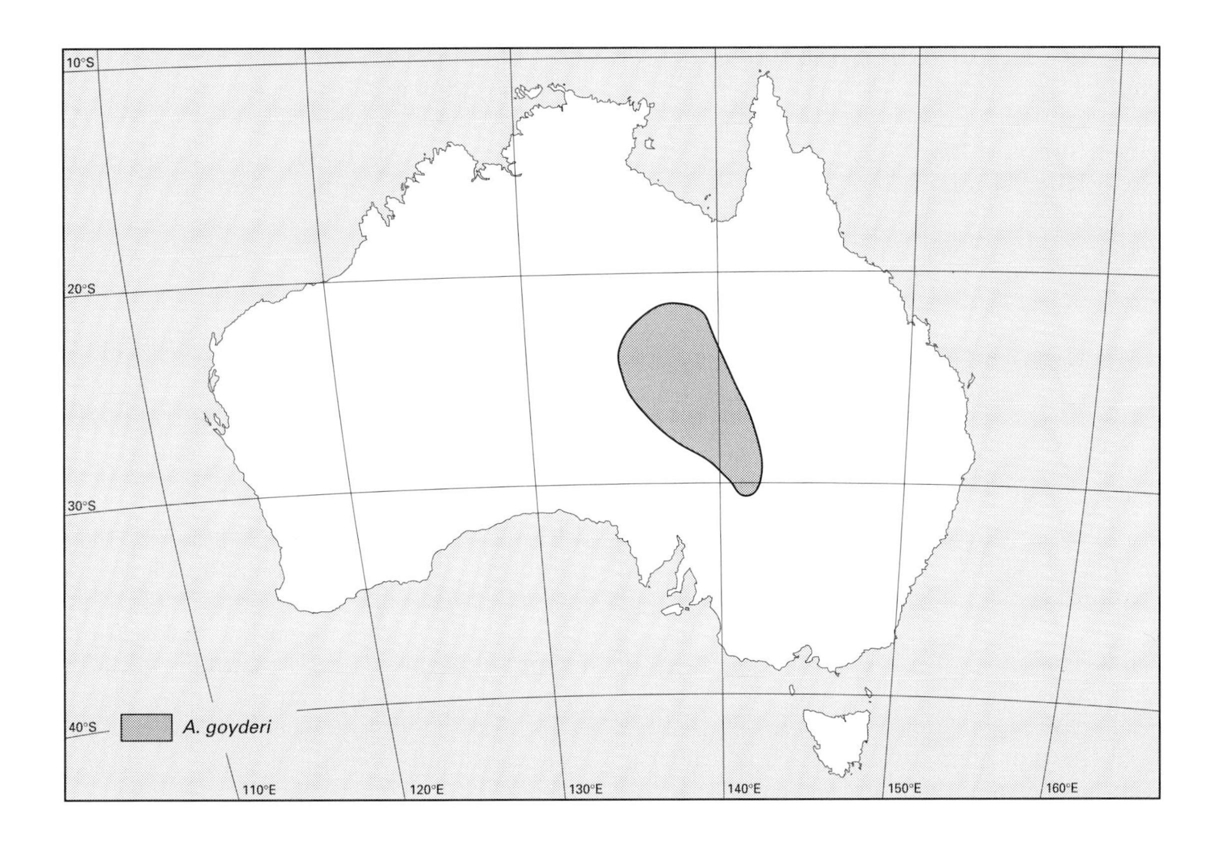

half-bounded from one tussock to the next with great rapidity. Depending on the degree with which they used their wings during these dashes, they touched the ground with their feet at intervals of 20–240 cm. This method of locomotion left tracks of paired footprints, with one print of each pair invariably a little ahead of the other. These tracks are so distinctive that we used them to ascertain the presence of the species on a dune before seeing the birds themselves.' (Parker *et al.* 1978, p. 367). Almost entirely terrestrial, forage on ground within or at edges of clumps of cane-grass and spinifex. Diet, seeds (mostly *Zygochloa* and *Aristida*), and insects, especially bugs and ants, also spiders, weevils, beetles, wasps, and antlions. Since they are no more granivorous than other grasswrens, their finch-like bills may have evolved to cope with seeds of cane-grass, much larger than small seeds of *Triodia* eaten by *A. striatus*. Encountered singly, in pairs, or small groups; no marked birds studied, so social organization unknown.

Displays and breeding behaviour

Nothing recorded.

Breeding and life cycle

No detailed studies of life history. Lives in most arid and variable climatic region of Australia; significant rainfall or none at all may occur in any month. Pairs or groups dispersed in territories of *c.* 1 ha (Schodde 1982*b*). Laying recorded early Aug–Sept for the few nests known (Morgan *et al.* 1961; Parker *et al.* 1978; Schodde 1982*b*); breeding at any time in response to rainfall; suspension of breeding during drought highly probable. Few details of breeding known. Nest compact, globular (110–140 mm × 80–100 mm), semi-domed to open cup with rear rim elevated; loosely woven of strips and stems of grass (not *Zygochloa*, probably *Aristida*); lined with softer grass, sometimes vegetable down and spiders' cocoons; wedged into stout stems in middle of *Zygochloa* tussocks, 100–800 mm above ground. Clutch 2–3, eggs (18–20.5 × 14–15.5 mm, n = 15, Schodde 1982*b*) broadly oval, dull, white, with spots and blotches of dull reddish brown and grey, concentrated at broader end. Incubation and nestling periods unknown, but probably similar to *A. striatus* (similar size, incubation period 14–15 days, nestling period 12–14 days, Hutton 1991). Nestlings have dark grey down on head and wings, yellow gape, and a bright yellow throat (Morgan *et al.* 1961). Co-operative breeding not established; small groups observed, but not known which birds feed nestlings.

Thick-billed Grasswren *Amytornis textilis* (Dumont, 1824)

Malurus textilis Dumont, 1824. *Dictionnaire des Sciences Naturelles,* **30**, 117

PLATE 7

Polytypic. Three subspecies. *Amytornis textilis textilis* (Dumont, 1824); *Amytornis textilis myall* (Mathews, 1916); *Amytornis textilis modestus* (North, 1902)

Description

A. t. textilis

ADULT MALE: grey-brown above, crown darker; from crown to rump, strongly marked, feathers with long white shaft streaks edged black; wings, dark grey-brown, flight feathers with paler margins and creamy shafts; tail, long, dark grey-brown with paler shafts and pale rufous fringes; face and lores, grey-brown with white streaks, brow obscure, rufous; no malar stripe; ventral surface, uniform pale fawn-brown, with white shaft-streaks on chin and breast; sometimes whitish on centre belly; flanks, tawny-fawn; eyes, dark brown; bill, black, blunt and finch-like, deeper than wide; feet, dark grey.

ADULT FEMALE: as male, except for chestnut patch at side of lower breast, just under wings. Tail shorter.

IMMATURE: as adult, but duller, streaks less distinct; at first similar, females acquire chestnut patch on flank after 1–2 months; bill and feet pinkish brown.

MOULT: adults moult once a year. In SE, adults moult Jan–May, after nesting; at Shark Bay, moult Apr–July (Brooker 1988). Tail feathers replaced as they wear; no eclipse plumage (Schodde 1982*b*).

History and subspecies

Three subspecies recognized (Schodde 1982*b*): *A. t. textilis* (Dumont, 1824), once widespread in WA, now restricted to Peron Peninsula and Shark Bay region. *A. t. myall* (Mathews, 1916), from Eyre Peninsula, SA: slightly smaller; a little darker above, strongly streaked; paler below; tail slightly shorter, longer in male. *A. t. modestus* (North, 1902), from inland E Australia; small, with short tail, equal in male and female, most dully streaked subspecies; drab-brown above with short, white streaks, edged dusky; ventral surface dull fawn-white, faint shaft streaks may be barely visible.

The taxonomy of *Amytornis textilis* has been marked by confusion, owing to variation across its large range, similarities with the Dusky Grasswren (*A. purnelli*), and misidentification. Eastern and western forms of *A. textilis* have at times been included in one species together with *A. purnelli*, and at other times treated as separate species. The first *Amytornis textilis* was collected from Shark Bay WA by Quoy and Gaimard in 1818, and described by Dumont in 1824 from drawings, after the specimens were lost in a shipwreck. The species was not seen in this locality again until Tom Carter collected specimens from Dirk Hartog Is. in 1917. Gould (1847) collected a grasswren on the Lower Namoi R., NSW, and thinking Dumont's grasswren came from E Australia, identified and illustrated his specimen as *Amytis textilis*. Subsequently, Gould received birds from Wongan Hills, in SW WA, and described them as a new species *Amytis macroura*. When A. J. North received five specimens collected by G. A. Keartland on the Horn Expedition, he identified these first grasswrens from central Australia as Dumont's *Amytis textilis*, at the same time realizing that Gould's *Amytis macroura* was also *Amytis textilis*. In fact, Keartland's five specimens included three that belonged to an undescribed species, the Dusky Grasswren (*A. purnelli*) (see p. 242). Prompted by Keartland, who always considered that his material included two species, North (1896) described a new species *Amytis modesta* based on a male Thick-billed Grasswren, and included in it Gould's early *textilis* specimens from NSW.

Later, collecting over a vast environmental range in WA led to the description of a number of very similar species; Mathews (1912*a*; 1913) recognized seven subspecies: from WA (*textilis*, *gigantura*, *morgani*, *macrourus*), central Australia (*modestus*), NSW (*inexpectatus*), and *merrotsyi* from the Flinders Ra., SA (now *A. striatus merrotsyi*); in 1916, he recognized a subspecies *myall* from the Gawler Ra., and in 1917, *A. t. carteri* from Dirk Hartog Is. Mathews (1914) regarded *purnelli* as a subspecies of *textilis*, but later separated it as a species (1918*b*). In 1923, he rearranged the Thick-billed Grasswrens, accepting *A. textilis* for all subspecies from WA, *A. modestus* from central Australia, and *A. inexpectatus* from NSW and Eyre Peninsula in SA. Within *A. inexpectatus*, he recognized *i. inexpectatus* from the interior of NSW and *i. myall* from Eyre Peninsula. He distinguished between *A. m. modestus* from the Macdonnell Ra. in central Australia and *A. m. obscurior* from north of Broken Hill in NSW. In 1926, the RAOU Checklist introduced some welcome simplicity, listing three species: Dusky Grasswren (*A. purnelli*), Western Grasswren (*A. textilis*), and Thick-billed Grasswren (*A. modestus*) (which included Mathews' *myall*). A. G. Campbell's review of the genus *Amytornis* in 1927 followed this arrangement, which was generally used for some years. Mathews (1930) persisted with his three species classification of *A. textilis* (with five subspecies), *A. modestus* (two subspecies), and *A. inexpectatus* (two subspecies), but Mayr and Serventy (1944) grouped in one what they considered geographical representatives of a single species: *A. goyderi, A. purnelli, A. textilis*, and *A. modestus*. Condon (1951, 1962, 1969*a,b*), and Keast (1958), however, giving no reason for doing so, went back to North's arrangement of 1902, accepting two species, *modestus* and *textilis*, and treating *purnelli* as a subspecies of *textilis*. The last three authors did, however, recognize the affinity of the birds from the Gawler Ra. (Mathews' subspecies *myall*) with the western rather than the central Australian birds.

Finally, Parker (1972) unravelled the history of this confusion, finding *textilis* and *modestus* as conspecific, and *purnelli* a separate species. For further details, the reader is referred to that classic of taxonomic detection. Schodde (1982*b*) followed this arrangement, recognizing that most of the subspecies previously described in eastern and western forms were stages in continuous variation over the geographical range. He envisaged a once continuous population of Thick-billed Grasswrens in chenopod shrublands right across southern Australia, split by the incursion of the sea from the head of Spencer Gulf N to Lake Eyre, which Serventy called the Eyrean Barrier (1953, 1972). The two divided populations diverged in isolation. At a later stage, some Western Grasswrens were isolated in the region of the Gawler Ra. and diverged further. Schodde (1982*b*) recognized three subspecies, the western *A. t. textilis*, *A. t. myall* from the Gawler Ra., and *A. t. modestus* from inland eastern Australia.

Measurements and weights

Amytornis textilis textilis: wing, M (n = 13) 66.3 ± 2.0, F (n = 12) 64.8 ± 2.2; tail, M (n = 13) 91.8 ± 2.0, F (n = 12) 85.6 ± 4.6; bill, M (n = 13) 10.7 ± 0.8, F (n = 12) 10.8 ± 0.4; tarsus, M (n = 13) 25.2 ± 0.8, F (n = 12) 25.5 ± 1.2; weight, M (n = 17) 21.6–27.6 (24.2 ± 0.8), F (n = 18) 19.6–25.6 (22.1 ± 1.4). (Weights from Brooker 1988.)

Amytornis textilis myall: wing, M (n = 6) 64.7 ± 1.5, F (n = 3) 63.7 ± 2.1; tail, M (n = 6) 85.2 ± 2.8, F (n = 3) 80.5 ± 3.5; bill, M (n = 6) 11.4 ± 0.4, F (n = 3) 10.7 ± 0.4; tarsus, M (n = 6) 25.6 ± 0.9, F (n = 3) 24.2 ± 0.2; weight, F (n = 1) 20.0.

Amytornis textilis modestus: wing, M, 61.3 ± 2.1, F, 60.1 ± 2.0; tail, M, 75.9 ± 5.4, F, 74.7 ± 6.4; bill, M, 10.7 ± 0.6, F, 9.9 ± 0.5; tarsus, M, 24.8 ± 1.0 , F, 23.5 ± 0.8. (Measurements from Schodde 1982*b*; no n given.)

Field characters

Grasswren of moderate size with no black markings on face and uniformly brown appearance; only grasswren to inhabit saltbush–bluebush shrublands on inland plains. May overlap with or occur near Dusky Grasswren (*A. purnelli*), Striated Grasswren (*A. striatus*), Eyrean Grasswren (*A.goyderi*), and Grey Grasswren (*A. barbatus*). Distinguished from *A. purnelli*, the other uniformly streaked grasswren, by dull brown plumage which lacks rufous on back or brow, thicker bill, and habitat; *A. purnelli* occurs on rocky, spinifex-clad hillsides. *A. textilis* lacks black malar stripe and rufous brow and is generally less rufous than *A. striatus*, which is always in habitat with spinifex. Eastern subspecies *A. t. modestus* is much less streaked than *A. s. striatus*. *A. barbatus* and *A. goyderi* both have black facial markings, unstreaked white undersurfaces, and occur in different habitats.

Voice

Tape-recorded (ANWC). Reputedly most silent grasswren (Carter 1917; Whitlock 1924). Songs of *A. t. textilis* from Shark Bay and *A. t. modestus* from Central Australia both include a series of short pips, down-slurred whistles, and short trill. *A. t. myall* not recorded. Two types of contact call: soft twittering trills and louder 'chet' given in twos or threes, with some frequencies above 8 kHz; alarm a short sharp 'tik' (Fig. 5.12(b,c)).

Range and status

Resident. Formerly widespread, ranging patchily over arid Australia S of Tropic of

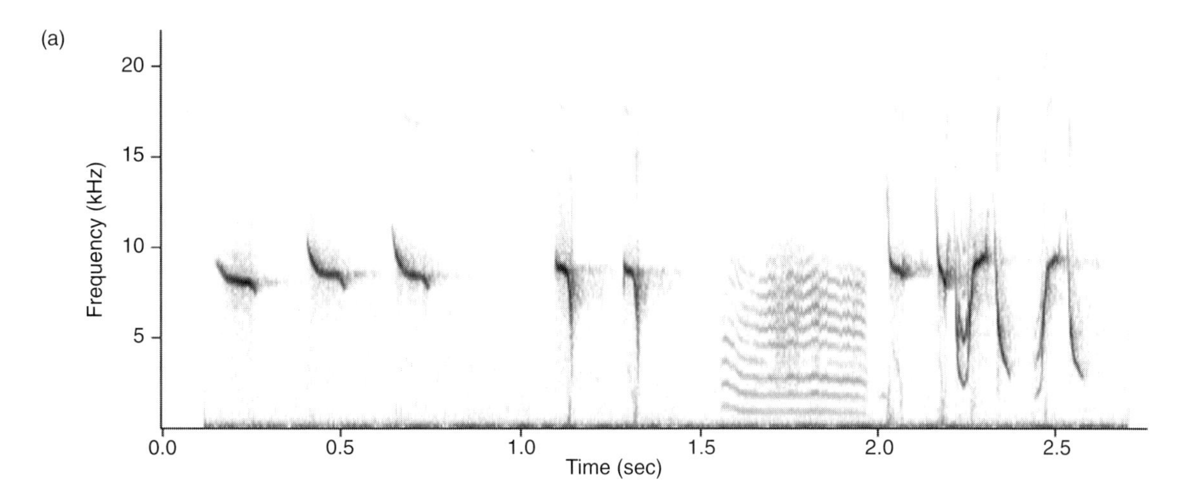

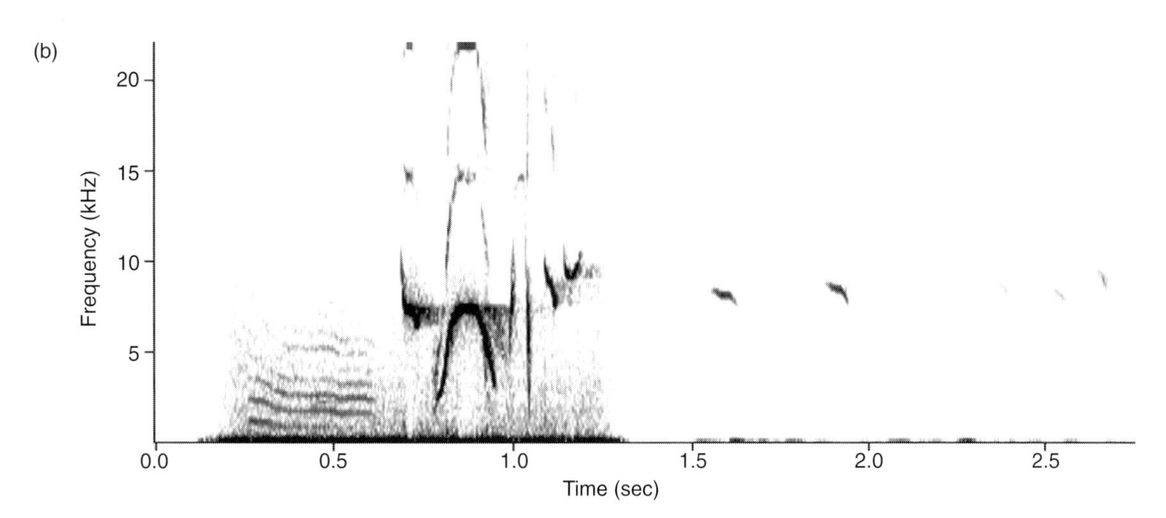

Thick-billed Grasswren (*Amytornis textilis*)
(a) and (b) *A. t. textilis*, two phrases of song. E. M. Russell, Peron National Park, Western Australia.

Capricorn wherever chenopod shrublands occurred.

A. t. textilis: former range, large area of S half of WA, from W coast between Shark Bay and Point Cloates in NW to Wongan Hills, Beverley, and Broome Hill in SW, Lake Way (Wiluna) in NE, and Kalgoorlie in SE (Storr 1985, 1986, 1991). SE extent of former range in some doubt. Rabbits spread west from Eucla after 1894 (Tomlinson 1979). 'Evidently occurring on the Nullarbor Plain before the arrival of the rabbit. 'Odd ones noted here and there right through, chiefly amongst the blue-bush' by C. G. Gibson (1909) during his trip from Kalgoorlie to Eucla in 1908, but not recorded by F. L. Whitlock (1922) between

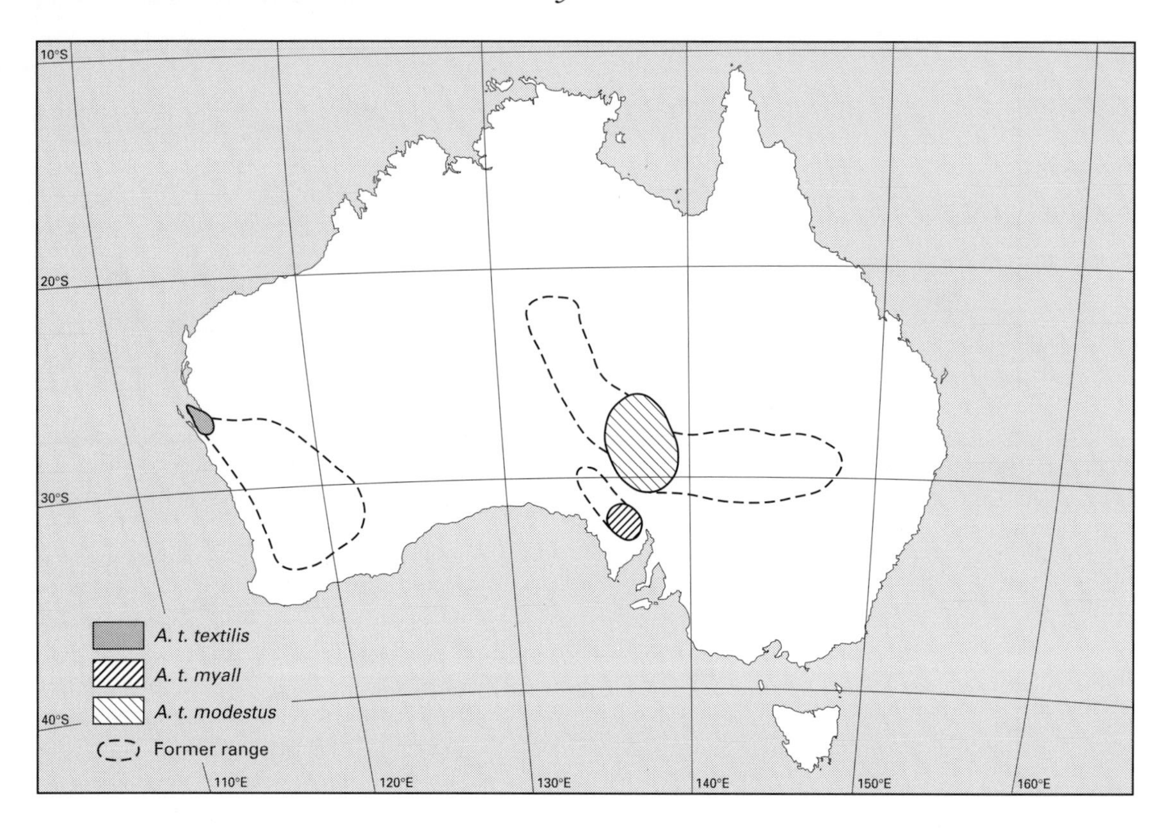

Naretha and Loongana in 1921, A. S. Le Soeuf (1927) at Rawlinna in 1927, or H. Collins (1934) between Limekilns and Haig in 1931–32.' (Storr 1987, p. 60). *A. t. textilis* not recorded 1967–78 on NW Nullabor Plain N of railway line between Naretha and Haig during extensive fieldwork (Brooker *et al.* 1979). Presently known only from arid plains in Shark Bay region of WA; abundant in N two-thirds of Peron Peninsula (Brooker 1988), with one recent record from S of Exmouth Gulf (Blakers *et al.* 1984). Classified as vulnerable (Garnett 1992*a*). Shark Bay area now listed as World Heritage Area and most of Peron Peninsula included in Francois Peron National Park declared in 1991.

A. t. myall: some specimens collected in 1909 near Ooldea, in SW SA, *c.* 200 km W of Tarcoola, no records since. All other specimens and sightings from N Eyre Peninsula, from Mt. Ive, just S of Lake Gairdner in NW to Port Augusta and Whyalla in SE (Parker 1972). Probably still common in suitable habitat, though hard to see, W of Port Augusta and Whyalla (Cox 1974). Status 'Insufficiently Known' (Garnett 1992*a*); none of known range in any conservation reserve.

A. t. modestus : formerly widespread in E Australia, from Lake Eyre basin NW to Macdonnell Ra. in NT, W to Coober Pedy, S to Leigh Creek, SA, E to Broken Hill, Mossgiel, and Lower Namoi–Barwon R., NSW. Present range much reduced following clearing and grazing; in NSW had declined significantly even by 1885 (Bennett in North 1904, p. 249); not seen on lower Namoi R. since Gould's description and plate in 1847. In NT, seen on Finke R. at Palm Valley in 1923 by Whitlock (1924). Since 1936, not seen outside the basins of Lake Eyre, Lake Torrens, and Lake Frome in SA (Garnett 1992*a*). Distribution patchy, but common in some parts of present range. Classified 'Vulnerable' (Garnett 1992*a*). Not present in any conservation reserves, although occurring in Woomera Prohibited Military Area.

Habitat and general habits

Primary habitat, arid and semi-arid shrublands with major shrub component semi-succulent saltbush (*Atriplex* spp.), bluebush (*Maireana* spp.), and similar genera of family Chenopodiaceae; trees, if any, sparse (<10% cover) in range of E subspecies. In W subspecies: a large part of former range covered by shrublands with an upper storey of *Acacia* spp. and *Eucalyptus* spp. growing as tall shrubs >2 m high and a mixed understorey of lower shrubs, including chenopods (Anon. 1990). Present range of *A. t. textilis* in WA includes tall *Acacia* shrublands, coastal dunes, and sandplain, with the following common factors: presence of chenopod species and shrubs 1–3 m tall with a recumbent growth form through which a dense tangle of grass, smaller shrubs, and twining species grows, protected from being grazed by large herbivores (Curry 1986; Brooker 1988). Generally elusive and hard to follow, often climb briefly to exposed vantage point as observer approaches, before disappearing into thick cover. Inquisitive rather than shy if allowed to approach a still, silent observer (Keartland 1904; Carter 1917; Whitlock 1924). Group of four watched for 5 min dust-bathing in the middle of the entrance road to a Shark Bay car park (B. Brooker, Rowley, and Russell, unpublished data). Forage largely on ground, around bases of shrubs, running quickly across open ground. Rarely fly if undisturbed. Diet: small berries, seeds of grasses and various dicotyledons; centipedes, termites, grasshoppers, bugs, beetles, and ants (Whitlock 1924; Lea and Gray 1935; Barker and Vestjens 1990). Encountered singly, in pairs, or small groups. Shark Bay population of *A. t. textilis* studied; 69 adults colour-banded (37 F, 32 M), 18 nests located. At three nests, an adult additional to breeding pair seen nearby; one of these (male) carried food to nest. Pairs or groups nested in same areas from year to year, clearly territorial; marked individuals did not wander from territory after breeding season; probably co-operative breeding when population densities high (Brooker 1988).

Displays and breeding behaviour

Nothing recorded of breeding behaviour or courtship displays. Typical malurid Rodent-run display when nest or one of group threatened (Carter 1917; Brooker 1988).

Breeding and life cycle

Pairs or groups dispersed in territories of *c.* 1 ha at Shark Bay (Brooker 1988). Laying season July–Oct if conditions suitable, may breed at any time in response to sufficient rainfall (Whitlock 1910, 1924; Beruldsen 1980; Schodde 1982*b*; Brooker 1988). Nesting frequency usually once, more often during prolonged abundant seasons (Beruldsen 1980). Nests (90–150 mm × 80–120 mm), built by female unaided attended by male, tightly or loosely woven of dry grass with some strips of bark, lined with finer grass and sometimes with mammal fur. Collection and carrying of nest material all done on foot (Russell, unpublished data). At Shark Bay, WA, nests all had incomplete hoods; 20–1100 mm above ground ($n = 25$, median 300 mm), in centre of substrate vegetation with 150–1100 mm foliage above, tended to face S with entrance towards densest part of bush (Brooker 1988). In other areas, nests varied from domed to cup with rudimentary hood (Whitlock 1921, 1924). At Shark Bay, nests in several species of shrub; in central Australia, usually in saltbush or bluebush or in cane-grass (*Zygochloa paradoxa*) (Whitlock 1910, 1924). Clutch 2, sometimes 3, laid on alternate days (B. Brooker, personal communication); eggs (20–22.5 × 14.5–16 mm in Central and E. Australia, $n = 32$; larger in WA, 22–25 × 16–17 mm, $n = 8$; Schodde 1982*b*; Brooker 1988) tapered oval, creamy-white, thickly freckled all over with irregular reddish-brown spots and blotches, denser at larger end. Parasitized by

Horsfield's Bronze-Cuckoo (*Chrysococcyx basalis*) (Brooker and Brooker 1989*a*). Males thought to share incubation and brooding, but extent of contribution not known (Whitlock 1910, 1924; Schodde 1982*b*; Brooker 1988). Incubation period 15–17 days; nestling period 10–12 days (B. Brooker, personal communication). Unlike naked nestlings of *Malurus* spp., hatchlings covered with charcoal grey down; rictal flanges white, inside mouth orange-yellow (Brooker and Brooker 1987). Both parents feed young; one report of third bird (male) also feeding at nest; probably co-operative breeding in appropriate demographic circumstances. After fledging, young cryptic 1–2 weeks. No long-term studies of survival.

Dusky Grasswren *Amytornis purnelli* (Mathews, 1914)

Diaphorillas textilis purnelli Mathews, 1914. *Australasian avian Record*, **2**, 99.

PLATE 7

Polytypic. Two subspecies. *Amytornis purnelli purnelli* (Mathews, 1914); *Amytornis purnelli ballarae* Condon, 1969.

Description

A. p. purnelli

ADULT MALE: crown, dusky, grading to dull reddish-brown on back and rump; from head to mid-back, feathers with white shaft-streaks, narrower than in any other grasswren, streaks edged black; rump plain; wings, dusky-brown, flight feathers with rufous edges and paler shafts; tail, slender, moderately long, dusky brown, shafts and edges rufous; face and lores, dusky-rufous with white streaks; no obvious brow or malar stripe; throat and breast, dull rufous with whitish streaks edged rufous, grading to plain, dark, rufous-brown belly; eye, brown; bill, relatively long and slender, dark grey-brown to black, as feet.

ADULT FEMALE: as male except for rich rufous patch at sides of breast.

IMMATURE: as adults, but duller, with less streaking, sexes initially similar.

MOULT: adults moult once a year, after breeding, generally Nov–May; tail feathers replaced as they wear; no eclipse plumages (Schodde 1982*b*).

History and subspecies

Two subspecies recognized: *A. p. purnelli* (Mathews, 1914), from central Australia; *A. p. ballarae* Condon, 1969, an isolated population from Selwyn Ra. near Mt. Isa, Qld, with adult male slightly smaller, brighter reddish-brown on back, mid-grey over belly, throat and breast straw coloured, thin shaft-streaks edged dusky; female as male except for dark rufous flanks.

The original specimens collected by G. A. Keartland from rocky habitat on the Horn Expedition to central Australia in 1894 were misidentified as *A.textilis* by North (1896). Prompted by Keartland, who was sure he had collected two different species, North (1902) described *textilis* specimens in the same collection from saltbush and cane-grass as a new species, *modesta*. Mathews (1914) named specimens from the Macdonnell Ra. in central Australia as *purnelli*, a subspecies of *textilis*; in 1918 he recognized it as a species (1918*b*). He continued to hold this view (1922–3, 1930) as did the 1926 Checklist and Campbell (1927). However, Condon (1951) and Keast (1958) treated *textilis* and *purnelli* as conspecific and *modestus* as a separate species, an arrangement that was generally followed. Thus, in 1966, when W. Horton discovered a population of *purnelli* -like birds near Mt. Isa, separated from the nominate race in central Australia, they were described as *A. textilis ballarae* by Condon (1969*a*). Keast (1958) separated Dusky Grasswrens from the Everard Ra. in SA as a subspecies *A. t. everardi* on the basis of bill size, but Harrison (1974) disagreed because birds from the Rawlinson Ra. in WA spanned the size range of *everardi* and *purnelli* proper. Parker (1972) unravelled the history of this confusion, and recognized *textilis* and *purnelli* as separate species, the latter with two subspecies *purnelli* and *ballarae*. Schodde followed this arrangement. Indeed, he considered the two subspecies so different that they 'surely verge on distinct species' (Schodde 1982*b*, p. 177).

Measurements and weights

Amytornis purnelli purnelli: wing, M (*n* = 52) 62.4 ± 1.7, F (*n* = 38) 60.6 ± 1.7; tail, M (*n* = 52) 76.5 ± 4.0, F (*n* = 38) 74.5 ± 2.9; bill, M (*n* = 52) 12.1 ± 0.6, F (*n* = 38) 11.7 ± 0.6; tarsus, M (*n* = 52) 24.9 ± 1.2, F (*n* = 38) 24.2 ± 1.1; weight, M (*n* = 2) 22–23 (22.5), F (*n* = 4) 19–21 (20.5 ± 1.5).
Amytornis purnelli ballarae: wing, M (*n* = 4) 60.8 ± 1.3, F (*n* = 7) 59.8 ± 1.2; tail, M (*n* = 4) 74.0 ± 0.8, F (*n* = 7) 71.5 ± 2.2; bill, M (*n* = 4) 12.1 ± 0.3, F (*n* = 7) 11.5 ± 0.4; tarsus, M (*n* = 4) 24.1 ± 0.3, F (*n* = 7) 24.0 ± 0.7; weight, M (*n* = 5) 20.0–25.5 (23.1 ± 1.7), F (*n* = 5) 19.0–22.4 (20.8 ± 1.6). (Weights from museum specimens.)

Field characters

Brownest of all the grasswrens, no obvious marks of any kind. May overlap or occur near Striated Grasswren (*A. striatus*) and Thick-billed Grasswren (*A. textilis*) in central Australia and Carpentarian Grasswren (*A. dorotheae*) near Mt. Isa, Qld. Distinguished from *A. textilis*, the other uniformly streaked grasswren, by darker appearance, more rufous back, slender bill, and habitat of rocky spinifex-clad hills. *A. textilis*, generally found in saltbush–bluebush shrublands, is generally less rufous and is paler underneath. Lack of black malar stripe, generally darker appearance, and rocky habitat distinguishes *A. purnelli* from *A. striatus*, a bird of spinifex plains. Gleaming white throat and breast contrasting with blackish head of *A. dorotheae* distinguishes it from *A. p. ballarae*.

Voice

Tape-recorded (ANWC). Song a series of varied short trills for 2–3 sec (sonograms), heard throughout year from male and less frequently female. Singing usually performed by bird perched in open, on rock; bird stands, tail cocked, moving head from side to side as it sings. Song of *A. p. ballarae* has components with higher frequencies and fewer repeated trill elements (sonogram). Contact call a high pitched ventriloquial single or repeated 'seet'; also a series of soft, high-pitched twittering trills. Alarm calls sharp, rather harsh, loud 'tchk tchk'.

Range and status

Resident; restricted to rocky, spinifex-covered ranges of central Australia and NW Qld. In central Australia, occurs in Macdonnell Ra., N to Ashburton Ra., E to Hart's and Jervois Ra., S to Everard, Musgrave, Mann, and Blackstone Ra., and W to Rawlinson and Kintore Ra. Population in NW Qld restricted to Selwyn Ra. SE of Mt. Isa, S to Mt. Unbunmarra, NE of Boulia. An outlier based on unconfirmed sighting at Thorntonia 100 km NW of Mt Isa (Carruthers *et al.* 1970) now doubtful, since range of *A. dorotheae* extended to nearby locality (Harris 1992). NW Qld and central Australian populations separated by extensive floodplain of Georgina R. Common in suitable habitat; greatest threats to NW Qld population are altered fire regime of too frequent fires that reduce size and cover of spinifex hummocks (Carruthers *et al.* 1970) and restricted range in a single climatic region; not represented in any National Park. Central Australian subspecies occurs in several small National Parks and reserves.

Habitat and general habits

Rocky slopes, gullies, and ridges of ranges and stony hills with open woodland of *Eucalyptus*

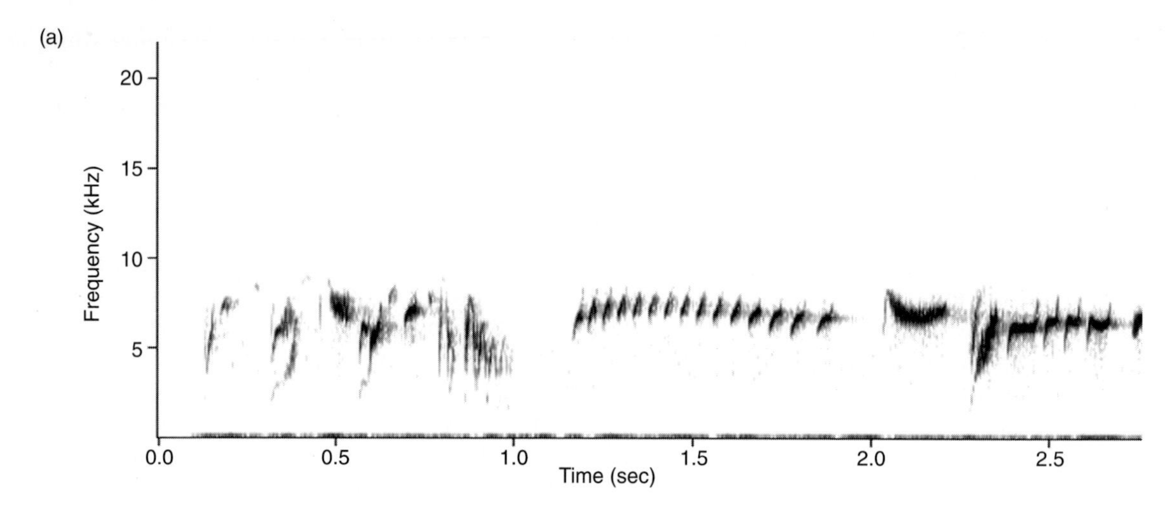

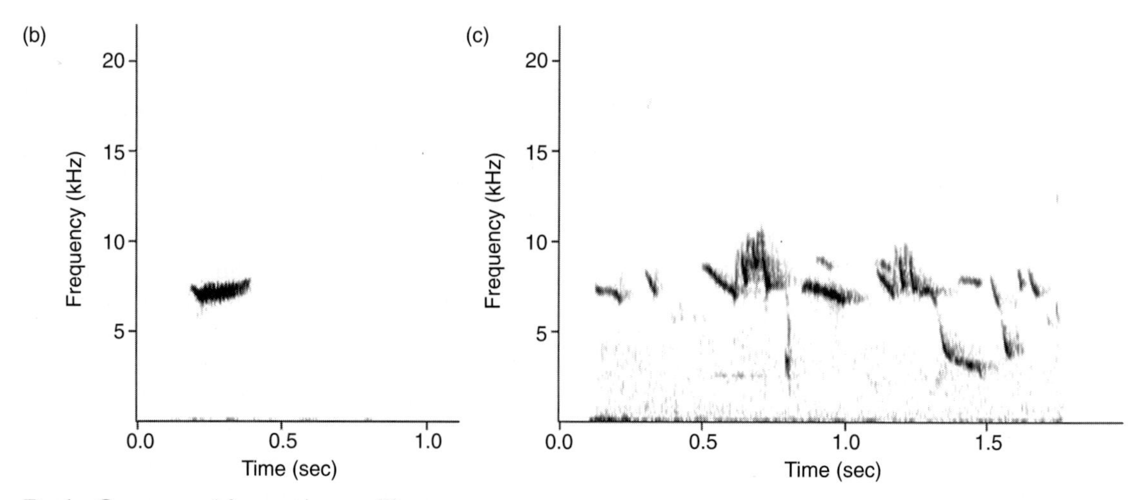

Dusky Grasswren (*Amytornis purnelli*)
(a) *A. p. purnelli*, song. (b) contact 'seep'. G. S. Chapman, Serpentine Gorge, Alice Springs, Northern Territory.
(c) *A. p. ballarae*, song. G. S. Chapman, near Mary Kathleen, S. E. of Mt. Isa, Queensland.

spp., occasional shrubs and ground cover of spinifex (*Triodia* and *Plectrachne*). Does not occur on flat valley floors, even where spinifex is dense. Erect fairy-wren-like stance, normally terrestrial, hops rather than flies, especially in rough country and moving uphill. In broken rock, runs through internal crevices. Flies short distances, generally downhill. When pursued, scurries rat-like with tail down at great speed through spinifex. Generally seen in pairs or small parties of 3–10 birds. Seeks shelter from sun in rock crevices. Forages on ground in rock crevices and under spinifex hummocks. Diet, seeds and insects in roughly equal proportions (Schodde 1982*b*; Barker and Vestjens 1990).

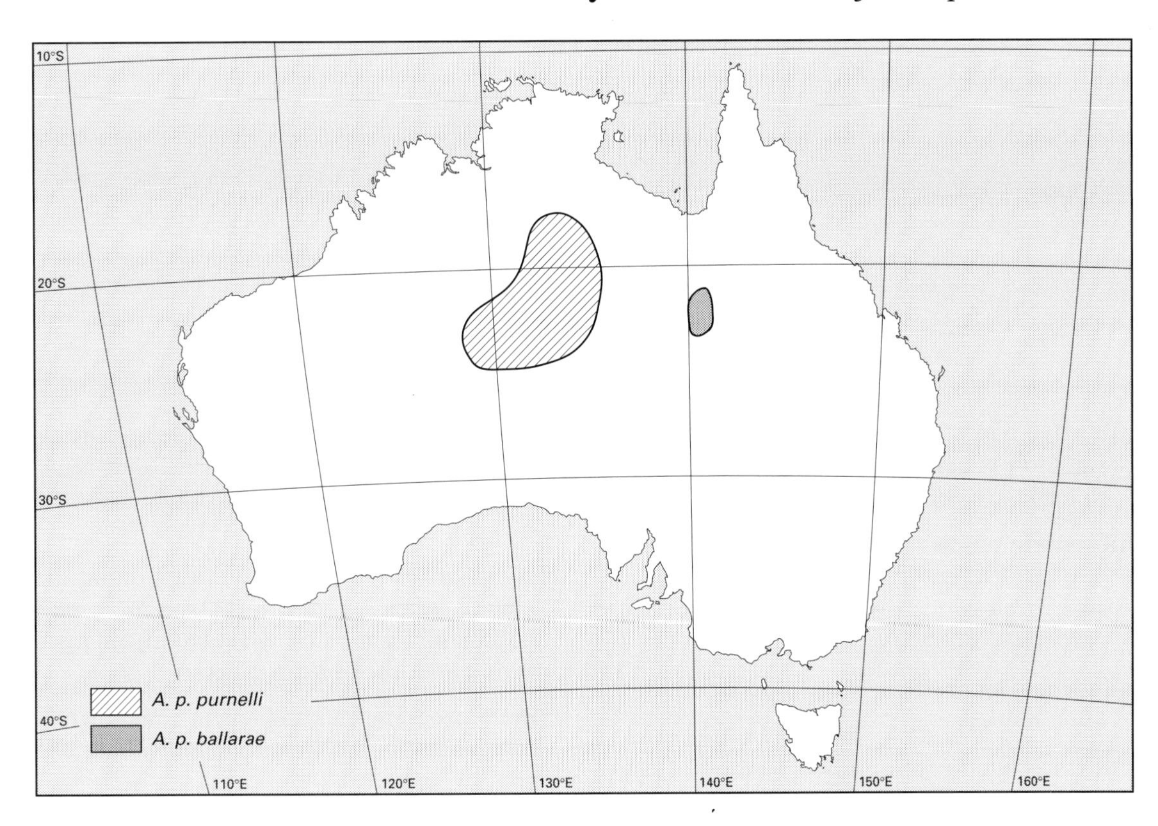

Displays and breeding

Nothing recorded of breeding behaviour or courtship displays. Typical malurid Rodent-run display when one of group threatened (Goodwin 1967).

Breeding and life cycle

No detailed studies of life history. Pairs or groups dispersed in territories of *c*. 1 ha (Schodde 1982*b*). Laying season July–Oct if conditions suitable (Whitlock 1924; Carruthers *et al*. 1970; Schodde 1982*b*), but may breed at other times in response to rainfall (Glass 1973); in some years, prolonged breeding may be possible (Cowle in North 1904; Schodde 1982*b*). No marked birds studied, so number of broods raised per year not known; nest bulky (140–160 mm × 100–130 mm), with thick projecting roof (NW Qld, Carruthers *et al*. 1970); in central Australia, vary from completely roofed to semi-dome (Whitlock 1924); loosely built of grass stems and fine twigs, lined with finer fibre; well concealed, 100–500 mm above ground near top of spinifex hummock. Clutches 2–3, eggs (19–22 × 14.5–16 mm, Schodde 1982*b*) tapered oval, creamy-white, heavily marked all over with brown and purplish-red spots and blotches, with some concentration at larger end. So far as is known, only female incubates and broods (Whitlock 1924; Schodde 1982*b*). Incubation and nestling periods not known, but probably similar to *A. striatus* (similar size, incubation period 14–15 days, nestling period 12–14 days, Hutton 1991). Both parents feed young and remove faecal sacs, approaching nest furtively and on foot (Carruthers *et al*. 1970). Probably a co-operative breeder, since groups of 3–10 birds seen at all times of year; Glass (1973) saw three birds at a nest with two unfeathered nestlings.

Glossary

adaptive radiation evolution from an ancestral form of several divergent forms adapted to distinct modes of life

allopatric species or subspecies whose geographical distributions do not overlap are said to be allopatric; contrasted with sympatric; also allopatry

allopreening preening of one bird by another; often mutual

allozyme electrophoresis a method of estimating paternity using polymorphic proteins, frequently from the blood, as genetic markers

barbs side branches from the shaft that when linked by barbules form the vane or webbing of most feathers

barbules projections along both edges of the feather barbs. They normally have hooks that interlock, binding the barbs together to form the webbing of the feather into a vane; reduced in malurid genera *Clytomyias* and *Stipiturus*

billabong an arm of a river made by water flowing from the main stream, usually in time of flood, to form a backwater, blind creek, anabranch, or, when the water level falls, a pool or a lagoon

bluebush shrubs of the genus *Maireana* (Chenopodiaceae), an important component of chenopod shrublands in arid and semi-arid areas of Australia

chenopod shrubland arid and semi-arid areas covered with low shrubs of the family Chenopodiaceae; salt-tolerant shrubs 0.5–1 metres tall, with greyish hairy, semi-succulent leaves; including bluebush, saltbush, and samphires

cloacal protuberance a swelling around the cloaca serving as a site for sperm storage and maturation before ejaculation; enlarged throughout the breeding season

cline a geographical gradient in some feature such as size or colour among individuals of a species

conspecific belonging to the same species

co-operative breeding where non-breeding birds, often themselves sexually mature, assist others in various aspects of the reproductive cycle, such as territory defence and rearing young

cordillera a high mountain chain

cryptic phase the post-fledging period (*c.* one week) when young remain hidden in cover and are fed there

culmen the ridge of the upper half of the bill (upper mandible)

DNA–DNA hybridization measures degrees of genealogical relationship among species by comparing their DNA (genetic material)

DNA fingerprinting a method of establishing identity and estimating paternity by comparing patterns of fragments of genetic material (DNA); an individual's unique DNA fingerprint includes elements from each parent

ear tufts lengthened feathers covering ear openings; often of contrasting colour, erectile for display in some malurids

eclipse specifically, a special temporary post-nuptial plumage in ducks, but now generally used for an alternate non-breeding plumage. In this book, we use it for the dull brown plumage of adult males not in nuptial plumage, which may nevertheless be reproductively active

egg-dumping laying of an egg by one female in the nest of a con-specific female

emarginate applied to the margin of a feather, usually a flight feather, where the vane narrows suddenly near the tip

faecal sac a jelly-like envelope enclosing nestling faeces; easily removed from nest by attendants

feather-tract a restricted area or line on skin from which feathers originate

flanks the sides of mid and lower ventral surface

fledging action of leaving the nest as nestlings mature

fledgling a young bird that has left the nest, is still dependent on its parents for food, and in most malurids is still completing growth of wings and tail

frequency modulation a periodic alternation of frequency in a sound wave; e.g. a trill involves rapid frequency modulation

gape the mouth, from corner to corner of the bill

group describes the situation in which more birds than one pair occupy a territory and attend a nest. In contrast to the looser association of a flock, the group has a definite membership over an extended period of time, and members are often closely related

group-year one group studied for one year; e.g. 10 groups studied over two years covers 20 group-years. Similarly, pair-years

helpers full-grown group members that assist pair to raise young; usually adult and frequently related

immature a young bird after growth of wings and tail is complete and before sexual maturity

juvenile a young bird that has not yet completed growth of wings and tail

Karri *Eucalyptus diversicolor*, a tall forest tree of south-western Australia

Lignum *Muehlenbeckia cunninghamii*, a tangled shrub with cane-like stems and reduced leaves, growing on flood-plains in arid Australia

lores the area between the base of the upper mandible and the eye

malar the area on the side of the throat, immediately below the base of the lower mandible

mallee describes eucalypts with a multi-stemmed habit, and the low open woodland vegetation formed by them

malurid any member of the family Maluridae

mandible upper and lower make the two halves of the bill

mantle feathers of the upper back

monotypic refers to any higher taxon comprising only one lower taxon, i.e. a genus of one species

mulga arid woodlands dominated by various *Acacia* species, the commonest being Mulga *A. aneura*

nuptial pertaining to the breeding season, especially to plumage and display

oscines one of two major sub-divisions of the order Passeriformes, originally based on the morphology of the syrinx, latterly supported by biochemical evidence; oscines are thought to have the more advanced syrinx, and include most of the Old World songbirds; suboscines include the Old World pittas and broadbills and the New World flycatchers, ant-birds, oven-birds, etc.

pair-year see group-year

philopatric remaining in or close to the natal area

polytypic refers to any higher taxon comprising more than one lower taxon, i.e. a species with several subspecies

primaries the outermost flight feathers (10 in malurids)

rectrix, rectrices tail feathers (10 or fewer in malurids)

rictal pertaining to gape, often applied to bristles in that area

saltbush shrubs of the genus *Atriplex* (Chenopodiaceae), important component of chenopod shrublands in arid and semi-arid areas of Australia

samphire low shrubfield of *Salicornia*, *Arthrocnemum*, etc. (Chenopodiaceae), growing at the margins of saline lakes

scapulars the feathers above the shoulders

sclerophyll a plant with hard, stiff leaves

secondaries the inner flight feathers (10 in malurids)

sexual selection the evolution of characters involved in competition for mates; this may involve either competition within one sex for members of the opposite sex or differential choice by members of one sex for members of the opposite sex

shaft the central rib of a feather

shaft-streak individual feathers, especially of grasswrens *Amytornis*, are strongly streaked, with an area close to the central shaft contrasting strongly with the rest of the vane; the shaft itself is frequently white and may be edged with black, while the vane is brown or rufous

sonogram a visual representation of a sound, usually with time on the horizontal axis, frequency on the vertical axis, and intensity represented by the density of the trace

spinifex a vernacular name used in Australia for the spiny hummock grasses *Triodia* and *Plectrachne*

suboscines one of two major subdivisions of the order Passeriformes, originally based on the morphology of the syrinx, latterly supported by biochemical evidence; oscines are thought to have the more advanced syrinx, and include most of the Old World songbirds; suboscines include the Old World pittas and broadbills, and the New World flycatchers, ant-birds, oven-birds, etc.

sympatric occurring in the same geographical area, contrasting with allopatric

tarsus birds walk on their toes; the tarsus is a fusion of bones between the toes and the ankle. It is elongated, unfeathered, and commonly referred to as the leg in birds. Most of the rest of the leg above the ankle joint is usually obscured by feathers

taxon, plural taxa any category used in classification e.g. species, subspecies

vibrato a rapid and continuous periodic rise and fall in pitch; a form of frequency modulation

xeromorphic applied to plants which grow naturally in dry regions and are structurally modified to withstand dry conditions

Bibliography

Ambrose, S. J. (1984). The response of small birds to extreme heat. *Emu*, **84**, 242–3.

Ambrose, S. J. and Murphy, D. P. (1994). Synchronous breeding of land birds on Barrow Island, Western Australia, after cyclonic summer rains. *Emu*, **94**, 55–8.

Ames, P. L. (1970). The morphology of the syrinx in passerine birds. *Yale University Peabody Museum of Natural History, Bulletin*, **36**, 1–194.

Andersson, M. (1994). *Sexual selection*. Princeton University Press, Princeton, New Jersey.

Anon. (1990). *Vegetation. Atlas of Australian resources*. Third Series. Vol. 6. Australian Surveying and Land Information Group, Canberra.

Archer, M. (1981). Results of the Archbold Expeditions 104. Systematic revision of the marsupial genus *Sminthopsis* Thomas. *Bulletin of the American Museum of Natural History*, **168**, 63–223.

Archer, M. and Fox, B. (1984). Background to vertebrate zoogeography in Australia. In *Vertebrate zoogeography and evolution in Australia*, (ed. M. Archer and G. Clayton), pp. 1–15. Hesperian Press, Perth.

Ashby, E. (1920). Birds of Mt. Compass district, South Australia. *Emu*, **19**, 299–303.

Ashby, E. (1924). Note on the discovery of *Malurus pulcherrimus* (Gould) in the state of South Australia. *South Australian Ornithologist*, **7**, 184–5.

Audley-Charles, M. G. (1987). Dispersal of Gondwanaland: relevance to evolution of the angiosperms. In *Biogeographical evolution of the Malay Archipelago*, (ed. T. C. Whitmore), pp. 5–25. Oxford University Press, Oxford.

Baker, M. C. (1995). A comparison of songs from four species of fairy-wrens (*Malurus*). *Emu*, **95**, 294–7.

Barker, R. D. and Vestjens, W. J. M. (1990). *The food of Australian birds. 2. Passerines*. CSIRO, Melbourne.

Barnard, H. G. (1914). The search for *Amytornis woodwardi* in the Northern Territory. *Emu*, **13**, 188–90.

Baverstock, P.R., Schodde, R., Christidis, L., Krieg, M., and Sheedy, C. (1991). Microcomplement fixation: preliminary results from the Australasian avifauna. *Acta XX Congressus Internationalis Ornithologici*, 611–8.

Beadle, N. C. W. (1981). *The vegetation of Australia*. Cambridge University Press, Cambridge.

Beecher, W. J. (1953). A phylogeny of the oscines. *Auk*, **70**, 270–333.

Beehler, B. M. (1978). *Upland birds of northeastern New Guinea*. Wau Ecology Institute, Handbook No. 4. Wau, New Guinea.

Beehler, B. M. and Finch, B. W. (1985). *Species check-list of the birds of New Guinea. Australasian Ornithological Monographs* No. 1, Royal Australasian Ornithologists' Union, Melbourne.

Beehler, B. M., Pratt, T. K., and Zimmermann, D. A. (1986). *Birds of New Guinea*. Princeton University Press, Princeton, New Jersey.

Bell, H. L. (1969). Field notes on the birds of the Ok Tedi River drainage, New Guinea. *Emu*, **69**, 193–211.

Bell, H. L. (1970). Field notes on the birds of Amazon Bay, Papua. *Emu*, **70**, 23–6.

Bell, H. L. (1971). Field-notes on birds of Mt. Albert Edward, Papua. *Emu*, **71**, 13–19.

Bell, H. L. (1982*a*). A bird community of lowland rainforest in New Guinea. 2. Seasonality. *Emu*, **82**, 65–74.

Bell, H. L. (1982*b*). Abundance and seasonality of the savanna avifauna at Port Moresby, Papua New Guinea. *Ibis*, **124**, 252–74.

Bell, H. L. (1982*c*). A bird community of lowland rainforest in New Guinea. 4. Birds of secondary vegetation. *Emu*, **82**, 217–24.

Bell, H. L. (1984). A bird community of lowland rainforest in New Guinea. 6. Foraging ecology and community structure of the avifauna. *Emu*, **84**,142–58.

Bell, H. L. (1985). Seasonal variation and the effects of drought on the abundance of arthropods on savanna woodland on the northern tablelands of New South Wales. *Australian Journal of Ecology*, **10**, 207–22.

Bell, H. L. and Ford, H. A. (1986). A comparison of the social organization of three syntopic species of Australian thornbill, *Acanthiza*. *Behavioral Ecology and Sociobiology*, **19**, 381–92.

Bell, H. L., Coates, B. J., and Layton, W. A. (1979). Notes on Wallace's Wren-warbler *Todopsis wallacii* Gray, with a description of the nest and eggs. *Emu*, **79**, 152–4.

Bennett, A. T. D. and Cuthill, I. C. (1994). Ultraviolet vision in birds: what is its function? *Vision Research*, **34**, 1471–8.

Berlioz, J. (1950). Systematique. In *Traité de zoologie*, Vol. 15. *Oiseaux*, (ed. P. P. Grassé), pp. 845–1055. Masson, Paris.

Beruldsen, G. (1980). *A field guide to nests and eggs of Australian birds*. Rigby, Adelaide.

Beruldsen, G. (1992). Another Queensland locality for the Carpentarian Grasswren. *Sunbird*, **22**, 49–50.

Blakers, M., Davies, S. J. J. F., and Reilly, P. N. (1984) *The atlas of Australian birds*. RAOU and Melbourne University Press, Melbourne.

BMR Palaeogeographic Group (1990). *Australia: evolution of a continent*. Bureau of Mineral Resources, Canberra.

Bock, W. J. (1960). The palatine process of the premaxilla in the Passeres. *Bulletin of the Museum of Comparative Zoology, Harvard*, **122**, 361–488.

Boehm, E. F. (1957). Perching birds (Passeriformes) of the Mount Mary Plains, South Australia. *Emu*, **57**, 311–24.

Boles, W. E. (1991). The origin and radiation of Australasian birds: perspectives from the fossil record. *Acta XX Congressus Internationalis Ornithologici*, 383–91.

Boles, W. E. (1993). A logrunner *Orthonyx* (Passeriformes: Orthonychidae) from the Miocene of Riversleigh, north-western Queensland. *Emu*, **93**, 44–9.

Boles, W. E. (1995). The world's oldest songbird. *Nature, London*, **374**, 21–2.

Boles, W. E. (1997). Fossil songbirds (Passeriformes) from the Early Eocene of Australia. *Emu*, **97**, 42–9.

Boles, W. E. and Dingley, M. (1977). A white-backed White-winged Wren. *Australian Birds*, **12**, 10.

Bonaparte, C. L. J. L. (1854). Notes ornithologique sur les collections rapportees en 1853, par M. A. Delattre, et classification parallelique des Passereaux chanteurs. *Comptes Rendus de l'Academie des Sciences, Paris*, **38**, 650–5.

Bradley, E. and Bradley, J. (1958). Notes on the behaviour and plumage of colour-ringed blue wrens. *Emu*, **58**, 313–26.

Bridges, L. (1994*a*). Breeding biology of a migratory population of the Rufous Whistler *Pachycephala rufuventris*. *Emu*, **94**, 106–15.

Bridges, L. (1994*b*). Territory and mate fidelity in a a migratory population of the Rufous Whistler *Pachycephala rufuventris*. *Emu*, **94**, 156–65.

Brooker, L. C. and Brooker, M. G. (1995). A model for the effects of fire and fragmentation on the population viability of the Splendid Fairy-wren. *Pacific Conservation Biology*, **1**, 344–58.

Brooker, M. G. (1988). Some aspects of the biology and conservation of the Thick-billed Grasswren *Amytornis textilis* in the Shark Bay area, Western Australia. *Corella*, **12**, 101–8.

Brooker, M. G. and Brooker, L. C. (1987). Description of some neonatal passerines in Western Australia. *Corella*, **11**, 116–8.

Brooker, M. G. and Brooker, L. C. (1989*a*). Cuckoo hosts in Australia. *Australian Zoological Reviews*, **No. 2**, 1–67.

Brooker, M. G. and Brooker, L. C. (1989*b*). The comparative breeding behaviour of two sympatric cuckoos, Horsfield's Bronze-Cuckoo *Chrysococcyx basalis* and the Shining Bronze-Cuckoo *C. lucidus*, in Western Australia: a new model for the evolution of egg morphology and host specificity in avian brood parasites. *Ibis*, **131**, 528–47.

Brooker, M. G. and Rowley, I. (1995). The significance of territory size and quality in the mating strategy of the Splendid Fairy-wren. *Journal of Animal Ecology*, **64**, 614–27.

Brooker, M. G., Ridpath, M. G., Estbergs, A. J., Bywater, J., Hart, D. S., and Jones, M. S. (1979). Bird observations on the north-western Nullabor Plain and neighbouring regions, 1967–1978. *Emu*, **79**, 176–90.

Brooker, M. G., Brooker, L. C., and Rowley, I. (1988). Egg deposition by the bronze-cuckoos *Chrysococcyx basalis* and *Ch. lucidus*. *Emu*, **88**, 107–9.

Brooker, M. G., Rowley, I, Adams, M., and Baverstock, P. R. (1990). Promiscuity: an inbreeding avoidance mechanism in a socially monogamous species? *Behavioral Ecology and Sociobiology*, **26**, 191–9.

Brosset, A. (1990). A long-term study of the rainforest birds in M'Passa (Gabon). In *Biogeography and ecology of forest bird communities*, (ed. A. Keast), pp. 259–74. SPB Academic Publishing bv, The Hague.

Brown, J. L. (1975). *The evolution of behavior*. W. W. Norton, New York.

Brown, J. L. (1987). *Helping and communal breeding in birds: ecology and evolution*. Princeton University Press, Princeton, New Jersey.

Brown, R. J., Brown, M. N., and Russell, E. M. (1990). Survival of four species of passerine in Karri forest in southwestern Australia. *Corella*, **14**, 69–78.

Burrett, C., Duhig, N., Berry, R., and Varne, R. (1991). Asian and south-western Pacific continental terranes derived from Gondwana, and their biogeographic significance. *Australian Systematic Botany*, **4**, 13–24.

Calder, W. A. (1984). *Size, function and life history*. Harvard University Press, Cambridge, Massachusetts.

Cale, P. (1994). Temporal changes in the foraging behaviour of insectivorous birds in a sclerophyll forest in Tasmania. *Emu*, **94**, 116–26.

Calver, M. C. and Wooller, R. D. (1981). Seasonal differences in the diets of small birds in the Karri forest understorey. *Australian Wildlife Research*, **8**, 653–7.

Calver, M. C. and Wooller, R. D. (1982). A technique for assessing the taxa, length, dry weight and energy content of the arthropod prey of birds. *Australian Wildlife Research*, **9**, 293–301.

Campbell, A. G. (1927). The genus *Amytornis*: a review. *Emu*, **27**, 23–35.

Campbell, A. J. (1899*a*). Provisional description of a new emu-wren. *Victorian Naturalist*, **15**, 116.

Campbell, A. J. (1899*b*). Description of a new emu-wren. *Ibis* series 7, **5**, 399.

Campbell, A. J. (1901*a*). *Nests and eggs of Australian birds including the geographical distribution of the species and popular observations thereon*. 2 volumes. A. J. Campbell, Melbourne.

Campbell, A. J. (1901*b*). On a new species of blue wren from King Island, Bass Strait. *Ibis* series 8, **1**, 10–1.

Campbell, A. J. (1901*c*). Description of a new wren or *Malurus*. *Victorian Naturalist*, **17**, 203–4.

Campbell, A. J. (1902). Notes on certain Maluri, with a description of a new species. *Emu*, **1**, 65–7.

Campbell, A. J. (1908). Description of a new emu-wren. *Emu*, **8**, 134–5.

Campbell, A. J. (1912). Western Emu-wren. *Emu*, **11**, 222.

Campbell, A. J. and Kershaw, J. A. (1913). Notes on a small collection of bird skins from the Northern Territory. *Emu*, **12**, 274–8.

Carnaby, I. C. (1954). Nesting seasons of Western Australian birds. *Western Australian Naturalist*, **4**, 149–56.

Carruthers, R. K., Horton, W., and Vernon, D. P. (1970). Distribution, habits and sexual dimorphism of the Western Grass-wren *Amytornis textilis ballarae* Condon in north-western Queensland. *Memoirs of the Queensland Museum*, **15**, 335–41.

Carter, T. (1903). Birds occurring in the region of the North-West Cape. *Emu*, **3**, 30–8.

Carter, T. (1916). On behalf of Mr. Tom Carter, Mr. G. M. Mathews sent the following descriptions of new subspecies of Australian birds. *Bulletin of the British Ornithologists' Club*, **37**, 6–7.

Carter, T. (1917). The birds of Dirk Hartog Island and Peron Peninsula, Shark Bay, Western Australia 1916–17. With nomenclature and remarks by Gregory M. Mathews. *Ibis* series 10, **5**, 564–611.

Castles, I. (1994). *Year book Australia*. Australian Bureau of Statistics, Canberra.

Catchpole, C. K. (1982). The evolution of bird sounds in relation to mating and spacing behavior. In *Acoustic Communication in Birds*. Vol 1. *Production, perception, and design features of sounds*, (ed. D. E. Kroodsma and E. H. Miller), pp. 297–319. Academic Press, New York.

Cayley, N. W. (1949). *The fairy wrens of Australia*. Angus and Robertson, Sydney.

Chapman, G. S. (1996). The Grasswrens—a brief pictorial. *Wingspan*, **6**, 20–8.

Chandler, L. G. (1940). Notes on the Striated Grass-wren. *Emu*, **39**, 245–6.

Chisholm, A. H. (1946). Observations and reflections on birds of the Victorian mallee. *Emu*, **46**, 168–86.

Chisholm, A. H. and Cayley, N. W. (1929). The birds of Port Stephens, N. S. W. *Emu*, **28**, 243–51.

Christidis, L. and Boles, W. E. (1994). The taxonomy and species of birds of Australia and its territories. *RAOU Monograph*, **2**, 1–112.

Christidis, L. and Schodde, R. (1991). Relationships of Australo-Papuan song-birds: protein evidence. *Ibis*, **133**, 277–85.

Christidis, L. and Schodde, R. (in press) Phylogeny of malurids. *Australian Journal of Zoology*.

Clarke, M. F. (1995). Co-operative breeding in Australasian birds: a review of hypotheses and evidence. *Corella*, **19**, 73–90.

Clarke, M. F. and Heathcote, C. F. (1990). Dispersal, survivorship and demography in the co-operatively breeding Bell Miner *Manorina melanophrys*. *Emu*, **90**, 15–23.

Coate, K. H. (1994). Another instance of Thick-billed Grass-wrens hiding in burrows. *Australian Bird Watcher*, **15**, 278–9.

Coates, B. J. (1973). No title. *New Guinea Bird Society Newsletter*, **83**, 3.

Coates, B. J. (1985). *The birds of Papua New Guinea*. Vol. 1. Dove Publications, Alderley, Qld.

Coates, B. J. (1990). *The birds of Papua New Guinea*. Vol. 2. Dove Publications, Alderley, Qld.

Cockburn, A. (1991). *An introduction to evolutionary ecology*. Blackwell, Oxford.

Cockburn, A. (1996). Why do so many Australian birds cooperate? Social evolution in the Corvida. In *Frontiers in population ecology* (ed. R. Floyd, A. Sheppard, and P. de Barro). pp. 21–42. CSIRO, Melbourne.

Cole, J. R. and Gibson, D. F. (1987). The Eyrean Grasswren *Amytornis goyderi* in the Northern Territory. *South Australian Ornithologist*, **30**, 57–9.

Collar, N. J., Crosby, M. J., and Stattersfield, A. J. (1994). *Birds to watch 2: The world list of threatened birds*. Birdlife conservation series Number 4. Birdlife International, Cambridge.

Collins, H. J. (1934). Some notes on the birds of the Nullabor Plain. *South Australian Ornithologist*, **12**, 198–201.

Condon, H. C. (1951). Notes on the birds of South Australia: occurrence, distribution and taxonomy. *South Australian Ornithologist*, **20**, 26–8.

Condon, H. C. (1962). A handlist of the birds of South Australia with annotations. *South Australian Ornithologist*, **23**, 85–151.

Condon, H. C. (1969*a*). A new subspecies of the Western Grass-wren *Amytornis textilis* (Dumont) in north-western Queensland. *Memoirs of the Queensland Museum*, **15**, 205–6.

Condon, H. C. (1969*b*). *A handlist of the birds of South Australia*. (2nd edn.) South Australian Ornithological Society, Adelaide.

Conover, M. R. and Miller, D. E. (1980). Rictal bristle function in Willow Flycatcher. *Condor*, **82**, 469–71.

Cook, J. (1784). *Voyage to the Pacific Ocean..... Performed under the direction of Captains Cook, Clerke and Gore, in His Majesty's Ships the Resolution and Discovery. In the years 1776, 1777, 1778, 1779 and 1780*. 3 volumes. Lords Commissioners at the Admiralty, Dublin.

Cooney, R. and Cockburn, A. (1995). Territorial defence is the major function of female song in the superb fairy-wren, *Malurus cyaneus*. *Animal Behaviour*, **49**, 1635–47.

Cooper, R. P. (1969). Multiple feeding habits of wrens. *Australian Bird Watcher*, **3**, 145–50.

Cox, J. B. (1974). Some birds of north-eastern Eyre Peninsula, S. A. *South Australian Ornithologist*, **26**, 142–4.

Cox, J. B. (1976). Grey Grasswrens and Grass Owls at Goyder's Lagoon, South Australia. *South Australian Ornithologist*, **27**, 96–100.

Cracraft, J. (1972). Continental drift and Australian avian biogeography. *Emu*, **72**,171–4.

Cracraft, J. (1981). Toward a phylogenetic classification of the recent birds of the world (Class Aves). *Auk*, **98**, 681–704.

Cracraft, J. (1986). Origin and evolution of continental biotas: speciation and historical congruence within the Australian avifauna. *Evolution*, **40**, 977–96.

Curry, P. J. (1986). Habitat characteristics of the Thick-billed Grass-wren *Amytornis textilis* in grazed shrubland in Western Australia. In *Rangelands: a resource under seige*, (ed. P. J. Joss, P. W. Lynch, and O. B. Williams), pp. 566–70. Australian Academy of Science, Canberra.

Darlington, P. J. (1957). *Zoogeography: the geographical distribution of animals*. John Wiley, New York.

Darwin, C. (1859). *On the origin of species by means of natural selection*. John Murray, London.

Davies, T. (1798). An account of a new species of *Muscicapa*, from New South Wales. *Transactions of the Linnean Society, London*, **4**, 207–10.

Diamond, J. M. (1972). Avifauna of the eastern highlands of New Guinea. *Publications of the Nuttall Ornithological Club*, **12**, 1–438.

Diamond, J. M. (1981). Distribution, habits and nest of *Chenorhamphus grayi*, a malurid endemic to New Guinea. *Emu*, **81**, 97–100.

Dow, D. D. (1980). Communally breeding Australian birds, with an analysis of distributional and environmental factors. *Emu*, **80**, 121–40.

Dumont, C. H. F. (1824). *Dictionnaire des sciences naturelles*, **30**, 117–8. Levrault, Paris.

Dunn, P. O., Cockburn, A., and Mulder, R. A. (1995). Fairy-wren helpers often care for young to which they are unrelated. *Proceedings of the Royal Society of London, Series B*, **259**, 339–43.

Eckert, J. (1977). The distribution of the emu-wrens *Stipiturus malachurus* and *S. ruficeps mallee* in South Australia. *South Australian Ornithologist*, **27**, 186–7.

Edwards, S. V. and Naeem, S. (1993). The phylogenetic component of cooperative breeding in perching birds. *American Naturalist*, **141**, 754–89.

Ellis, W. W. (1782). *An authentic narrative of a voyage performed by Captain Cook and Capt. Clerke in His Majesty's Ships Resolution and Discovery during the years 1776, 1777, 1778, 1779 and 1780.* Vol. 1. G. Robinson, J. Sewell and J. Debrett, London.

Emison, W. B., Beardsell, C. M., Norman, F. I., and Loyn, R. H. (1987). *Atlas of Victorian birds* . Department of Conservation, Forests and Lands and RAOU, Melbourne.

Emlen, S. T. (1978). Cooperative breeding. In *Behavioural ecology: an evolutionary approach*, (ed. J. R. Krebs and N. B. Davies), pp. 245–81. Blackwell, Oxford.

Emlen, S. T. (1982). The evolution of helping. I. An ecological constraints model. *American Naturalist*, **119**, 29–39.

Emlen, S. T. (1991). Evolution of cooperative breeding in birds and mammals. In *Behavioural ecology: an evolutionary approach*. 3rd edn. (ed. J. R. Krebs and N. B. Davies), pp. 301–37. Blackwell, Oxford.

Ewing, T. J. (1841). A catalogue of the birds of Tasmania. *Tasmanian Journal of Natural Science*, **1**, 52–8.

Favaloro, N. J. (1931). Notes on a trip to the Macpherson Range, south-eastern Queensland. *Emu*, **31**, 48–59.

Favaloro, N. J. and McEvey, A. (1968). A new species of Australian grass-wren. *Memoirs of the National Museum of Victoria*, **28**, 1–9.

Feduccia, A. (1995). Explosive radiation in Tertiary birds and mammals. *Science*, **267**, 637–8.

Filewood, L. W. (1970). Observations. *New Guinea Bird Society Newsletter*, **56**, 1.

Filewood, L. W. (1971). A New Guinea jungle banding station. *Australian Bird Bander*, **9**, 3–7.

Finch, B. W. (1982). Birds of Port Moresby district. *New Guinea Bird Society Newsletter*, **36**, 197–8.

Fitzpatrick, J. W. (1980). Foraging behavior of neotropical tyrant flycatchers. *Condor*, **82**, 43–57.

Fitzpatrick, J. W. (1985). Form, foraging behavior and adaptive radiation in the Tyrannidae. In *Neotropical ornithology*, (ed. P. A. Buckley, M. S. Foster, E. S. Morton, R. S. Ridgely, and F. C. Buckley), pp. 447–70. Ornithological Monographs No. 36, American Ornithologists' Union, Washington, D. C.

Fleming, M. and Strong, B. (1990). The discovery of the Carpentarian Grasswren—the original location. *South Australian Ornithologist*, **31**, 50–3.

Fletcher, J. A. (1913). Field notes on the emu-wren (*Stipiturus malachurus*). *Emu*, **12**, 168–70.

Fletcher, J. A. (1915). Further field notes on the emu-wren (*Stipiturus malachurus*). *Emu*, **14**, 213–7.

Ford, H. A., Noske, S., and Bridges, L. (1986). Foraging birds in eucalypt woodland in north-eastern New South Wales. *Emu*, **86**, 168–79.

Heinsohn, R. (1995). Hatching asynchrony and brood reduction in cooperatively breeding White-winged Choughs. *Emu*, **95**, 252–8.

Hill, G. F. (1913). Ornithological notes, Barclay expedition. *Emu*, **12**, 238–62.

Hindwood, K. A. (1931). Historical associations and early records of the emu-wren. *Emu*, **31**, 99–110.

Hindwood, K. A. (1945). The Eyrean Grass-wren (*Amytornis goyderi*). A third specimen. *Emu*, **44**, 321–3.

Hindwood, K. A. (1948). The use of flower petals in courtship display. *Emu*, **47**, 389–91.

Hindwood, K. A. (1970). The 'Watling' drawings, with incidental notes on the 'Lambert' and the 'Latham' drawings. *Proceedings of the Royal Zoological Society of New South Wales* for **1968-69**, 16–32.

Hitchcock, W. B. and Jarman, H. E. A. (1944). Bird observations in the Alice Springs district, Northern Territory. *South Australian Ornithologist*, **17**, 12–7.

Hope, J. (1982). Late Cainozoic vertebrate faunas and the development of aridity in Australia. In *Evolution of the flora and fauna of arid Australia*, (ed. W. R. Barker and P. J. M. Greenslade), pp. 85–100 . Peacock Publications, Adelaide.

Howard, E. and Howard, M. (1984). Nesting and observations of the Striated Grasswren at Wittenoom. *Western Australian Naturalist*, **16**, 21.

Howard, M. (1986). Notes on birds of the Wittenoom area, Western Australia. *Australian Bird Watcher*, **11**, 247–57.

Howe, F. E. (1933). The Mallee Emu-wren (*Stipiturus mallee*). *Emu*, **32**, 266–9.

Hoyt, D. F. (1979). Practical methods of estimating volume and fresh weight of bird eggs. *Auk*, **96**, 73–7.

Hutton, R. (1991). *Australian softbill management.* Singil Press, Austral, NSW.

Iredale, T. (1939). The eclipse plumage of the Elfin Wren (*Ryania melanocephala*). *Emu*, **39**, 39–40.

Iredale, T. (1956). *Birds of New Guinea*. Vol. 2. Georgian House, Melbourne.

IUCN. (1994). *IUCN Red list categories*. IUCN, Gland, Switzerland.

Izzard, J., Jenkins, V., and Miller, R. (1973). Further notes on the Striated Grass-Wren in New South Wales. *Australian Birds*, **8**, 51–2.

Jacobs, S. W. L. (1982). Relationships, distribution and evolution of *Triodia* and *Plectrachne* (Gramineae). In *Evolution of the flora and fauna of arid Australia*, (ed. W. R. Barker and P. J. M. Greenslade), pp. 287–90. Peacock Publications, Adelaide.

Jamieson, I. G. and Craig, J. L. (1987). Critique of helping behavior in birds: a departure from functional explanations. In *Perspectives in ethology*, Vol. 7, (ed. P. Bateson and P. Klopfer), pp. 79–98. Plenum Press, New York.

Johnstone, R. E. and Smith, L. A. (1981). Birds of Mitchell Plateau and adjacent coasts and lowlands. In *Biological survey of Mitchell Plateau and Admiralty Gulf, Western Australia*, pp. 171–212. Western Australian Museum, Perth.

Joseph, L. (1982). A further population of the Grey Grasswren. *Sunbird*, **12**, 51–3.

Jurisevic, M. A. and Sanderson, K. J. (1994). Alarm vocalisations in Australian birds: convergent characteristics and phylogenetic differences. *Emu*, **94**, 69–77.

Karr, J. R., Nichols, J. D., Klimkiewicz, M. K., and Brawn, J. D. (1990). Survival rates of birds of tropical and temperate forests: will the dogma survive? *American Naturalist*, **136**, 277–91.

Keartland. G. A. (1904). Ornithological notes—the grass wren. *Victorian Naturalist*, **20**, 133–9.

Keast, A. (1957). Variation in the Australian emu-wrens (*Stipiturus*). *Proceedings of the Royal Zoological Society of New South Wales* 1955–56. 47–53.

Keast, A. (1958). Speciation in the genus *Amytornis* Stejneger (Passeres: Muscicapidae, Malurinae) in Australia. *Australian Journal of Zoology*, **6**, 33–52.

Keast, A. (1959). Australian birds: their zoogeography and adaptations to an arid continent. In *Biogeography and ecology in Australia*, (ed. A. Keast, R. L. Crocker, and C. S. Christian), pp. 115–35. W. Junk, The Hague.

Keast, A. (1961). Bird speciation on the Australian continent. *Bulletin of the Museum of Comparative Zoology, Harvard*, **123**, 303–495.

Keast, A. (1975). Zonal feeding in the birds of Culeenup Island, Yunderup. *Western Australian Naturalist*, **13**, 25–9.

Keast, A. (1994). Temporal vocalisation patterns in members of a eucalypt forest bird community: the effects of weather on song production. *Emu*, **94**, 172–80.

Keast, A. (1996). Wing shapes in insectivorous passerines inhabiting New Guinea and Australian rainforests and eucalypt forest/eucalypt woodlands. *Auk*, **113**, 94–104.

Kinghorn, J. R. and Iredale, T. (1924). Rediscovery of the White-backed Wren *Malurus leuconotus*, Gould. *Emu*, **24**, 59–60.

Kinnear, N. B. (1949). Gregory M. Mathews C. B. E. *Ibis*, **91**, 521–4.

Koenig, W. D. and Mumme, R. L. (1987). *Population ecology of the cooperatively breeding acorn woodpecker*. Princeton University Press, Princeton, New Jersey.

Koenig, W. D., Pitelka, F. A., Carmen, W. J., Mumme, R. L., and Stanback, M. T. (1992). The evolution of delayed dispersal in cooperative breeders. *Quarterly Review of Biology*, **67**, 111–50.

Komdeur, J. (1992). Importance of habitat saturation and territory quality for evolution of cooperative breeding and the Seychelles warbler. *Nature, London*, **358**, 493–5.

Kroodsma, D. E. (1977). Correlates of song organisation among North American wrens. *American Naturalist*, **111**, 995–1008.

Lack, D. (1968). *Ecological adaptations for breeding in birds*. Methuen, London.

Langmore, N. E. and Mulder, R. A. (1992). A novel context for bird song: predator calls prompt male singing in the kleptogamous Superb Fairy-wren *Malurus cyaneus*. *Ethology*, **90**, 143–53.

Latham, J. (1783). *A general synopsis of birds*. Vol. 2. Benj. White, London.

Latham, J. (1801). *Supplementum indicis ornithologici*. London.

Laurance, W. F. and Grant, J. D. (1994). Photographic identification of ground-nest predators in Australian tropical rainforest. *Wildlife Research*, **21**, 241–8.

Lavery, H. J., Seton, D., and Bravery, J. A. (1968). Breeding seasons of birds in north-eastern Australia. *Emu*, **68**, 133–47.

Lea, A. M. and Gray, J. T. (1935). The food of Australian birds. An analysis of stomach contents. Part 3. *Emu*, **35**, 145–78.

LeCroy, M. and Diamond, J. (1995). Plumage variation in the Broad-billed Fairy-wren *Malurus grayi*. *Emu*, **95**, 185–93.

Le Soeuf, A. S. (1927). Birds of the Nullabor Plain. *Emu*, **27**, 195–7.

Lesson, R. P. (1831). *Traité d'ornithologie*. 2 Volumes and 8 livraisons. (Genus *Stipiturus*, livr.6, p. 414). Levrault, Paris.

Lewin, J. W. (1808). *Birds of New Holland and their natural history*. Vol. 1. J. White and S. Bagster, London.

Ligon, J. D. (1993). The role of phylogenetic history in the evolution of contemporary avian mating and parental care systems. In *Current ornithology*, Vol. 10, (ed. D. M. Power), pp. 1–46. Plenum Press, New York.

Ligon, J. D., Ligon, S. H., and Ford, H. A. (1991). An experimental study of the bases of male philopatry in the cooperatively breeding Superb Fairy-wren *Malurus cyaneus*. *Ethology*, **87**, 134–48.

Loaring, W. H. (1948). Splendid Wren with flower petal. *Emu*, **48**, 163–4.

Macdonald, J. D. and Colston, P. R. (1966). J. R. Elsey and his bird observations on Gregory's overland expedition, Australia, 1856. *Emu*, **65**, 255–78.

Macgillivray, W. D. (1914). Notes on some north Queensland birds. *Emu*, **13**, 132–86.

Mack, G. (1934). A revision of the genus *Malurus*. *Memoirs of the National Museum of Victoria*, **8**, 100–25.

Major, R. E. (1991*a*). Breeding biology of the White-fronted Chat *Ephthianura albifrons* in a saltmarsh near Melbourne. *Emu*, **91**, 236–49.

Major, R. E. (1991*b*). Identification of nest predators by photography, dummy eggs and adhesive tape. *Auk*, **108**, 190–5.

Marchant, S. (1974). Analysis of nest-records from the Willie Wagtail. *Emu*, **74**, 149–60.

Marchant, S. (1980). Incubation and nestling periods of some Australian birds. *Corella*, **4**, 30–2.

Marchant, S. (1984). Nest-records of the Eastern Yellow Robin *Eopsaltria australis*. *Emu*, **84**, 167–74.

Marchant, S. (1985). Breeding of the Eastern Yellow Robin, *Eopsaltria australis*. In *Birds of eucalypt forest and woodlands: ecology, conservation and management*, (ed. A. Keast, H. F. Recher, H. Ford, and D. A. Saunders), pp. 231–40. Surrey Beatty, Sydney.

Marchant, S. (1986). Commentary. Long laying intervals. *Auk*, **103**, 247.

Marchant, S. (1992). A bird observatory at Moruya, N.S.W., 1975–84. *Eurobodalla Natural History Society, Occasional Publication* No. **1**, 1–99.

Marchant, S. and Fullagar, P. (1983). Nest records of the Welcome Swallow. *Emu*, **83**, 66–74.

Martin, T. E. (1995). Avian life history evolution in relation to nest sites, nest predation, and food. *Ecological Monographs*, **65**, 101–27.

Mathews, G. M. (1908*a*). Handlist of the birds of Australia. *Emu*, **7**, Supplement, 1–108.

Mathews, G. M. (1908*b*). *Malurus dulcis* sp.n. *Bulletin of the British Ornithologists' Club*, **21**, 100–1.

Mathews, G. M. (1909). *Malurus dulcis* (Lavender-flanked wren) sp. nov. *Emu*, **8**, 113–4.

Mathews, G. M. (1910). *Amytornis whitei,* subsp. n. *Bulletin of the British Ornithologists' Club*, **25**, 34.

Mathews, G. M. (1911*a*). No title. *Bulletin of the British Ornithologists' Club*, **27**, 48.

Mathews, G. M. (1911*b*). No title. *Bulletin of the British Ornithologists' Club*, **27**, 99–101.

Mathews, G. M. (1912*a*). A reference-list to the birds of Australia. *Novitates zoologicae*, **18**, 171–656.

Mathews, G. M. (1912*b*). Additions and corrections to my Reference List to the Birds of Australia. *Austral Avian Record*, **1**, 25–52.

Mathews, G. M. (1912*c*). New generic names for Australian birds. *Austral Avian Record*, **1**, 105–17.

Mathews, G. M. (1913). *A list of the birds of Australia*. Witherby and Co., London.

Mathews, G. M. (1914). Additions and corrections to my list of the birds of Australia. *Austral Avian Record*, **2**, 83–107.

Mathews, G. M. (1916). List of additions of new subspecies to, and changes in, my list of the birds of Australia. *Austral Avian Record*, **3**, 25–68.

Mathews, G. M. (1917). The re-discovery of two lost birds. *Austral Avian Record*, **3**, 79–90.

Mathews, G. M. (1918*a*). Birds of the north and north-west of Australia. *South Australian Ornithologist*, **3**, 174–80.

Mathews, G. M. (1918*b*).Additions and corrections to my 1913 list. *Austral Avian Record*, **3**, 159–60.

Mathews, G. M. (1919). No title. *Bulletin of the British Ornithologists' Club*, **40**, 44–45.

Mathews, G. M. (1922–3). *The birds of Australia*. Vol. 10. H. F. and G. Witherby, London.

Mathews, G. M. (1928). *Sipodotus* gen. nov. *Bulletin of the British Ornithologists' Club*, **48**, 83.

Mathews, G. M. (1930). *Systema avium australasianarum*, Part 2. British Ornithologists' Union, London.

May, I. A. (1977*a*). Recent re-discovery of the Eyrean Grasswren. *Emu*, **77**, 230–1.

May, I. A. (1977*b*). Sighting of the Rufous-crowned Emu-wren in the Simpson Desert. *South Australian Ornithologist*, **27**, 172.

May, I. (1982). Bird note. In Bird notes (B. Glover, compiler), *South Australian Ornithological Association Newsletter*, **102**, 11.

Mayr, E. (1941). *List of New Guinea birds*. American Museum of Natural History, New York.

Mayr, E.(1944*a*). Timor and the colonization of Australia by birds. *Emu*, **44**, 113–30.

Mayr, E.(1944*b*). The birds of Timor and Sumba. I. *Bulletin of the American Museum of Natural History*, **83**, 123–94.

Mayr, E. and Amadon, D. (1951). A classification of recent birds. *American Museum Novitates*, **1496**, 1–42.

Mayr, E. and Cottrell, G. W. (ed.) (1986). *Peters' checklist of birds of the world*. Vol. 11. Museum of Comparative Zoology, Cambridge, Massachusetts.

Mayr, E. and Gilliard, E. T. (1937). Birds of central New Guinea. Results of the American Museum of Natural History expeditions to New Guinea in 1950 and 1952. *Bulletin of the American Museum of Natural History*, **103**, 311–74.

Mayr, E. and Meyer de Schauensee, R. (1939). Zoological results of the Denison–Crockett South Pacific expedition for the Academy of Natural Sciences of Philadelphia 1937–1938. Part IV. Birds from North-west New Guinea. *Proceedings of the Academy of Natural Sciences of Philadelphia*, **91**, 97–144.

Mayr, E. and Rand, A. L. (1935). Results of the Archbold expeditions. 6. Twenty-four apparently undescribed birds from New Guinea and the d'Entrecasteaux Archipelago. *American Museum Novitates*, **814**, 1–17.

Mayr, E. and Rand, A. L. (1937). Results of the Archbold expeditions. 14. Birds of the 1933–1934 Papuan expedition. *Bulletin of the American Museum of Natural History*, **73**, 1–248.

Mayr, E. and Serventy, D. L. (1944). The number of Australian bird species. *Emu*, **44**, 33–40.

McAllan, I. A. W. (1987). Early records of the Thick-billed Grasswren *Amytornis textilis* and Striated Grasswren *Amytornis striatus* in New South Wales. *Australian Birds*, **21**, 33–43.

McLennan, C. H (1906). Nature notes. *The Argus*, Melbourne, 27 April, 1906.

McCulloch, E. M. (1975). Tongues of some passerine birds. *Australian Bird Watcher*, **6**, 1–3.

McGilp, J. N. (1921). Emu-wrens breeding at Mt. Compass, South Australia. *South Australian Ornithologist*, **6**, 41–2.

McGilp, J. N. (1942). No title. *South Australian Ornithologist*, **16**, 44.

McKean, J. and Martin, K. C. (1989). Distribution and status of the Carpentarian Grasswren *Amytornis dorotheae*. *Northern Territory Naturalist*, **11**, 12–19.

Mees, G. F. (1961). An annotated catalogue of a collection of bird-skins from West Pilbara, Western Australia. *Journal of the Royal Society of Western Australia*, **44**, 97–143.

Mellor, J. W. (1913). Description of a new grass-wren. *Emu*, **12**, 166–7.

Mellor, J. W. (1921). Description of a new wren. *South Australian Ornithologist*, **5**, 10.

Meyer, A. B. (1874). No title. *Sitzungsberichte der Mathematisch-Naturwissenschaftlichen Classe der Kaiserlichen Akademie der Wissenschaften, Wien*, **69**, 496.

Miller, R. (1973). The rediscovery of the Striated Grass-wren in New South Wales. *Australian Birds*, **8**, 9–11.

Milligan, A. W. (1902). Report on the birds collected. In *Report on exploration of the north-west Kimberley, 1901*, (ed. F. S. Brockman), pp. 52–9. West Australian Parliamentary Paper No. 2, Perth.

Milligan, A. W. (1903). Notes on a trip to the Stirling Range. *Emu*, **3**, 9–19.

Morgan, D. G., Morgan, L. R., Robinson, L. N., Robinson, P. A., and Ashton, D. H. (1961). The Eyrean Grass-wren (*Amytornis goyderi*). *Australian Bird Watcher*, **1**, 161–71.

Morris, A. K., McGill, A. K., and Holmes, G. (1981). *Handlist of birds in New South Wales*. New South Wales Field Ornithologists' Club, Sydney.

Mulder, R. A. (1992). Evolutionary ecology of the mating system of Superb Fairy-wrens. Unpublished Ph.D. thesis. Australian National University, Canberra.

Mulder, R. A. (1995). Natal and breeding dispersal in a cooperative, extra-group-mating bird. *Journal of Avian Biology*, **26**, 234–40.

Mulder, R. A. and Cockburn, A. (1993). Sperm competition and the reproductive anatomy of male Superb Fairy-wrens. *Auk*, **110**, 588–93.

Mulder, R. A. and Langmore, N. E. (1993). Dominant males punish short-term defection by helpers in superb fairy-wrens. *Animal Behaviour*, **45**, 830–3.

Mulder, R. A. and Magrath, M. J. L. (1994). Timing of pre-nuptial molt as a sexually selected indicator of male quality in Superb Fairy-wrens. *Behavioral Ecology*, **5**, 393–400.

Mulder, R. A., Dunn, P. O., Cockburn, A., Lazenby-Cohen, K. A., and Howell, M. J. (1994). Helpers liberate female fairy-wrens from constraints on extra-pair mate choice. *Proceedings of the Royal Society of London, Series* B, **255**, 223–9.

Nevill, S. (1994). Kimberley birding. *Wingspan*, **15**, 20–2.

Nias, R. C. (1984). Territory quality and group size in the Superb Fairy-wren *Malurus cyaneus*. *Emu*, **84**, 178–80.

Nias, R. C. (1987). Co-operative breeding in the Superb Fairy-wren *Malurus cyaneus*. Unpublished Ph.D. thesis. University of New England, Armidale, Australia.

Nias, R. C. and Ford, H. A. (1992). The influence of group size and habitat on reproductive success in the Superb Fairy-wren *Malurus cyaneus*. *Emu*, **92**, 238–43.

Nicholson, C. J. and Coates, B. J. (1975) Observations on the nesting of the White-shouldered Wren. *New Guinea Bird Society Newsletter*, **106**, 3–4.

Nix, H. A. (1976). Environmental control of breeding, post-breeding dispersal and migration of birds in the Australian region. *Proceedings of the 16th International Ornithological Congress, Canberra, 1974*, 272–305.

Noriega, J. I. and Chiappe, L. M. (1993). An early Miocene passeriform from Argentina. *Auk*, **110**, 936–8.

North, A. J. (1888). Notes on the nidification of *Rhipidura preissei*, Cabanis, and *Malurus pulcherrimus*, Gould. *Proceedings of the Linnean Society of New South Wales*, series 2, **3**, 414–5.

North, A. J. (1896). Aves. In *Report on the work of the Horn scientific expedition to central Australia*, Part 2, (Zoology), (ed. W. B. Spencer), pp. 53–111. Dulau and Co., London.

North, A. J. (1901*a*). Description of a new species of the genus *Malurus*. *Victorian Naturalist*, **18**, 29–30.

North, A. J. (1901*b*). Notes on *Malurus cyaneus* and *M. superba*, and on cuckoos. *Proceedings of the Linnean Society of N. S. W.*, **26**, 632.

North, A. J. (1902). On three apparently undescribed species of Australian birds. *Victorian Naturalist*, **19**, 101–4.

North, A. J. (1904). *Nests and eggs of birds found breeding in Australia and Tasmania*. Vol. 1, part 4. Australian Museum, Sydney.

Noske, R. (1991*a*). A demographic comparison of cooperatively breeding and non-cooperative treecreepers. *Emu*, **91**, 73–86.

Noske, R. (1991*b*). Good news on the White-throated Grasswren. *Wingspan*, **December 1991**, 14–5.

Noske, R. (1992). The status and ecology of the White-throated Grasswren *Amytornis woodwardi*. *Emu*, **92**, 39–51.

Oberholser, H. C. (1899). Some untenable names in ornithology. *Proceedings of the Academy of Natural Sciences, Philadelphia*, **51**, 201–16.

Officer, H. R. (1970). Rufous-crowned Emu-wren, *Stipiturus ruficeps*, a new Queensland species. *Sunbird*, **1**, 64–6.

Ogilvie-Grant, W. R. (1909). New species of birds. *Bulletin of the British Ornithologists' Club*, **23**, 72–4.

Olson, S. L. (1988). Aspects of global avifaunal dynamics during the Cenozoic. *Acta XIX Congressus Internationalis Ornithologici*, 2023–9.

Oustalet, J. F. E. (1878). No title. *Bulletin de l'Association Scientifique de la France*, n. **533**, 248.

Osborne, W. S. and Green, K. (1992). Seasonal changes in composition, abundance and foraging behaviour of birds in the Snowy Mountains. *Emu*, **92**, 93–105.

Paijmans, K. 1976. Vegetation. In *New Guinea vegetation*, (ed. K. Paijmans), pp. 23–105. Australian National University Press, Canberra.

Parker, S. A. (1972). Remarks on distribution and taxonomy of the grass wrens *Amytornis textilis, modestus* and *purnelli*. *Emu*, **72**, 157–66.

Parker, S. A. (1975). Maluridae. In *Interim list of Australian songbirds*, (ed. R. Schodde), pp. 1–46. RAOU, Melbourne.

Parker, S. A. (1980). Birds and conservation parks in the north-east of South Australia. *South Australian Parks and Conservation*, **3**, 11–4.

Parker, S. A. (1982*a*). Notes on *Amytornis striatus merrotsyi* Mellor, a subspecies of the Striated Grasswren inhabiting the Flinders Ranges. *South Australian Ornithologist*, **29**, 13–6.

Parker, S. A. (1982*b*). Remarks on the tympanic cavity of *Malurus*, *Stipiturus* and *Amytornis* (Passeriformes, Maluridae). *South Australian Ornithologist*, **29**, 17–22.

Parker, S. A. and Reid, N. (1978). Remarks on the status of some Australian passerines. In *The status of endangered Australian wildlife*, (ed. M. Tyler), pp. 109–15. Royal Zoological Society of South Australia, Adelaide.

Parker, S. A., May, I. A., and Head, W. (1978). Some observations on the Eyrean Grass-wren *Amytornis goyderi* (Gould, 1875). *Records of the South Australian Museum*, **17**, 361–71.

Parsons, F. E. (1920). New subspecies of emu-wren, *Stipiturus malachurus halmaturina*. Kangaroo Island Emu-wren. *South Australian Ornithologist*, **5**, 15–7.

Parsons, F. E. (1968). *Pterylography. The feather tracts of Australian birds with notes and observations. Occasional Papers in Zoology*, Number 1. Libraries Board of South Australia, Adelaide.

Parsons, F. E. and Cleland, J. B. (1926). Untitled. *South Australian Ornithologist*, **8**, 137–8.

Parsons, F. E. and McGilp, J. N. (1934). Notes taken on a trip to Panitya, Victoria. *South Australian Ornithologist*, **12**, 192–8.

Payne, R. B., Payne, L. L., and Rowley, I. (1985). Splendid wren *Malurus splendens* response to cuckoos: an experimental test of social organization in a communal bird. *Behaviour*, **94**, 108–27.

Payne, R. B., Payne, L. L., and Rowley, I. (1988). Kin and social relationships in splendid fairy-wrens: recognition by song in a cooperative bird. *Animal Behaviour*, **36**, 1341–51.

Payne, R. B., Payne, L. L., Rowley, I., and Russell, E. M. (1991). Social recognition and response to song in cooperative Red-winged Fairy-wrens. *Auk*, **108**, 811–9.

Phillip. A. (1789). *The voyage of Governor Phillip to Botany Bay*. Stockdale, London.

Pickett, M. (1995). Some observations on petal carrying and display by the Superb Fairy-wren *Malurus cyaneus*. *South Australian Ornithologist*, **32**, 64.

Pringle, J. D. (1982). *The wrens and warblers of Australia*. Angus and Robertson, Sydney.

Pruett-Jones, S.G. and Lewis, M. J. (1990). Habitat limitation and sex ratio promote delayed dispersal in Superb Fairy-wrens. *Nature, London*, **348**, 541–2.

Quoy, J. R. C. and Gaimard, J. P. (1830). *Voyage de découvertes de l'Astrolabe exécuté par ordre du roi, pendant les années 1826–1827–1828–1829, sous le commandement de M. J. Dumont d'Urville. Zoologie.* Vol. 1. Tastu, Paris.

Rahn, H., Paganelli, C. V., and Ar, A. (1975). Relation of avian egg weight to body weight. *Auk*, **92**, 750–65.

Ralls, K., Harvey, P. H., and Lyles, A. M. (1986). Inbreeding in natural populations of birds and mammals. In *Conservation biology: the science of scarcity and diversity*, (ed. M. Soulé), pp. 35–56. Sinauer, New York.

Rand, A. L. (1940). Results of the Archbold expeditions. 25. New birds from the 1938–39 expedition. *American Museum Novitates*, **1072**, 1–14.

Rand, A. L. (1942). Results of the Archbold expeditions. 42. Birds of the 1936–1937 New Guinea expedition. *Bulletin of the American Museum of Natural History*, **79**, 289–366.

Rand, A. L. and Gilliard, E. T. (1967). *Handbook of New Guinea birds*. Weidenfeld and Nicholson, London.

RAOU (1913). Official check-list of the birds of Australia. *Emu* **12**, Supplement, 1–116.

RAOU (1926). *The official checklist of the birds of Australia*. (2nd edn). RAOU, Melbourne.

RAOU (1978). Recommended English names for Australian birds. *Emu*, **77**, Supplement, 245–313.

Recher, H. F., Holmes, R. T., Schulz, M., Shields, J., and Kavanagh, R. (1985). Foraging patterns of breeding birds in eucalypt forest in southeastern Australia. *Australian Journal of Ecology*, **10**, 399–419.

Reichenow, A. (1897). No title. *Ornithologische Monatsbuch*, **5**, 25.

Reichenow, A. (1920). *Chenorhamphus pileatus* Rchw. n. sp. *Journal für Ornithologie (Leipzig)*, **68**, 399.

Reid, N., Paton, J. B., and Paton, D. C. (1977). Critical range limits of the Turquoise and Black-backed wrens in S. A. *South Australian Ornithologist*, **27**, 216–21.

Rich, P. V. (1975). Antarctic dispersal routes, wandering continents, and the origin of Australia's non-passeriform avifauna. *Memoirs of the National Museum of Victoria*, **36**, 63–124.

Richardson, K. C. and Wooller, R. D. (1986). The structures of the gastrointestinal tract of honeyeaters and other small birds in relation to their diets. *Australian Journal of Zoology*, **34**, 119–24.

Ripley, S. D. (1964). A systematic and ecological study of the birds of New Guinea. *Bulletin of the Peabody Museum of Natural History, Yale University*. No. 19, pp. 1–87.

Robinson, A. H. (1955). Nesting seasons of Western Australian birds—a further contribution. *Western Australian Naturalist*, **4**, 187–92.

Robinson, D. (1990*a*). The nesting ecology of sympatric Scarlet Robin *Petroica multicolor* and Flame Robin *P. phoenicea* in open eucalypt forest. *Emu*, **90**, 40–52.

Robinson, D. (1990*b*). The social organization of the Scarlet Robin *Petroica multicolor* and Flame Robin *P. phoenicea* in southeastern Australia: a comparison between sedentary and migratory flycatchers. *Ibis*, **132**, 78–94.

Robinson, L. (1973). The Grey Grass-wren. *Australian Bird Watcher*, **4**, 251–6.

Rosenberg, (1863). No title. *Naturkundig Tijdschrift voor Nederlands-Indie*, **25**, 231–3.

Rothschild, W. and Hartert, E. (1903). Notes on Papuan birds. *Novitates Zoologicae*, **10**, 435–80.

Rothschild, W. and Hartert, E. (1907). List of collections of birds made by A. S. Meek in the mountains on the upper Aroa River and on the Angabunga River, British New Guinea. *Novitates Zoologicae*, **14**, 447–83.

Rowley, I. (1957). Cooperative feeding of young by Superb Blue Wrens. *Emu*, **57**, 356–7.

Rowley, I. (1962). 'Rodent-run' distraction display by a passerine, the Superb Blue Wren *Malurus cyaneus* (L.). *Behaviour*, **19**, 170–6.

Rowley, I. (1963). The reaction of the Superb Blue Wren, *Malurus cyaneus*, to models of the same and closely related species. *Emu*, **63**, 207–14.

Rowley, I. (1965). The life history of the Superb Blue Wren. *Emu*, **64**, 251–97.

Rowley, I. (1968). Communal species of Australian birds. *Bonner Zoologische Beitrage*, **19**, 361–8.

Rowley, I. (1976). Cooperative breeding in Australian birds. *Proceedings of the 16th International Ornithological Congress*. 657–66.

Rowley, I. (1978). Communal activities among White-winged Choughs *Corcorax melanorhamphos*. *Ibis*, **120**, 178–97.

Rowley, I. (1981*a*). The communal way of life in the Splendid Fairy-wren *Malurus splendens*. *Zeitschrift für Tierpsychologie*, **55**, 228–67.

Rowley, I. (1981*b*). A relict population of Blue-breasted Wrens *Malurus pulcherrimus* in the central wheatbelt. *West Australian Naturalist*, **15**, 1–8.

Rowley, I. (1988). The Purple-crowned Fairy-wren *Malurus coronatus*: an RAOU Conservation Statement. *RAOU Report*, No. **34**, pp. 1–12.

Rowley, I. (1991). Petal-carrying by Fairy-wrens of the genus *Malurus*. *Australian Bird Watcher*, **14**, 75–81.

Rowley, I. (1993). The Purple-crowned Fairy-wren *Malurus coronatus*. I. History, distribution and present status. *Emu*, **93**, 220–34.

Rowley, I. and Brooker, M. G. (1987). The response of a small insectivorous bird to fire in heathlands. In *Nature conservation: the role of remnants of native vegetation*, (ed. D. A. Saunders, G. W. Arnold, A. Burbidge, and A. J. M. Hopkins), pp. 211–8. Surrey Beatty, Sydney.

Rowley, I. and Russell, E. (1990). Philandering—a mixed mating strategy in the Splendid Fairy-wren *Malurus splendens*. *Behavioral Ecology and Sociobiology*, **27**, 431–7.

Rowley, I. and Russell, E. (1991). Demography of passerines in the temperate Southern Hemisphere. In *Bird population studies: relevance to conservation and management*, (ed. C. M. Perrins, J.-D. Lebreton, and G. S. Hirons), pp. 22–44. Oxford University Press, Oxford.

Rowley, I. and Russell, E. (1993). The Purple-crowned Fairy-wren *Malurus coronatus*. II. Breeding biology, social organisation, demography and management. *Emu*, **93**, 235–50.

Rowley, I. and Russell, E. (1995). The breeding biology of the White-winged Fairy-wren *Malurus leucopterus leuconotus* in a Western Australian coastal heathland. *Emu*, **95**, 175–84.

Rowley, I., Russell, E. M., and Brooker, M. G. (1986). Inbreeding: benefits may outweigh costs. *Animal Behaviour*, **34**, 939–41.

Rowley, I., Russell, E. M., Brown, R. J., and Brown, M. N. (1988). The ecology and breeding biology of the Red-winged Fairy-wren *Malurus elegans*. *Emu*, **88**, 161–76.

Rowley, I., Russell, E., Payne, R. B., and Payne, L. L. (1989). Plural breeding in the Splendid Fairy-wren, *Malurus splendens* (Aves: Maluridae), a cooperative breeder. *Ethology*, **83**, 229–47.

Rowley, I., Brooker, M. G., and Russell, E. M. (1991). The breeding biology of the Splendid Fairy-wren *Malurus splendens* : the significance of multiple broods. *Emu*, **91**, 197–221.

Russell, E. M. (1989). Co-operative breeding—a Gondwanan perspective. *Emu*, **89**, 61–2.

Russell, E. M. and Rowley, I. (1988). Helper contributions to reproductive success in the Splendid Fairy-wren (*Malurus splendens*). *Behavioral Ecology and Sociobiology*, **22**, 131–40.

Russell, E. M. and Rowley, I. (1993*a*). Philopatry or dispersal: competition for territory vacancies in the splendid fairy-wren, *Malurus splendens*. *Animal Behaviour*, **45**, 519–39.

Russell, E. M. and Rowley, I. (1993*b*). The demography of the cooperatively-breeding Splendid Fairy-wren *Malurus splendens*. *Australian Journal of Zoology*, **41**, 475–505.

Russell, E. M. and Rowley, I. (1996). Partnerships in promiscuous fairy-wrens. In *Partnerships in birds*, (ed. J. M. Black), pp. 162–73. Oxford University Press, Oxford.

Russell, E. M. Rowley, I., Brown, R. J., and Brown, M. N. (1991). Acquisition of nuptial plumage in the Red-winged Fairy-wren *Malurus elegans*. *Corella*, **15**, 125–33.

Salvadori, T. (1881). *Ornitologia della Papuasia e della Molucche*, Vol. 2. T. Salvadori, Turin.

Saunders, D. A. and Ingram, J. A. (1995). *Birds of south-western Australia: an atlas of changes in the distribution and abundance of the wheatbelt avifauna.* Surrey Beatty, Sydney.

Schmidt-Nielsen, K. (1984). *Scaling:why is animal size so important?* Cambridge University Press, Cambridge.

Schodde, R. (1975). *Interim list of Australian songbirds.* R.A.O.U., Melbourne.

Schodde, R. (1982*a*). Origin, adaptation and evolution of birds in arid Australia. In *Evolution of the flora and fauna of arid Australia,* (ed. W. R. Barker and P. J. M. Greenslade), pp. 191–224. Peacock Publications, Adelaide.

Schodde, R. (1982*b*). *The fairy-wrens: a monograph of the Maluridae.* Lansdowne, Melbourne.

Schodde, R. (1984*a*). First specimens of Campbell's Fairy-wren, *Malurus campbelli,* from New Guinea. *Emu,* **84**, 249–50.

Schodde, R. (1984*b*). Remarkable bird life of Papua New Guinea. In *Vertebrate zoogeography and evolution in Australia,* (ed. M. Archer and G. Clayton), pp. 1143–6. Hesperian Press, Perth.

Schodde, R. (1991). Concluding remarks: origins and evolution of the Australasian avifauna. *Acta XX Congressus Internationalis Ornithologici,* 413–6.

Schodde, R. and Calaby, J. H. (1972). The biogeography of the Australo-Papuan bird and mammal faunas in relation to Torres Strait. In *Bridge and barrier: the natural and cultural history of Torres Strait,* (ed. D. Walker), pp. 257–99. Australian National University, Canberra.

Schodde, R. and Christidis, L. (1987). Genetic differentiation and subspeciation in the Grey Grasswren *Amytornis barbatus* (Maluridae). *Emu,* **87**, 188–92.

Schodde, R. and Faith, D. P. (1991). The development of modern avifaunulas. *Acta XX Congressus Internationalis Ornithologici,* 404–12.

Schodde, R. and Hitchcock, W. B. (1968). Contributions to Papuasian ornithology. 1. Report on the birds of the Lake Kutubu area, Territory of Papua New Guinea. *C. S. I. R. O. Division of Wildlife Research Technical Paper,* No. 13, pp. 1–73.

Schodde, R. and Hitchcock, W. B. (1972). Birds. In *Encyclopaedia of Papua New Guinea,* (ed. P. Ryan), pp. 67–86. Melbourne University Press, Melbourne.

Schodde, R. and Mason, I. J. (1975). Occurrence, nesting and affinities of the White-throated Grasswren *Amytornis woodwardi* and White-lined Honeyeater *Meliphaga albilineata. Emu,* **75**, 12–8.

Schodde, R. and Mason, I. J. (1980). *Nocturnal birds of Australia.* Lansdowne, Australia.

Schodde, R. and Tidemann, S. C. (ed.) (1988). *Readers Digest complete book of Australian birds.* (2nd edn). Readers Digest, Sydney.

Schodde, R. and Weatherley, R. G. (1981). A new subspecies of the Southern Emu-wren *Stipiturus malachurus* from South Australia, with notes on its affinities. *South Australian Ornithologist,* **28**, 169–70.

Schodde, R. and Weatherley, R. G. (1983). Contributions to Papuasian ornithology. 8. Campbell's Fairy-wren (*Malurus campbelli)* a new species from New Guinea. *Emu,* **82**, 308–9.

Serventy, D. L. (1950). Taxonomic trends in Australian ornithology—with special reference to the work of Gregory Mathews. *Emu,* **49**, 257–67.

Serventy, D. L. (1951). The evolution of the chestnut-shouldered wrens (*Malurus*). *Emu,* **51**, 113–20.

Serventy, D. L. (1953). Some speciation problems in Australian birds: with particular reference to relations between Bassian and Eyrean 'species pairs'. *Emu,* **53**, 231–45.

Serventy, D. L. (1972). Causal ornithogeography of Australia. *Proceedings 15th International Ornithological Congress, London,* 374–84.

Serventy, D. L. and Marshall, A. J. (1957). Breeding periodicity in Western Australian birds: with an account of unseasonal nestings in 1953 and 1955. *Emu,* **57**, 99–126.

Serventy, D. L. and Marshall, A. J. (1964). A natural history reconnaissance of Barrow and Montebello Islands, 1958. *C.S.I.R.O Divison of Wildlife Technical Paper,* No. 6, pp. 1–23.

Serventy, D. L. and Whittell, H. M. (1948). *The birds of Western Australia.* (1st edn). Paterson Brokensha, Perth, Western Australia.

Serventy, D. L. and Whittell, H. M. (1967). *The birds of Western Australia.* (4th edn). Paterson Brokensha, Perth, Western Australia.

Serventy, D. L. and Whittell, H. M. (1976). *The birds of Western Australia.* (5th edn). University of Western Australia Press, Perth.

Sharpe, R. B. (1874–95). *Catalogue of the birds in the British Museum.* Vol. 1–27. Trustees of the British Museum, London.

Sharpe, R. B. (1879*a*). On *Pseudogerygone rubra*, a remarkable new species of flycatcher, from the Arfak Mountains, North-western New Guinea. *Notes Leyden Museum*, **1**, 31.

Sharpe, R. B. (1879*b*). *Catalogue of the birds in the British Museum*. Vol. 4, Cinclomorphae, part 1. Trustees of the British Museum, London.

Sharpe, R. B. (1881). No title. *Proceedings of the Zoological Society of London*, **1881**, 788.

Sharpe, R. B. (1883). *Catalogue of the birds in the British Museum*. Vol. 7, Cinclomorphae, part 4. Trustees of the British Museum, London.

Sharpe, R. B. (1899–1909). *A handlist of the genera and species of birds*. 6 volumes. British Museum, London.

Sharpe, R. B. (1903). *A handlist of the genera and species of birds*, Vol. 4. British Museum, London.

Shaw, G. (1789). *The naturalist's miscellany*. Vol. 1. London.

Shaw, G. (1790). In *Journal of a voyage to New South Wales with sixty-five plates of nondescript animals, birds lizards, serpents, curious cones of trees and other natural productions*. (ed. J. White). J. Debrett, London.

Shaw, G. (1798). An appendix to *Account of a new species of* Muscicapa *from New South Wales* by Major-General Thomas Davies. *Transactions of the Linnean Society of London*, **4**, 242.

Sheldon, F. H. and Bledsoe, A. H. (1993). Avian molecular systematics, 1970s to 1990s. *Annual Review of Ecology and Systematics*, **24**, 243–78.

Sibley, C. G. (1970). A comparative study of the egg-white proteins of passerine birds. *Bulletin of the Peabody Museum of Natural History*, Yale University, No. **32**, pp. 1–131.

Sibley, C. G. (1976). Protein evidence of the relationships of some Australian passerine birds. *Proceedings of the 16th International Ornithological Congress*, 557–70.

Sibley, C. G. and Ahlquist, J. E. (1982). The relationships of the Australo-Papuan fairy-wrens as indicated by DNA–DNA hybridization. *Emu*, **82**, 251–5.

Sibley, C. G. and Ahlquist, J. E. (1985). The phylogeny and classification of the Australo-Papuan passerine birds. *Emu*, **85**, 1–14.

Sibley, C.G. and Ahlquist, J. E. (1990). *Phylogeny and classification of birds: a study in molecular evolution*. Yale University Press, New Haven.

Skutch, A. F. (1935). Helpers at the nest. *Auk*, **52**, 257–73.

Skutch, A. F. (1961). Helpers among birds. *Condor*, **63**, 198–226.

Skutch, A. F. (1985). Clutch size, nesting success, and predation on nests of neotropical birds, reviewed. In *Neotropical ornithology*, (ed. P. A. Buckley, M. S. Foster, E. S. Morton, R. S. Ridgely, and F. C. Buckley), pp. 575–94. American Ornithologists' Union, Kansas.

Stacey, P. B. and Ligon, J. D. (1987).Territory quality and dispersal options in the Acorn Woodpecker, and a challenge to the habitat saturation model of cooperative breeding. *American Naturalist*, **130**, 654–76.

Stacey, P. B. and Ligon, J. D. (1991). The benefits-of-philopatry hypothesis for the evolution of cooperative breeding: variation in territory quality and group size effects. *American Natrualist*, **137**, 831–46.

Stejneger, L. (1885). *Amytornis* gen. nov. In *The standard natural history*, (ed. J. S. Kingsley), Vol. **4**, p. 499.

Stephens, J. F. (1826). *General zoology; or Systematic natural history*. 14 volumes, 1800–1826. London. Volumes 1–8 by Shaw, Volumes 9-14 by J. F. Stephens. *Malurus gularis* in Vol. 13, p. 224

Storer, R. W. (1971). Adaptive radiation of birds. In *Avian Biology*, Vol. 1, (ed. D. S. Farner and J. R. King), pp. 149–88. Academic Press, New York.

Storr, G. M. (1973). List of Queensland birds. *Special Publications of the West Australian Museum*, No. **5**, pp. 1–177.

Storr, G. M. (1984). Birds of the Pilbara region of Western Australia. *Records of the Western Australian Museum*, Supplement No. **16**, pp. 1–63.

Storr, G. M. (1985). Birds of the Gascoyne Region, Western Australia. *Records of the Western Australian Museum*, Supplement No. **21**, pp. 1–66.

Storr, G. M. (1986). Birds of the south-eastern interior of Western Australia. *Records of the Western Australian Museum*, Supplement No. **26**, pp. 1–60.

Storr, G. M. (1987). Birds of the Eucla Division of Western Australia. *Records of the Western Australian Museum*, Supplement No. **27**, pp. 1–81.

Storr, G. M. (1991). Birds of the South-west Division of Western Australia. *Records of the Western Australian Museum*, Supplement No. **35**, pp. 1–150.

Stringer, M. (1979). Taxonomical ordering of birds by the Waffa. *Papua New Guinea Bird Society Newsletter*, **156**, 4–7.

Strong, M. and Cuffe, E. (1985). Petal display by the Variegated Wren. *Sunbird*, **15**, 71.

Sutton, J. (1929). A trip to the Murray Mallee. *South Australian Ornithologist*, **10**, 22–45.

Swainson, W. and Richardson, J. (1831). *Fauna boreali-americana*. Part 2. *The birds*. J. Murray, London.

Tarr, H. E. (1948). Birds of Dunk Island, North Queensland. *Emu*, **48**, 8–13.

Tarr, H. E. (1949). Concerning the Lovely Wren. *Emu*, **49**, 143.

Tidemann, S. C. (1980). Notes on breeding and social behaviour of the White-winged Fairy-wren *Malurus leucopterus*. *Emu*, **80**, 158–61.

Tidemann, S. C. (1983). The behavioural ecology of three coexisting fairy-wrens (Maluridae: *Malurus*). Unpublished Ph.D. thesis. Australian National University.

Tidemann, S. C. (1986). Breeding in three species of fairy-wrens (*Malurus*): do helpers really help? *Emu*, **86**, 131–8.

Tidemann, S. C. (1989). Acquisition of nuptial plumage in White-winged Fairy-wrens *Malurus leucopterus*. *Corella*, **13**, 15–7.

Tidemann, S. C. (1990). Factors affecting territory establishment, size and use by three co-existing species of fairy-wrens. *Emu*, **90**, 7–14.

Tidemann, S. C. and Marples, T. G. (1987). Periodicity of breeding behaviour of three species of fairy-wren (*Malurus* spp.). *Emu*, **87**, 73–7.

Tidemann, S. C., Green, B., and Newgrain, K. (1989). Water turnover and estimated food consumption in three species of fairy-wren (*Malurus* spp.) *Australian Wildlife Research*, **16**, 187–94.

Tomlinson, A. R. (1979). Living enemies of Agriculture II. Vermin (vertebrate pests). In *Agriculture in Western Australia 1829–1979*, (ed. G. H. Burvill), pp. 181–7. University of Western Australia Press, Perth.

Tordoff, H. B. (1954). A systematic study of the avian family Fringillidae, based on the structure of the skull. *Miscellaneous Publications, Museum of Zoology, University of Michigan*, No. **81**, 1–42.

Truswell, E. M., Kershaw, A. P., and Sluiter, I. R. (1987). The Australian–south-east Asian connection: evidence from the paleobotanical record. In *Biogeographical evolution of the Malay archipelago*, (ed. T. C. Whitmore), pp. 32–49. Oxford University Press, Oxford.

Tullis, K. J., Calver, M. C., and Wooller, R. D. (1982). The invertebrate diets of small birds in *Banksia* woodland near Perth, W. A., during winter. *Australian Wildlife Research*, **9**, 303–9.

Tuttle, E. M., Pruett-Jones, S., and Webster, M. S. (1996). Cloacal protruberances and extreme sperm production in Australian fairy-wrens. *Proceedings of the Royal Society of London*, Series B, **263**, 1359–64.

Veevers, J. G. (1991). Phanerozoic Australia in the changing configuration of Proto-Pangea through Gondwanaland and Pangea to the present dispersed continents. *Australian Systematic Botany*, **4**, 1–11.

Vevers, G. (1985). Article 'Colour'. In *A dictionary of birds*, (ed. B. Campbell and E. Lack), pp. 99–100. T. and A. D. Poyser, Calton.

Vernon, D. P. (1972). The Rufous-crowned Emu-wren, *Stipiturus ruficeps*, in Queensland. *Sunbird*, **3**, 34–5.

Vieillot, L. P. (1816). *Analyse d'une nouvelle ornithologie elementaire*. Deterville, Paris.

Vieillot, L. P. (1818). In *Noveau dictionnaire d'histoire naturelle appliquee aux arts, principalement a l'agriculture at a l'economie rurale et domestique par une societe de naturalistes at d'agriculteurs*. Nouvelle edition. Volume 20. Deterville, Paris.

Vigors, N. A. and Horsfield, T. (1827). A description of the Australian birds in the collection of the Linnean Society; with an attempt at arranging them according to their natural affinities. *Transactions of the Linnean Society, London*, **15**, 170–331.

Vuilleumier, F., LeCroy, M., and Mayr, E. (1992). New species of birds described from 1981 to 1990. *Bulletin of the British Ornithologists' Club Centenary Supplement*, **112A**, 267–309.

Wallace, A.R. (1862). Narrative of search after birds of paradise. *Proceedings of the Zoological Society of London*, **1862**, 153–66.

Warham, J. (1954). The behaviour of the Splendid Blue Wren. *Emu*, **54**, 135–40.

Webster, D. B. and Webster, M. (1975). Auditory systems of Heteromyidae: functional morphology and evolution of the inner ear. *Journal of Morphology*, **146**, 343–76.

Weiner, J. (1994). *The beak of the finch.* Jonathan Cape, London.

Wesolowski, T. (1994). On the origin of parental care and the early evolution of male and female parental roles in birds. *American Naturalist*, **143**, 39–58.

Wetmore, A. (1960). A classification for the birds of the world. *Smithsonian Miscellaneous Collections*, **139**, 1–37.

Whitaker, J. (1987). Some observations on the Carpentarian Grasswren. *Northern Territory Naturalist*, **10**, 14–5.

White, H. L. (1910). Description of two new nest and eggs from north-west Australia, with field notes by the collector, G. F. Hill. *Emu*, **10**, 132–4.

White, H. L. (1914). Description of new Australian birds' eggs. *Emu*, **14**, 57–9.

White, H. L. (1917). Description of the nest and eggs of the Rufous-crowned Emu-wren. *Emu*, **17**, 39.

White, S. R. (1946). Notes on the bird life of Australia's heaviest rainfall region. *Emu*, **46**, 81–122.

White, S. R. (1949). Notes on the Blue-and-white Wren (*Malurus leuconotus*). *Western Australian Naturalist*, **1**,162–8.

Whitlock, F. L. (1910). On the East Murchison. Four months' collecting trip. *Emu*, **9**, 181–219.

Whitlock, F. L. (1921). Notes on Dirk Hartog Island and Peron Peninsula, Shark Bay, Western Australia. *Emu*, **20**, 168–86.

Whitlock, F. L. (1922). Notes from the Nullabor Plain. *Emu*, **21**, 170–80.

Whitlock, F. L. (1924). Journey to central Australia in search of the Night Parrot. *Emu*, **23**, 248–81.

Wigan, M. L. (1932). The Black-backed Wren. *Emu*, **32**, 115–6.

Wilkes, J. (1817). *Encyclopaedia londinensis*. Vol. 16. London.

Wilson, F. E. (1912). Oologists in the Mallee. *Emu*, **12**, 30–9.

Woinarski, J. C. Z. (1985). Breeding biology and life history of small insectivorous birds in Australian forests: response to a stable environment. *Proceedings of the Ecological Society of Australia*, **14**, 159–68.

Woinarski, J. C. Z. (1990). Effects of fire on the bird communities of tropical woodlands and open forests in northern Australia. *Australian Journal of Ecology*, **15**, 1–22.

Woinarski, J. C. Z. and Cullen, J. M. (1984). Distribution of invertebrates on foliage in forests of south-eastern Australia: response to a stable environment? *Australian Journal of Ecology*, **9**, 207–32.

Wolfson, A. (1954). Notes on the cloacal protruberance, seminal vesicles, and a possible copulatory organ in male passerine birds. *Bulletin of the Chicago Academy of Sciences*, **10**, 1–23.

Woolfenden, G. E. and Fitzpatrick, J. W. (1984). *The Florida scrub jay: demography of a cooperatively breeding bird.* Princeton University Press, Princeton, New Jersey.

Wooller, R. D. (1984). Bill size and shape in honeyeaters and other small insectivorous birds in Western Australia. *Australian Journal of Zoology*, **32**, 657–62.

Wooller, R. D. and Calver, M. C. (1981*a*). Feeding segregation within an assemblage of small birds in the karri forest understorey. *Australian Wildlife Research*, **8**, 401–10.

Wooller, R. D. and Calver, M. C. (1981*b*). Diet of three insectivorous birds on Barrow Island, WA. *Emu*, **81**, 48–50.

Wyndham, E. (1986). Length of birds' breeding seasons. *American Naturalist*, **128**, 155–64.

Yom-Tov, Y. (1987). The reproductive rates of Australian passerines. *Australian Wildlife Research*, **14**, 319–30.

Yom-Tov, Y. (1994). Clutch size of passerines at mid-latitudes: the possible effect of competition with migrants. *Ibis*, **136**, 161–5.

Yom-Tov, Y., Christie, M. L., and Iglesias, G. J. (1994). Clutch size in passerines of southern South America. *Condor*, **96**, 170–7.

Index

Species accounts can be found on the page numbers set in bold type.